Early Life Nutrition and Future Health

Early Life Nutrition and Future Health

Special Issue Editor
Kristin Connor

MDPI • Basel • Beijing • Wuhan • Barcelona • Belgrade

Special Issue Editor
Kristin Connor
Carleton University
Canada

Editorial Office
MDPI
St. Alban-Anlage 66
4052 Basel, Switzerland

This is a reprint of articles from the Special Issue published online in the open access journal *Nutrients* (ISSN 2072-6643) in 2019 (available at: https://www.mdpi.com/journal/nutrients/special_issues/ Early_Life_Nutrition_Future_Health).

For citation purposes, cite each article independently as indicated on the article page online and as indicated below:

LastName, A.A.; LastName, B.B.; LastName, C.C. Article Title. *Journal Name* **Year**, *Article Number*, Page Range.

ISBN 978-3-03928-250-0 (Pbk)
ISBN 978-3-03928-251-7 (PDF)

© 2020 by the authors. Articles in this book are Open Access and distributed under the Creative Commons Attribution (CC BY) license, which allows users to download, copy and build upon published articles, as long as the author and publisher are properly credited, which ensures maximum dissemination and a wider impact of our publications.
The book as a whole is distributed by MDPI under the terms and conditions of the Creative Commons license CC BY-NC-ND.

Contents

About the Special Issue Editor vii

Preface to "Early Life Nutrition and Future Health" ix

Paul McJarrow, Hadia Radwan, Lin Ma, Alastair K.H. MacGibbon, Mona Hashim,
Hayder Hasan, Reyad Shaker Obaid, Farah Naja, Hamid Jan Jan Mohamed,
Hessa Al Ghazal and Bertram Y. Fong
Human Milk Oligosaccharide, Phospholipid, and Ganglioside Concentrations in Breast Milk
from United Arab Emirates Mothers: Results from the MISC Cohort
Reprinted from: *Nutrients* **2019**, *11*, 2400, doi:10.3390/nu11102400 1

Andrea Ziesmann, Ruhi Kiflen, Vanessa De Rubeis, Brendan T. Smith, Jonathon L. Maguire,
Catherine S. Birken, Laura N. Anderson and on behalf of the TARGet Kids Collaboration
The Association between Early Childhood and Later Childhood Sugar-Containing Beverage
Intake: A Prospective Cohort Study
Reprinted from: *Nutrients* **2019**, *11*, 2338, doi:10.3390/nu11102338 15

Audrée Lebrun, Anne-Sophie Plante, Claudia Savard, Camille Dugas,
Bénédicte Fontaine-Bisson, Simone Lemieux, Julie Robitaille and Anne-Sophie Morisset
Tracking of Dietary Intake and Diet Quality from Late Pregnancy to the Postpartum Period
Reprinted from: *Nutrients* **2019**, *11*, 2080, doi:10.3390/nu11092080 27

Gwenola Le Dréan, Anne-Lise Pocheron, Hélène Billard, Isabelle Grit, Anthony Pagniez,
Patricia Parnet, Eric Chappuis, Malvyne Rolli-Derkinderen and Catherine Michel
Neonatal Consumption of Oligosaccharides Greatly Increases L-Cell Density without
Significant Consequence for Adult Eating Behavior
Reprinted from: *Nutrients* **2019**, *11*, 1967, doi:10.3390/nu11091967 42

Sebastian A. Srugo, Enrrico Bloise, Tina Tu-Thu Ngoc Nguyen and Kristin L. Connor
Impact of Maternal Malnutrition on Gut Barrier Defense: Implications for Pregnancy Health
and Fetal Development
Reprinted from: *Nutrients* **2019**, *11*, 1375, doi:10.3390/nu11061375 68

Geetha Gopalsamy, Elissa Mortimer, Paul Greenfield, Anthony R. Bird, Graeme P. Young and
Claus T. Christophersen
Resistant Starch Is Actively Fermented by Infant Faecal Microbiota and Increases
Microbial Diversity
Reprinted from: *Nutrients* **2019**, *11*, 1345, doi:10.3390/nu11061345 92

Aurore Camier, Manik Kadawathagedara, Sandrine Lioret, Corinne Bois, Marie Cheminat,
Marie-Noëlle Dufourg, Marie Aline Charles and Blandine de Lauzon-Guillain
Social Inequalities in Prenatal Folic Acid Supplementation: Results from the ELFE Cohort
Reprinted from: *Nutrients* **2019**, *11*, 1108, doi:10.3390/nu11051108 108

Dan Liu, Dong-mei Yu, Li-yun Zhao, Hong-yun Fang, Jian Zhang, Jing-zhong Wang,
Zhen-yu Yang and Wen-hua Zhao
Exposure to Famine During Early Life and Abdominal Obesity in Adulthood: Findings from
the Great Chinese Famine During 1959–1961
Reprinted from: *Nutrients* **2019**, *11*, 903, doi:10.3390/nu11040903 119

Jessica F. Briffa, Rachael O'Dowd, Tania Romano, Beverly S. Muhlhausler, Karen M. Moritz and Mary E. Wlodek
Reducing Pup Litter Size Alters Early Postnatal Calcium Homeostasis and Programs Adverse Adult Cardiovascular and Bone Health in Male Rats
Reprinted from: *Nutrients* **2019**, *11*, 118, doi:10.3390/nu11010118 . . . . **130**

Chien-Ning Hsu and You-Lin Tain
The Good, the Bad, and the Ugly of Pregnancy Nutrients and Developmental Programming of Adult Disease
Reprinted from: *Nutrients* **2019**, *11*, 894, doi:10.3390/nu11040894 . . . . **149**

About the Special Issue Editor

Kristin Connor is an Assistant Professor of Developmental Origins of Health and Disease in the Department of Health Sciences at Carleton University in Ottawa, Canada. She is a molecular geneticist and nutritionist by first training (University of Guelph) and obtained her doctorate in reproductive and developmental physiology in the Faculty of Medicine at the University of Toronto, where she conducted her research internationally. Dr. Connor was a Research Fellow and investigator at the Liggins Institute and the National Research Centre for Growth and Development (Gravida) in Auckland, New Zealand, and a senior Research Fellow at the Lunenfeld-Tanenbaum Research Institute at Mount Sinai Hospital in Toronto. Her research programme aims to understand how developmental trajectories are established and modified in early life (specifically in the contexts of malnutrition, metabolic diseases, and pro-inflammatory environments) to influence an individual's resilience and risk for chronic disease. Dr. Connor is funded through a number of agencies including the Molly Towell Perinatal Research Foundation, the Natural Sciences and Engineering Research Council of Canada, and the Canadian Institutes for Health Research. She serves as an Associate Editor for the Journal of Developmental Origins of Health and Disease.

Preface to "Early Life Nutrition and Future Health"

Inequity starts before birth and is programmed in part by nutritional exposures. If these exposures occur around the time of conception, during pregnancy, and/or in infancy or childhood (all critical periods of development) they may alter a child's health trajectory and impact risk for chronic diseases and disorders across the child's lifespan. Adverse health outcomes associated with suboptimal nutrition in early life include poor growth, impaired cognition and learning, cardiometabolic diseases, immune and inflammatory-mediated diseases and disorders, and neuropsychiatric illnesses. This Special Issue on "Early Life Nutrition and Future Health" has the following aims: 1) understand the origins of offspring health inequities from an early nutritional perspective; 2) uncover new insights into the environmental, biological, and social mechanisms that underpin these health outcomes in offspring; and 3) present novel targets and approaches to optimise health trajectories and prevent chronic diseases and disorders in later life and across generations. The research projects included herein highlight novel mechanistic, epidemiologic, and intervention studies that target key windows where nutrition has greatest influence on future health (preconception, prenatal, and postnatal periods) and that explore vulnerable populations and animal models of early life nutritional programming.

Kristin Connor
Special Issue Editor

Article

Human Milk Oligosaccharide, Phospholipid, and Ganglioside Concentrations in Breast Milk from United Arab Emirates Mothers: Results from the MISC Cohort

Paul McJarrow [1,*,†], Hadia Radwan [2,†], Lin Ma [1], Alastair K.H. MacGibbon [1], Mona Hashim [2,3], Hayder Hasan [2], Reyad Shaker Obaid [2], Farah Naja [4], Hamid Jan Jan Mohamed [3], Hessa Al Ghazal [5] and Bertram Y. Fong [1]

[1] Fonterra Research and Development Centre, Dairy Farm Road, Private Bag 11029, Palmerston North 4442, New Zealand; kevin.ma@fonterra.com (L.M.); alastair.macgibbon@fonterra.com (A.K.H.M.); bertram.fong@fonterra.com (B.Y.F.)
[2] Department of Clinical Nutrition and Dietetics, College of Health Sciences, Research Institute of Medical and Health Sciences (RIMHS), University of Sharjah, Sharjah 27272, UAE; hradwan@sharjah.ac.ae (H.R.); mhashim@sharjah.ac.ae (M.H.); haidarah@sharjah.ac.ae (H.H.); robaid@sharjah.ac.ae (R.S.O.)
[3] School of Health Sciences, Universiti Sains Malaysia, Kubang Kerian 16150, Malaysia; hamidjan@usm.my
[4] Department of Nutrition and Food Sciences, American University of Beirut, Beirut 1107 2020, Lebanon; fn14@aub.edu.lb
[5] Family Health Promotion Centre, Sharjah 27272, UAE; Hessa.ALGhazal@scf.shj.ae
* Correspondence: paul.mcjarrow@fonterra.com; Tel.: +64-6-3504649
† Both authors contributed equally to this manuscript.

Received: 1 August 2019; Accepted: 18 September 2019; Published: 8 October 2019

Abstract: Human milk oligosaccharides (HMOs), phospholipids (PLs), and gangliosides (GAs) are components of human breast milk that play important roles in the development of the rapidly growing infant. The differences in these components in human milk from the United Arab Emirates (UAE) were studied in a cross-sectional trial. High-performance liquid chromatography-mass spectrometry was used to determine HMO, PL, and GA concentrations in transitional (5–15 days) and mature (at 6 months post-partum) breast milk of mothers of the United Arab Emirates (UAE). The results showed that the average HMO (12 species), PL (7 species), and GA (2 species) concentrations quantified in the UAE mothers' transitional milk samples were (in mg/L) 8204 ± 2389, 269 ± 89, and 21.18 ± 11.46, respectively, while in mature milk, the respective concentrations were (in mg/L) 3905 ± 1466, 220 ± 85, and 20.18 ± 9.75. The individual HMO concentrations measured in this study were all significantly higher in transitional milk than in mature milk, except for 3 fucosyllactose, which was higher in mature milk. In this study, secretor and non-secretor phenotype mothers showed no significant difference in the total HMO concentration. For the PL and GA components, changes in the individual PL and GA species distribution was observed between transitional milk and mature milk. However, the changes were within the ranges found in human milk from other regions.

Keywords: human milk; human milk oligosaccharides; phospholipids; sphingomyelin; gangliosides; LC–MS

1. Introduction

Human milk (HM) is the complete food source for infants, providing all the nutrients required for growth in the early stages of life. There has been long interest in the composition of human milk and its changes with lactation and differences between mothers and between populations in different geographical locations or with different diets. Changes in the macronutrient composition of HM with respect to fat, protein, and lactose in different population cohorts have been investigated for many years [1].

As a group, human milk oligosaccharides (HMOs) are a major component of HM, forming the third most dominant component (12–18.6 g/L) after lactose (55–70 g/L) and fat 20–60 g/L) [2,3], and on average having a higher concentration than protein. However, whether they are a macronutrient in sensu stricto is open to interpretation [4]. Phospholipids (PLs) and gangliosides (GAs), which are constituents of the milkfat globule membrane (MFGM) that surrounds the lipid droplet, are minor components. Increasing evidence indicates that HMOs, PLs, and GAs play critical roles in the development of the growing infants and have been of increasing interest in recent years as infant formula manufacturers look to humanise their formulations.

HMOs are shown to play significant biological roles as prebiotics, antimicrobials, and immune modulators [5–16] for the growing infant. To date, there are over 200 oligosaccharides [17] reported in HM. One impact on the HMO profile from an individual mother is the absence (non-secretor) or presence (secretor) of a fully functional α-(1,2)-fucosyltransferase 2, which is coded by the FUT2 gene; milk produced by non-secretor mothers contains low concentrations of 2'-fucosyllactose (2'FL) as well as lacto-N-fucopentaose I (LNFP I) [5,18].

In addition to their essential roles in cell membrane structural integrity, PLs play critical roles in lung and brain development in the growing infant [19–23]. The PLs in HM are mainly found in the tri-layer of the MFGM and consist of sphingomyelins (SMs) and the glycerophospholipids phosphatidyl choline (PC) and its lyso species (L-PC), phosphatidyl ethanolamine (PE) and its lyso species (L-PE), phosphatidyl inositol (PI), and phosphatidyl serine (PS).

GAs are also important in neurological development, memory formation, and synaptic signal transduction, and are implicated in regulating the immune system and supporting gut maturation in the new born [24–26].

There are recent studies that report the changes in HMOs [27–30], PLs [31–33], and GAs [32–35] across lactation for various population cohorts; however, to our knowledge, there is no specific study reported for the UAE or other Middle Eastern populations. In this study, the HMO, PL, and GA concentrations in transitional (5–15 days) and mature milk (at 6 months post-partum) were measured in a cross section of UAE mothers' breast milk samples, collected from a wider mother–infant study cohort (MISC) [36]. The information obtained from this study allows comparison with other similar studies on HMO, PL, and GA levels in HM from other population cohorts to better address the hypothesis that geography and ethnicity impact the levels of these HM bioactives [28].

2. Materials and Methods

2.1. Study Setting and Population

A randomly selected subsample of breast milk samples (transitional milk period, days 5 to 15 post-partum, $n = 41$, and mature milk at 6 months post-partum, $n = 40$) were made from a large cohort collected as part of the MISC in the UAE to comprehensively investigate maternal and infant factors in relation to child health outcomes, as well as early-life determinants of non-communicable diseases, through integration of sociodemographic, dietary, lifestyle, anthropometric, and biological and cognitive data [36].

Arab participants were recruited from antenatal clinics in three main public governmental hospitals, and seven primary health care clinics and mother and child centres in the Emirates of Sharjah, Dubai, and Ajman. The inclusion criteria were pregnant women who were Emirati or Arab expatriate; aged 19–40 years; singleton pregnancy; within the third trimester of pregnancy (27–42 weeks of gestation); free of chronic diseases (including diabetes, hypertension, kidney disease, cancer, and others), autoimmune disorders, infections with the human immunodeficiency virus, or hepatitis in preconception; receiving antenatal care in any of the above-mentioned clinics; and expected to give birth in a participating public hospital and remaining in the UAE during the timeline of the study. The exclusion criteria were as follows: multiple pregnancy, high risk pregnancy or pre-eclampsia, and history of chronic diseases [36].

The study was approved by the Research and Ethics Committee, University of Sharjah (REC/14/01/1505) and by Al Qassimi Clinical Research Centre Ethical Research Committee (REC Reference Number: 215 12015-03), by the Ministry of Health Ethical Research Committee (R02), and by Dubai Health Authority (DSREC-0/2016).

2.2. Phospholipid, Ganglioside, and Oligosaccharide Analysis

Extraction of HM samples and analysis using HPLC-MS was as described previously for PLs [31] and GAs [34]. Briefly, 0.25 mL of HM samples was extracted using the modified Svennerholm and Fredman [37] method described by Fong et al. [38]. The non-polar (lower) phase was diluted to 5 mL with choroform/methanol (1:2) and an aliquot was used for quantification of PC, PS, PI, PE, SM, L-PC, and L-PE. Separation and quantification were as described previously [38] with individual PL species separated using an ACQUITY UPLC system (Waters, Milford, MA, USA) equipped with an APS-2 Hypersil column (150 mm × 2.1 mm, 3 Um, Thermo Electron Corporation, Waltham, MA, USA). The eluate from the HPLC was directed into a TSQ Quantum mass spectrometry (Thermo Electron Corporation, Waltham, MA, USA) with individual PL species detected either by precursor ion or neutral loss experiments [38]. The polar (upper) phase from the extraction was diluted to 10 mL with methanol/water (1:1), and then GA classes (disialoganglioisde 3 (GD3) and monosialoganglioside 3 (GM3)) were separated on an Agilent 1100 series HPLC system (Santa Clara, CA, USA) equipped with an APS-2 Hypersil column (150 mm × 2.1 mm, 3 mm), interfaced to a SCIEX 6500 QTrap mass spectrometer (AB SCIEX, Framingham, MA, USA) and quantified as described by Ma et al. [26] using multiple reaction monitoring in negative mode.

HMOs (3'-sialyllactose (3'SL), 6'-sialyllactose (6'SL), 6'-sialyllactosamine (6'SLN), disialyllactose (DSL), 3'-sialyl-3-fucosyllactose (3'S3FL), LS-tetrasaccharide a/b (LSTa/b), LS-tetrasaccharide (LSTc), lacto-N-neotetraose (LNnT), lacto-N-tetraose (LNT), Lacto-N-fucopentaose (LNFP total against LNFP I as standard), 2'-fucosyllactose (2'FL), and 3-fucosyllactose (3FL)); were analysed for, and in all cases, the sialic acid is N-acetyl neuraminic acid. HMOs were separated using a Luna hydrophilic interaction liquid chromatography column (150 mm × 2.1 mm, 3 um, Phenomenex, Torrence, CA, USA) on an Agilent 1100 series HPLC [30]. The HPLC system was coupled to a SCIEX 6500 QTrap mass spectrometer, operated in negative ion mode, and HMOs were quantified as described [30] using 3'-sialyllactosamine as an internal standard.

2.3. Statistical Analysis

Statistical analysis was conducted using one-way analysis of variance (ANOVA), where difference among the means is significant when $p < 0.05$ (MiniTab Release 17.2.1 2016, MiniTab Inc., State College, PA, USA).

3. Results

3.1. HMO

Seven acidic (3'SL, 6'SL, 6'SLN, 3'S3FL, DSL, LSTa/b, and LSTc) and five neutral (LNnT, LNT, LNFP (total), 2'FL, and 3FL) oligosaccharides were measured.

The average HMO concentration in transitional milk (8204 ± 2389 mg/L; Table 1) was significantly higher than that measured in mature milk (3905 ± 1466 mg/L; Table 1). This trend was also reflected at the individual HMO level (Table 1). When secretor and non-secretor milk samples were considered separately (based on a non-secretor milk having a 2'FL concentration <50 mg/L), the total HMO concentrations for both groups were not significantly different for either transitional or mature milk (Table 1).

The proportion of the acidic HMOs in transitional milk (18%) was higher than that measured in mature milk (7%); concomitantly, the neutral HMOs in transitional milk were 82% of total HMOs measured, compared with 93% in mature milk. When considered separately, secretors and non-secretors were not significantly different in this aspect. The HMO species for which there was a significant difference in concentration between secretor and non-secretor milk samples were 2'FL, 3FL, and 3'S3FL (Table 1). The secretor milk samples were higher in 2'FL and lower in 3FL and 3'S3FL compared with non-secretor milk samples.

3.2. Phospholipids

The average total PL concentration (± SD) measured in the transitional milk (269.0 ± 89.2 mg/L) was significantly ($p < 0.05$) higher than that measured for mature milk (219.6 ± 85.0 mg/L, Table 2). The concentration of PE, SM, and L-PE did not change significantly ($p > 0.05$) between transitional and mature milk, but PI, PC, PS, and L-PC decreased significantly in concentration in mature milk (Table 2).

The relative distribution of the five different PL classes (PI, PC, PE, PS, and SM) and L-PC and L-PE is presented in Table 2. Because of the changes in individual PL species concentration, significant changes in the relative distribution of PL classes were observed, with PE increasing from 25% in transitional milk to 36% in mature milk (6 months post-partum), while PC and PS decreased from 25% to 14% and 11% to 7%, respectively. SM remained the dominant PL class during both lactation milk time points, and only increased slightly (Table 1). Little change in the PI distribution was observed over the two-time points, decreasing from 4% to 3%.

3.3. Gangliosides

The average total GA (TGA) concentration measured in the UAE mothers' transitional milk (21.2 ± 11.46 mg/L) and mature milk (20.2 ± 9.8 mg/L) was not significantly different ($p > 0.05$) across the two-time points. However, the relative distribution of ganglioside classes, GD3 and GM3, changed across the two-time points from 56% and 44%, respectively, for transition milk, to 9% and 91% respectively, for mature milk (Table 2), with both GM3 and GD3 showing significant changes in concentration.

Table 1. Human milk oligosaccharide (HMO) concentration in transitional and mature human milk from United Arab Emirates (UAE) mothers. 3'SL, 3'-sialyllactose; 6'SL, 6'-sialyllactose; 6'SLN, 6'-sialyllactosamine; DSL, disialyllactose; 3'S3FL, 3'-sialyl-3-fucosyllactose; LSTa/b, LS-tetrasaccharide a/b; LSTc, LS-tetrasaccharide; LNnT, lacto-N-neotetraose; LNT, lacto-N-tetraose LNFP, lacto-N-fucopentaose I; 2'FL, 2'-fucosyllactose; 3FL, 3-fucosyllactose.

Sample	3'SL	6'SL	6'SLN	DSL	3'S3FL	LSTa/b	LSTc	2'FL	3FL	LNnT	LNT	LNFP	Total HMO
Transitional (all) ($n = 41$)	226 ± 107 [a]	621 ± 212 [a]	15 ± 15 [a]	2.2 ± 2.3 [a]	19 ± 21 [a]	104 ± 46 [a]	488 ± 224 [a]	2021 ± 1776 [a]	581 ± 868 [a]	765 ± 350 [a]	1429 ± 693 [a]	1932 ± 762 [a]	8204 ± 2388 [a]
Non-secretor	256 ± 144	562 ± 232	20 ± 23	1.8 ± 1.9	48 ± 49	124 ± 52	456 ± 243	4.3 ± 8.6	1599 ± 524	990 ± 524	1917 ± 973	1490 ± 528	7466 ± 1812
Secretor	216 ± 91	643 ± 204	13 ± 11	2.4 ± 2.4	8.1 ± 7.0	98 ± 42	500 ± 220	2761 ± 1497 #	208 ± 112 #	682 ± 220	1250 ± 460	2094 ± 777 #	8475 ± 2541
Mature (all) ($n = 40$)	134 ± 69 [b]	91 ± 108 [b]	5 ± 1 [b]	0.2 ± 0.4 [b]	10 ± 14 [a]	31 ± 25 [b]	11 ± 8 [b]	997 ± 885 [a]	1194 ± 106 [a]	250 ± 188 [b]	504 ± 337 [b]	650 ± 416 [b]	3876 ± 1403 [b]
Non-secretor	181 ± 98	81 ± 52	4.9 ± 2.2	0.1 ± 0.2	25 ± 20	28 ± 18	6.7 ± 5.3	4.0 ± 2.8	2526 ± 113	187 ± 113	420 ± 276	548 ± 300	4009 ± 1104
Secretor	116 ± 44	95 ± 124	4.4 ± 0.6	0.3 ± 0.5	4.6 ± 2.4 #	32 ± 28	13 ± 8 #	1374 ± 746 #	688 ± 398 #	273 ± 207	536 ± 357	689 ± 451	3826 ± 1515

Values (in mg/L) are means ± standard deviation. Different symbol pairings signify statistically significant values ($p < 0.05$); when comparing transitional and mature milk individual HMO and total, differences in groups are represented by [a],[b]. # is used if the difference between secretor/non-secretor groups is significant.

Table 2. Phospholipid (PL) and ganglioside (GA) concentrations in transitional and mature breast milk. SM, sphingomyelins; PC, phosphatidyl choline (PC); L-PC, PC lyso species; PE, phosphatidyl ethanolamine; L-PE, PE lyso species; PI, phosphatidyl inositol; PS, phosphatidyl serine; GD3, disialoganglioside 3; GM3, monosialoganglioside 3.

Time Point	Phospholipids									Gangliosides		
	PI	PE	PC	SM	PS	L-PE	L-PC	Total PL	GM3	GD3	Total GA
Transitional ($n = 41$)	11.2 ± 5.5 [a] (4%)	66.3 ± 27.16 [a] (25%)	66.4 ± 32.87 [a] (25%)	91.2 ± 26.38 [a] (34%)	28.5 ± 13.29 [a] (7%)	3.7 ± 2.37 [a] (1.4%)	1.7 ± 0.98 [a] (0.6%)	269.0 ± 89.2 [a]	9.47 ± 8.37 [a] (45%)	11.71 ± 9.46 [a] (55%)	21.18 ± 11.46 [a]
Mature ($n = 40$)	6.5 ± 3.61 [b] (3%)	80.0 ± 35.35 [a] (36%) #	30.2 ± 22.07 [b] (14%) #	82.9 ± 29.21 [a] (38%)	16.1 ± 6.99 [b] (7%)	3.1 ± 1.99 [a] (1.4%)	0.9 ± 0.63 [b] (0.4%)	219.6 ± 85.0 [b]	18.62 ± 9.69 [b] (92%) #	1.57 ± 2.24 [b] (8%) #	20.18 ± 9.75 [a]

Values (in mg/L) are means ± standard deviation; the relative distribution (%) of the individual phospholipid classes and ganglioside classes are in parenthesis. In a column, values with the same superscript indicate no significant difference ($p > 0.05$) between the transitional and mature milk samples; when comparing transitional and mature milk individual phospholipids and gangliosides and total, differences in concentration are represented by [a],[b]. # is used for when the differences in the relative percentages of the individual phospholipids and gangliosides is significant between the transitional and mature milk.

4. Discussion

HM is considered the "gold standard" for an infant's nutrition, to which infant formula manufacturers strive to emulate in both nutrient composition and performance. HM composition varies considerably between individual mothers and over lactation, and so could be considered to be personalised to each infant that is breastfed. Various factors such as diet, geography, ethnicity, milk collection time, and genetics have been implicated to have a significant influence on the HMO, PL, and GA composition in HM, but most of the data obtained to date indicate that the stage of lactation is perhaps the primary factor that has the greatest influence on HM composition [27–29,39]. There are several recent studies reporting the composition of HM, trying to gain a better understanding of the changes in the HMOs and complex lipids (PLs and GAs) through lactation of different geographical population cohorts [28,31,32,34,35,39,40]. However, this is the first study that looks at the transitional and mature milk from UAE mothers, helping to address geographical variation in HMOs, PLs, and GAs.

4.1. Human Milk Oligosaccharides

The UAE mothers' transitional breast milk samples had significantly higher average total HMO concentrations (8204 ± 2389 mg/L) compared with the mature milk samples (3876 ± 1403 mg/L). The largest decrease in the HMOs over these two lactation timepoints was observed with the acidic oligosaccharide, LSTc, which decreased by 98% from 488 ± 224 mg/mL in transitional milk to 11 mg/L in the six months post-partum mature milk (Table 1). The only HMO to increase across this period was 3FL (Table 1). This trend is in line with lactational trend data reported in the literature [27,29,30] (Figure 1). While only five neutral oligosaccharides (2′FL, 3FL, LNT, LNnT, and total LNFP) were measured in this study, they made up a significant proportion (82% and 93% for transitional and mature milk, respectively) of the total HMOs measured in this study, with the acidic HMOs making up 18% and 7% for transitional and mature milk, respectively. This finding is consistent with those reported by Ma et al. [30] for their Malaysian and Chinese cohort of 89%–91% and 8.5%–11% of neutral and acidic HMOs, respectively, and other similar studies [27–29], despite differences in the range of HMOs and respective concentrations being different.

The individual HMO levels measured in this study for the UAE breast milk samples were also in a range similar to that reported by Ma et al. [30] for Chinese and Malaysian mothers (Table 3 and Figure 1); and Larsson et al. [41], Coppa et al. [42], Bao et al. [43], and Austin et al. [29] for the common HMOs measured for the corresponding time points, except for 3′SL, LNT, and LNnT where Austin et al. [29] reported lower HMO lactational results (Table 3 and Figure 1).

HMOs have been implicated not just to provide anti-infective protection for the infant, but also as being involved as immune modulators, and may play a key role in gut maturation of the rapidly growing infant. Higher HMO concentrations in colostrum and transitional milk may be the consequence of increased protection required for the vulnerable infant during the early few days of life. The changes in HMO levels over the course of lactation [29,30] may reflect changes in the development stages of the growing infant, and a requirement for specific compositions of these HMOs.

On the basis of the 2′FL concentrations, 26% of the UAE mothers in this cohort expressed a non-secretor phenotype, having a 2′FL concentration <50 mg/L in their breast milk samples [29,44]. 2′FL (and LNFP I) are products of α-(1,2)-fucosyltransferase 2, which is coded by the FUT2 gene that is supposedly non-functional in non-secretor mothers. However, in the study of Austin et al. [20], 2′FL was found to be not completely absent in the non-secretor mothers, as was the case in this study. The secretor/non-secretor frequency is known to vary with geography and racial difference, with 22.5% non-secretor phenotype reported for the Han Chinese population, Eastern China region [29]; 37% reported for the Chinese cohort (Guangzhou); and 17% reported for Malay mothers [30]. The typical frequency of non-secretor phenotype reported by Azad et al. [39] was 28% for Caucasian mothers, while Asian mothers had higher non-secretor frequency at 40%; however, the Asian mothers' sub-ethnic groups were not defined. There is no current information as to the non-secretor frequency for the UAE population. Azad et al. [30] also reported that the non-secretor group had significantly less HMOs than the secretor group. For the HMOs measured in this study, however, there was no significant difference ($p > 0.05$) in the average total HMOs between secretor and non-secretor mothers for either their transitional milk (8292 ± 2516 mg/L versus 6994 ± 1905 mg/L, respectively) or their mature milk (4289 ± 1791 mg/L versus 4317 ± 1857 mg/L, respectively) (Table 1).

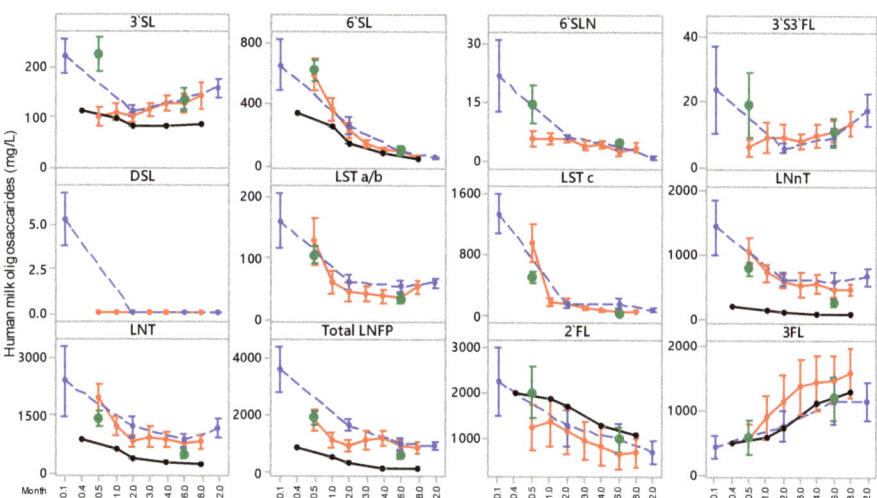

Figure 1. Human milk oligosaccharide (HMO) concentrations in the United Arab Emirates (UAE) mothers' milk (●) across the transitional milk and mature milk periods, plotted against the lactational trend reported by Ma et al. [30] for a Chinese cohort (●) and Malaysian cohort (●), and by Austin et al. [29] for a Chinese cohort (●). 3'SL, 3'-sialyllactose; 6'SL, 6'-sialyllactose; 6'SLN, 6'-sialyllactosamine; DSL, disialyllactose; 3'S3FL, 3'-sialyl-3-fucosyllactose; LSTa/b, LS-tetrasaccharide a/b; LSTc, LS-tetrasaccharide; LNnT, lacto-N-neotetraose; LNT, lacto-N-tetraose LNFP, lacto-N-fucopentaose I; 2'FL, 2'-fucosyllactose; 3FL, 3-fucosyllactose.

One important consideration in assessing crude population figures for HMO species is the impact that the percentage of non-secretors in each study population has on the average or mean 2'FL and 3FL concentrations that would be reported for the full cohort; the higher the percentage of non-secretors in a population, the lower the average 2'FL and higher the average 3FL concentrations for the cohort as a whole. Furthermore, it is not only the crude 2'FL and 3FL figures that are impacted, but also the figures for other oligosaccharides such as LNFP I and 3'S3FL. For example, in this study, the mature milk results for the full cohort showed that 2'FL and 3FL concentrations are relatively similar, which was also shown by Ma et al. [30] and Austin et al. [29], but if the non-secretor percentage was lower, then 3FL would be lower than 2'FL at later points in lactation, as evidenced in the limited data of Larsson et al. [41], which is the analysis of secretors alone, and the results of this study. Figure 1 also emphasises the impact of lactation on interpreting which oligosaccharides are in highest concentration, because, at six months in all four cohorts (two Chinese, one Malaysian, and one UAE), the concentrations of 3FL and 2'FL are very similar.

Table 3. Human milk oligosaccharide mean concentrations reported in the literature for both transitional and mature milk about six months *post partum* (mg/L, except for Austin et al. [20], which is mg/kg).

Reference	Milk (Post Partum)	n	3'SL	6'SL	6'SLN	DSL	3'S3FL	LSTa/b	LSTc	2'FL	3FL	LNnT	LNT	Total LNFP
Larsson et al. [41] [a,b] Denmark	Mature (5 month)	15	492	156				149	33	2989	229	578	703	2692
Coppa et al. [42] Italy	Transition (4 days)	18								3930	340	2040	840	1650
Bao et al. [43] USA	Transition (9–21 days)	90	76	396				74 [c]	148					
Austin et al. [20] (Table 4) China	Transition (5–11 days)	90	110	340						2600	510	180	880	1157
	Mature (4–8 months)	90	83	45						1300	1300	59	250	199
Ma et al. [21] China	Transition (14 days)	20	100	593	6	3	6	127	941	1281	543	1033	1979	1870
	Mature (6 months)	20	127	83	2	0	11	33	47	704	1476	446	785	945
Ma et al. [21] Malaysia	Mature (6 months)	21	135	84	4	0	9	84	145	1003	1146	571	867	1036
This study UAE	Transition (6–14 days)	41	226	621	15	2	19	104	488	2021	581	765	1429	1932
This study UAE	Mature (6 months)	40	134	91	5	0	10	31	11	997	1194	250	504	650

[a] secretors only. [b] calculated from nmol/mL data. [c] LSTb alone.

4.2. Phospholipids

The average total PL concentrations observed with HM samples of UAE mothers were within the typical ranges reported for human breast milk for other geographical population cohorts (Table 4). In this study, the UAE transitional milk had an average total PL concentration that was significantly higher ($p < 0.05$) than that measured for the six-month mature milk. While this trend is consistent with the majority of the published data [31,33,45–48], the trend reported from some studies showed the total PL concentration in colostrum and transitional milk was either much lower than [49–51] or the same as [52] that in mature milk (Table 4).

At the individual PL class level, changes in the relative distribution of the individual PL classes were observed over the transitional milk and mature milk periods (Figure 1). There was a significant increase in the relative amount of PE from 25% in transitional milk to 36% in the UAE mature milk samples, while PC and PS both decreased, the former from 25% to 14% and the latter from 11% to 7%. SM increased only slightly, while PI decreased slightly. Similarly, changes in the relative distribution between transition milk and mature milk was observed for a Malaysian HM cohort [31] (Figure 2). In contrast, however, the PL distribution was relatively constant for Spanish [34] and Chinese [24] breast milk cohorts (Figure 2).

Table 4. Human milk phospholipid concentration reported in the literature for both transitional and mature milk (modified from Ma et al. [31]) [a].

Total PL Ranges (mg/L)		Country	Reference
Transitional Milk	**Mature Milk**		
390 ± 50–440 ± 73 (8–15, 20)	370 ± 106–405 ± 80 (22–36, 40)	Germany	Harzer et al. [52]
310 ± 30 (11, 5)	270 ± 30 (23, 5 pooled)	USA	Bitman et al. [47]
158 (11, 17)	114 (23, 19)	Spain	Sala-Vila et al. [48]
148 (7, 6)	133–227 (20–84, 6)	USA	Bitman et al. [49]
973 (14, 10)	1023–1298 (42–112, 10)	USA	Clark et al. [50]
185 (6–15, 45)	182 (>16, 45)	Denmark	Zou et al. [51]
550 ± 260 (6–10, 7)	450 ± 260 (30, 16)	France	Gracia et al. [45]
	230 ± 49–242 ± 82 (30–120, 50)	Singapore	Thakkar et al. [32]
437 ± 23–535 ± 26 (6–15, 44)	260 ± 3–422 ± 13 (16–360, 44)	Spain	Claumarchirant et al. [46]
266 ± 57 (6–14, 12)	170 ± 80–219 ± 92 (60–365, 132)	Malaysia	Ma et al. [31]
285 ± 144 (6–15, 81)	242 ± 114 (16–240, 345)	China	Guiffrida et al. [33]
269 ± 89 (6–14, 41)	220 ± 85 (180, 40)	UAE	This study

[a] Lactational period, in days, and number of samples are given in parentheses. Data from Sala-Vila et al. [48] recalculated from total phospholipid data provided using average molecular masses of 758.4, 864.4, 787.1, 737.6, and 754.4 for PC, PI, PS, PE, and SM, respectively; data from Bitman et al. [49] recalculated from graph using termed mothers only.

However, across the mature milk period, three recent studies [31,32,46] and two earlier studies [49,52] showed that the individual PL class distribution remained relatively constant through the mature milk period, despite changes in their absolute concentrations.

Variation in absolute PL concentrations may be attributable to a variety of factors, such as time of sampling protocols (full breast expression, time of sampling, and breast variation [53]), diet, geographic, and even metabolic stage and gestational age at birth, in addition to different analytical methods used [54–56]. However, the fact that the relative distribution of the individual PL class remains constant through the mature milk period indicates that some metabolic controls are maintained over the biosynthesis of these PL classes, perhaps to maintain the integrity of the MFGM structure.

Changes in the individual PL distribution observed between early milk and mature milk may be the consequence of the changing structure. It is reported that colostrum and transitional milk has much larger fat droplet size than that of mature milk [57]. In fact, Cohen et al. [58] reported the phospholipid composition of the mammary epithelial cell regulated the lipid droplet size, rather than the cellular triglyceride content; this phospholipid composition is critical in maintaining membrane structure integrity as the lipid droplet size changes.

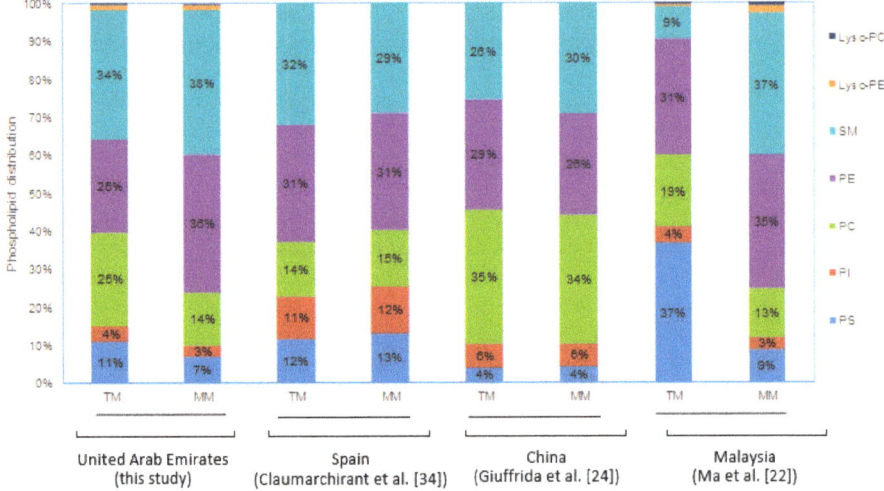

Figure 2. Distribution of phospholipid classes in transitional milk (TM) and mature milk (MM; six months post-partum) in UAE mothers' milk, and from other geographic cohorts. SM, sphingomyelins; PC, phosphatidyl choline (PC); L-PC, PC lyso species; PE, phosphatidyl ethanolamine; L-PE, PE lyso species; PI, phosphatidyl inositol; PS, phosphatidyl serine.

4.3. Gangliosides

In HM, GAs are present predominantly as the GM3 and GD3 classes (which have the same polar head group, but different sphingosine and fatty acid). Typically, GD3 is present at relatively high concentrations in colostrum and transitional HM, making up approximately 30%–80% of the TGAs, but decreasing to 8%–25% by four to six months post-partum. Conversely, the relative proportion of GM3 is higher in mature milk [33,35,59–61].

However, there is no clear consistent lactational trend observed for TGA concentration. While a few studies show colostrum and transitional milk to contain the highest TGA concentration [34,59], other studies show the opposite [60,62]. Similarly, the TGA lactation trend over the mature milk period is also not clear. While the studies of Thakkar et al. [32] and Ma et al. [34] showed a gradual increase in the TGA level over the mature milk period (of four months and eight months duration, respectively), from the data published for the Malaysian cohort [34], this gradual increase did not appear to be sustained: the average TGA levels dropped from ~25.3 ± 15.7 mg/L at 6 months to 16.6 ± 8.5 mg/L at around 12 months post-partum.

In this study, there was no significant difference between the average TGA results measured between the transitional milk (21.18 ± 11.46) and the mature milk period (20.18 ± 9.75), with GD3 making up 55% of the TGAs in transition milk, decreasing to 8% at six months post-partum. The change in the GA class distribution is consistent across other published cohort studies [34,35] (Table 5), suggesting that the biosynthesis of GAs is under metabolic control, and different classes may be required for different stages of the development and growth for the breast fed infant. Interestingly, Thakkar et al. [32] reported that HM from mothers with male infants had higher energy content and lipids compared with that mothers of female infants; these differences may be because of differences in nutritional requirements to support specific growth and development patterns between the two sexes [23]. However, in the current study, there were no significant differences in the GA concentrations of either the transitional milk or mature milk breast milk from mothers who had male or female infants. The average TGA levels observed with the UAE cohort were similar to those reported for the Malaysian and Chinese cohorts (25.3 ± 15.7 mg/L and 22.9 ± 9.9 mg/L, respectively [34,35]), but higher than those reported by Thakkar et al. [32] for their Singapore cohort (4.6–5.6 mg/L) and Giuffrida et al. [33] for their Chinese cohort (11.0 ± 5.0 mg/L).

Table 5. Human milk mean GM3, GD3, and total ganglioside concentrations reported in the literature for both transitional and mature milk (mg/L ± standard deviation) [a].

Reference	Milk	n	GM3	GD3	Total GA
Giuffrida et al. [40] China	Colostrum/Transition (0–11 days)	450	3.8 ± 0.4 (47)	4.3 ± 0.9 (53)	8.1
Ma et al. [34] Malaysia	Transition	12	8.3 ± 4.8 (44)	10.6 ± 4.3 (56)	18.9 ± 6.6
This Study UAE	Transition (5–15 days)	41	9.5 ± 8.4 (45)	11.7 ± 9.5 (55)	21.2 ± 11.5
Ma et al. [34] Malaysia	Mature (6 months)	42	21.4 ± 13 (85)	4.3 ± 5.5 (15)	25.3 ± 15.7
Ma et al. [35] China	Mature (6 months)	20	21.4 ± 9.5 (93)	1.5 ± 1.4 (7)	22.9 ± 9.9
This Study UAE	Mature (6 Months)	40	18.6 ± 9.7 (92)	1.6 ± 2.2 (8)	20.2 ± 9.8

[a] % of total GA is given in parenthesis.

5. Conclusions

This study provides new information about the HMO, PL, and GA concentrations in breast milk specific to UAE mothers. Despite reports indicating that human milk composition varies across different geographies and ethnicities, the average HMO, PL, and GA concentrations measured in the MISC study from UAE were within the typical ranges reported for other ethnic cohorts, especially data obtained by the same methodologies. There appear to be more similarities than differences for HMOs, PLs, and GAs in HM from different geographical locations and ethnicities. The similarity in the range of concentrations of HMO, PL, and GA between different cohorts suggests they each have specific biological and functional roles linked to timing of lactation.

Author Contributions: Study conceptualisation, B.Y.F., H.R., H.J.J.M., A.K.H.M., and P.M.; methodology, H.R., H.J.J.M., B.Y.F., and L.M.; formal analysis, B.Y.F., A.K.H.M., L.M., and P.M.; sample collection and selection investigation, M.H., H.R., F.N., H.H., H.J.J.N., R.S.O., and H.A.G.; resources and data curation M.H., H.R., F.N., H.H., H.J.J.M., R.S.O., and H.A.G.; writing—original draft preparation, B.Y.F., A.K.H.M., and P.M.; writing—review and editing, P.M., A.K.H.M., B.Y.F., L.M., H.H., and H.R.; visualisation, H.H., F.N., and R.S.O.; supervision, H.R. and M.H.; project administration, B.Y.F., H.R., and P.M.; funding acquisition, H.R.

Funding: We would like to note that the MISC research study has received funding from Al Jalila Foundation (AJF 201510) and University of Sharjah Collaborative Grant (1501057003-P). The analysis of the samples was supported by the New Zealand Primary Growth Partnership post-farm gate dairy programme, funded by Fonterra Co-operative Group and the New Zealand Ministry for Primary Industries.

Acknowledgments: The authors would like to acknowledge all the mothers who kindly supplied milk samples for this study, Tim Coolbear (Fonterra) for reviewing and editing, and Angela Rowan (Fonterra) for support and guidance for this project.

Conflicts of Interest: P.M., L.M., A.K.H.M., and B.Y.F. are employees of Fonterra Co-operative Group Limited, a company that manufactures infant formula and ingredients used for infant formula manufacture. Other authors declare no conflict of interest.

References

1. Michaelsen, K.F.; Skafte, L.; Badsberg, J.H.; Jorgensen, M. Variation in Macronutrients in Human bank milk-influencing factors and implications for human-milk banking. *J. Pediatr. Gastroenterol. Nutr.* **1990**, *11*, 229–239. [CrossRef] [PubMed]
2. Kunz, C.; Rudloff, S.; Baier, W.; Klein, N.; Strobel, S. Oligosaccharides in human milk: Structural, functional, and metabolic aspects. *Annu. Rev. Nutr.* **2000**, *20*, 699–722. [CrossRef] [PubMed]
3. Newburg, D.S.; Neubauer, S.H. Carbohydrates in milk: Analysis, quantities, and Significance. In *Handbook of Milk Composition*; Jensen, R.G., Ed.; Academic Press: London, UK, 1995; pp. 273–349.
4. Gopal, P.K.; Gill, H.S. Oligosaccharides and glycoconjugates in bovine milk and colostrum. *Br. J. Nutr.* **2000**, *84* (Suppl. S81), S69–S74. [CrossRef]
5. Bode, L. Human milk oligosaccharides: Every baby needs a sugar mama. *Glycobiology* **2012**, *22*, 1147–1162. [CrossRef] [PubMed]

6. Bode, L. The functional biology of human milk oligosaccharides. *Early Hum. Dev.* **2015**, *91*, 619–622. [CrossRef] [PubMed]
7. Lin, A.E.; Autran, C.A.; Szyszka, A.; Escajadillo, T.; Huang, M.; Godula, K.; Prudden, A.R.; Boons, G.J.; Lewis, A.L.; Doran, K.S.; et al. Human milk oligosaccharides inhibit growth of group B Streptococcus. *J. Biol. Chem.* **2017**, *292*, 11243–11249. [CrossRef]
8. Yu, H.; Lau, K.; Thon, V.; Autran, C.A.; Jantscher-Krenn, E.; Xue, M.; Li, Y.; Sugiarto, G.; Qu, J.; Mu, S.; et al. Synthetic disialyl hexasaccharides protect neonatal rats from necrotizing enterocolitis. *Angew. Chem. Int. Ed. Engl.* **2014**, *53*, 6687–6691. [CrossRef]
9. Yu, H.; Yan, X.; Autran, C.A.; Li, Y.; Etzold, S.; Latasiewicz, J.; Robertson, B.M.; Li, J.; Bode, L.; Chen, X. Enzymatic and Chemoenzymatic Syntheses of Disialyl Glycans and Their Necrotizing Enterocolitis Preventing Effects. *J. Org. Chem.* **2017**, *82*, 13152–13160. [CrossRef]
10. Ackerman, D.L.; Craft, K.M.; Doster, R.S.; Weitkamp, J.H.; Aronoff, D.M.; Gaddy, J.A.; Townsend, S.D. Antimicrobial and Antibiofilm Activity of Human Milk Oligosaccharides against Streptococcus agalactiae, Staphylococcus aureus, and Acinetobacter baumannii. *ACS Infect. Dis.* **2018**, *4*, 315–324. [CrossRef]
11. Ackerman, D.L.; Doster, R.S.; Weitkamp, J.H.; Aronoff, D.M.; Gaddy, J.A.; Townsend, S.D. Human Milk Oligosaccharides Exhibit Antimicrobial and Antibiofilm Properties against Group B Streptococcus. *ACS Infect. Dis.* **2017**, *3*, 595–605. [CrossRef]
12. Craft, K.M.; Thomas, H.C.; Townsend, S.D. Interrogation of Human Milk Oligosaccharide Fucosylation Patterns for Antimicrobial and Antibiofilm Trends in Group B Streptococcus. *ACS Infect. Dis.* **2018**, *4*, 1755–1765. [CrossRef] [PubMed]
13. Craft, K.M.; Thomas, H.C.; Townsend, S.D. Sialylated variants of lacto-N-tetraose exhibit antimicrobial activity against Group B Streptococcus. *Org. Biomol. Chem.* **2019**, *17*, 1893–1900. [CrossRef] [PubMed]
14. Kunz, C. Complex oligosaccharides in infant nutrition. *Mon. Kinderheilkd.* **1998**, *146*, S49–S56. [CrossRef]
15. Kunz, C.; Rudloff, S. Biological Functions of Milk Oligosaccharides. *Acta Paediatr.* **1994**, *83*, 1042. [CrossRef]
16. Craft, K.M.; Gaddy, J.A.; Townsend, S.D. Human Milk Oligosaccharides (HMOs) Sensitize Group B Streptococcus to Clindamycin, Erythromycin, Gentamicin, and Minocycline on a Strain Specific Basis. *ACS Chem. Biol.* **2018**, *13*, 2020–2026. [CrossRef] [PubMed]
17. Urashima, T.; Kitaoka, M.; Terabayashi, T.; Fukuda, K.; Ohnishi, M.; Kobata, A. Milk Oligosaccharides. In *Oligosaccharides: Sources, Properties and Application*; Gordon, N.S., Ed.; Nova Science Publishers, Inc.: New York, NY, USA, 2011; pp. 1–58.
18. Tonon, K.; de Morais, M.; Abrao, A.; Miranda, A.; Morais, T. Maternal and Infant Factors Associated with Human Milk Oligosaccharides Concentrations According to Secretor and Lewis Phenotypes. *Nutrients* **2019**, *11*, 1358. [CrossRef]
19. Mozzi, R.; Buratta, S. Brain Phosphatidylserine: Metabolism and Functions. In *Handbook of Neurochemistry and Molecular Neurobiology*; Lajtha, A., Tettamanti, G., Goracci, G., Eds.; Springer: Boston, MA, USA, 2010; pp. 39–58. [CrossRef]
20. Contarini, G.; Povolo, M. Phospholipids in Milk Fat: Composition, Biological and Technological Significance, and Analytical Strategies. *Int. J. Mol. Sci.* **2013**, *14*, 2808–2831. [CrossRef]
21. Hirabayashi, Y.; Furuya, S. Roles of L-serine and sphingolipid synthesis in brain development and neuronal survival. *Prog. Lipid Res.* **2008**, *47*, 188–203. [CrossRef]
22. Küllenberg, D.; Taylor, L.A.; Schneider, M.; Massing, U. Health effects of dietary phospholipids. *Lipids Health Dis.* **2012**, *11*. [CrossRef]
23. Farrell, P.M. The development biochemistry of lung phospholipid metabolism. In *Lung Development: Biological and Clinical Perspectives: Biochemistry and Physiology*; Farrell, P.M., Ed.; Academic Press: Cambridge, MA, USA, 1982; Volume 1, pp. 223–235.
24. McJarrow, P.; Schnell, N.; Jumpsen, J.; Clandinin, T. Influence of dietary gangliosides on neonatal brain development. *Nutr. Rev.* **2009**, *67*, 451–463. [CrossRef]
25. Rueda, R. Gangliosides, immunity, infection and inflammation. In *Diet, Immunity and Inflammation*; Calder, P.C., Yaqoob, P., Eds.; Woodhead Publishing Ltd.: Cambridge, UK, 2013.
26. Sonnino, S.; Mauri, L.; Ciampa, M.G.; Prinetti, A. Gangliosides as regulators of cell signaling: Ganglioside-protein interactions or ganglioside-driven membrane organization? *J. Neurochem.* **2013**, *124*, 432–435. [CrossRef] [PubMed]

27. Thurl, S.; Munzert, M.; Henker, J.; Boehm, G.; Muller-Werner, B.; Jelinek, J.; Stahl, B. Variation of human milk oligosaccharides in relation to milk groups and lactational periods. *Br. J. Nutr.* **2010**, *104*, 1261–1271. [CrossRef] [PubMed]
28. McGuire, M.K.; Meehan, C.L.; McGuire, M.A.; Williams, J.E.; Foster, J.; Sellen, D.W.; Kamau-Mbuthia, E.W.; Kamundia, E.W.; Mbugua, S.; Moore, S.E.; et al. What's normal? Oligosaccharide concentrations and profiles in milk produced by healthy women vary geographically. *Am. J. Clin. Nutr.* **2017**, *105*, 1086–1100. [CrossRef] [PubMed]
29. Austin, S.; De Castro, C.A.; Benet, T.; Hou, Y.F.; Sun, H.N.; Thakkar, S.K.; Vinyes-Pares, G.; Zhang, Y.M.; Wang, P.Y. Temporal Change of the Content of 10 Oligosaccharides in the Milk of Chinese Urban Mothers. *Nutrients* **2016**, *8*, 346. [CrossRef] [PubMed]
30. Ma, L.; McJarrow, P.; Mohamed, H.; Liu, X.Y.; Welman, A.; Fong, B.Y. Lactational changes in the human milk oligosaccharide concentration in Chinese and Malaysian mothers' milk. *Int. Dairy J.* **2018**, *87*, 1–10. [CrossRef]
31. Ma, L.; MacGibbon, A.K.H.; Jan Mohamed, H.J.B.; Loy, S.; Rowan, A.; McJarrow, P.; Fong, B.Y. Determination of phospholipid concentrations in breast milk and serum using a high performance liquid chromatography–mass spectrometry–multiple reaction monitoring method. *Int. Dairy J.* **2017**, *71*, 50–59. [CrossRef]
32. Thakkar, S.K.; Giuffrida, F.; Cristina, C.-H.; De Castro, C.A.; Mukherjee, R.; Tran, L.-A.; Steenhout, P.; Lee, L.Y.; Destaillats, F. Dynamics of human milk nutrient composition of women from Singapore with a special focus on lipids. *Am. J. Hum. Biol. Off. J. Hum. Biol. Counc.* **2013**, *25*, 770–779. [CrossRef]
33. Giuffrida, F.; Cruz-Hernandez, C.; Bertschy, E.; Fontannaz, P.; Elmelegy, I.M.; Tavazzi, I.; Marmet, C.; Sanchez-Bridge, B.; Thakkar, S.K.; De Castro, C.A.; et al. Temporal changes of human breast milk lipids of Chinese mothers. *Nutrients* **2016**, *8*, 715. [CrossRef]
34. Ma, L.; MacGibbon, A.K.H.; Jan Mohamed, H.J.B.; Loy, S.; Rowan, A.; McJarrow, P.; Fong, B.Y. Determination of ganglioside concentrations in breast milk and serum from Malaysian mothers using a high performance liquid chromatography-mass spectrometry-multiple reaction monitoring method. *Int. Dairy J.* **2015**, *49*, 62–71. [CrossRef]
35. Ma, L.; Liu, X.; MacGibbon, A.K.H.; Rowan, A.; McJarrow, P.; Fong, B.Y. Lactational changes in concentration and distribution of ganglioside molecular species in human breast milk from Chinese mothers. *Lipids* **2015**, *50*, 1145–1154. [CrossRef]
36. Radwan, H.; Hashim, M.; Obaid, R.S.; Hasan, H.; Naja, F.; Al Ghazal, H.; Mohamed, H.; Rizk, R.; Al Hilali, M.; Rayess, R.; et al. The Molther-Infant Study Cohort (MISC): Methodology, challenges, and baseline characteristics. *PLoS ONE* **2018**, *13*, e0198278. [CrossRef] [PubMed]
37. Svennerholm, L.; Fredman, P. A procedure for the quantitative isolation of brain gangliosides. *Biochim. Biophys. Acta* **1980**, *617*, 97–109. [CrossRef]
38. Fong, B.; Ma, L.; Norris, C. Analysis of phospholipids in infant formulas using high performance liquid chromatography-tandem mass spectrometry. *J. Agric. Food Chem.* **2013**, *61*, 858–865. [CrossRef] [PubMed]
39. Azad, M.B.; Robertson, B.; Atakora, F.; Becker, A.B.; Subbarao, P.; Moraes, T.J.; Mandhane, P.J.; Turvey, S.E.; Lefebvre, D.L.; Sears, M.R.; et al. Human milk oligosaccharide concentrations are associated with multiple fixed and modifiable maternal characteristics, environmental factors, and feeding practices. *J. Nutr.* **2018**, *148*, 1733–1742. [CrossRef] [PubMed]
40. Giuffrida, F.; Elmelegy, I.M.; Thakkar, S.K.; Marmet, C.; Destaillats, F. Longitudinal Evolution of the Concentration of Gangliosides GM3 and GD3 in Human Milk. *Lipids* **2014**, *49*, 997–1004. [CrossRef] [PubMed]
41. Larsson, M.W.; Lind, M.V.; Laursen, R.P.; Yonemitsu, C.; Larnkjaer, A.; Molgaard, C.; Michaelsen, K.F.; Bode, L. Human Milk Oligosaccharide Composition Is Associated with Excessive Weight Gain During Exclusive Breastfeeding-An Explorative Study. *Front. Pediatr.* **2019**, *7*, 297. [CrossRef] [PubMed]
42. Coppa, G.V.; Pierani, P.; Zampini, L.; Carloni, I.; Carlucci, A.; Gabrielli, O. Oligosaccharides in human milk during different phases of lactation. *Acta Paediatr. Suppl.* **1999**, *88*, 89–94. [CrossRef] [PubMed]
43. Bao, Y.W.; Zhu, L.B.; Newburg, D.S. Simultaneous quantification of sialyloligosaccharides from human milk by capillary electrophoresis. *Anal. Biochem.* **2007**, *370*, 206–214. [CrossRef] [PubMed]
44. Hong, Q.T.; Ruhaak, L.R.; Totten, S.M.; Smilowitz, J.T.; German, J.B.; Lebrilla, C.B. Label-Free Absolute Quantitation of Oligosaccharides Using Multiple Reaction Monitoring. *Anal. Chem.* **2014**, *86*, 2640–2647. [CrossRef] [PubMed]

45. Garcia, C.; Millet, V.; Coste, T.C.; Mimoun, M.; Ridet, A.; Antona, C.; Simeoni, U.; Armand, M. French Mothers' Milk Deficient in DHA Contains Phospholipid Species of Potential Interest for Infant Development. *J. Pediatr. Gastroenterol. Nutr.* **2011**, *53*, 206–212. [CrossRef] [PubMed]
46. Claumarchirant, L.; Cilla, A.; Matencio, E.; Sanchez-Siles, L.M.; Castro-Gomez, P.; Fontecha, J.; Alegria, A.; Lagarda, M.J. Addition of milk fat globule membrane as an ingredient of infant formulas for resembling the polar lipids of human milk. *Int. Dairy J.* **2016**, *61*, 228–238. [CrossRef]
47. Bitman, J.; Freed, L.M.; Neville, M.C.; Wood, D.L.; Hamosh, P.; Hamosh, M. Lipid-composition of prepartum human mammary secretion and postpartum milk. *J. Pediatr. Gastroenterol. Nutr.* **1986**, *5*, 608–615. [CrossRef] [PubMed]
48. Sala-Vila, A.; Castellote, A.I.; Rodriguez-Palmero, M.; Campoy, C.; Lopez-Sabater, M.C. Lipid composition in human breast milk from Granada (Spain): Changes during lactation. *Nutrition* **2005**, *21*, 467–473. [CrossRef] [PubMed]
49. Bitman, J.; Wood, L.; Hamosh, M.; Hamosh, P.; Mehta, N.R. Comparison of the lipid composition of breast milk from mothers of term and preterm infants. *Am. J. Clin. Nutr.* **1983**, *38*, 300–312. [CrossRef]
50. Clark, R.M.; Ferris, A.M.; Fey, M.; Brown, P.B.; Hundrieser, K.E.; Jensen, R.G. Changes in the lipids of human milk from 2 to 16 weeks postpartum. *J. Pediatr. Gastroenterol. Nutr.* **1982**, *1*, 311–315. [CrossRef]
51. Zou, X.Q.; Guo, Z.; Huang, J.H.; Jin, Q.Z.; Cheong, L.Z.; Wang, X.G.; Xu, X.B. Human Milk Fat Globules from Different Stages of Lactation: A Lipid Composition Analysis and Microstructure Characterization. *J. Agric. Food Chem.* **2012**, *60*, 7158–7167. [CrossRef]
52. Harzer, G.; Haug, M.; Dieterich, I.; Gentner, P.R. Changing patterns of human milk lipids in the course of the lactation and during the day. *Am. J. Clin. Nutr.* **1983**, *37*, 612–621. [CrossRef]
53. Jensen, R.G. F-Miscellaneous Factors Affecting Composition and Volume of Human and Bovine Milks. In *Handbook of Milk Composition*; Jensen, R.G., Ed.; Academic Press: San Diego, CA, USA, 1995; pp. 237–271. [CrossRef]
54. Da Cunha, J.; da Costa, T.H.M.; Ito, M.K. Influences of maternal dietary intake and suckling on breast milk lipid and fatty acid composition in low-income women from Brasilia, Brazil. *Early Hum. Dev.* **2005**, *81*, 303–311. [CrossRef]
55. Jensen, R.G. Lipids in human milk. *Lipids* **1999**, *34*, 1243–1271. [CrossRef]
56. Jensen, R.G. *Handbook of Milk Composition*; Academic Press: London, UK, 1995.
57. Michalski, M.C.; Briard, V.; Michel, F.; Tasson, F.; Poulain, P. Size distribution of fat globules in human colostrum, breast milk, and infant formula. *J. Dairy Sci.* **2005**, *88*, 1927–1940. [CrossRef]
58. Cohen, B.C.; Shamay, A.; Argov-Argaman, N. Regulation of Lipid Droplet Size in Mammary Epithelial Cells by Remodeling of Membrane Lipid Composition-A Potential Mechanism. *PLoS ONE* **2015**, *10*. [CrossRef] [PubMed]
59. Takamizawa, K.; Iwamori, M.; Mutai, M.; Nagai, Y. Gangliosides of Bovine Buttermilk-Isolation and Characterization of a Novel Monosialoganglioside with a New Branching Structure. *J. Biol. Chem.* **1986**, *261*, 5625–5630.
60. Nakano, T.; Sugawara, M.; Kawakami, H. Sialic acid in human milk: Composition and functions. *Acta Paediatr. Taiwanica* **2001**, *42*, 11–17.
61. Rueda, R.; Puente, R.; Hueso, P.; Maldonado, J.; Gil, A. New data on content and distribution of gangliosides in human milk. *Biol. Chem. Hoppe-Seyler* **1995**, *376*, 723–727. [CrossRef] [PubMed]
62. Pan, X.L.; Izumi, T. Chronological changes in the ganglioside composition of human milk during lactation. *Early Hum. Dev.* **1999**, *55*, 1–8. [CrossRef]

© 2019 by the authors. Licensee MDPI, Basel, Switzerland. This article is an open access article distributed under the terms and conditions of the Creative Commons Attribution (CC BY) license (http://creativecommons.org/licenses/by/4.0/).

Article

The Association between Early Childhood and Later Childhood Sugar-Containing Beverage Intake: A Prospective Cohort Study

Andrea Ziesmann [1], Ruhi Kiflen [1], Vanessa De Rubeis [1], Brendan T. Smith [2,3], Jonathon L. Maguire [3,4,5,6], Catherine S. Birken [3,5,7], Laura N. Anderson [1,7,*]
and on behalf of the TARGet Kids Collaboration

[1] Department of Health Research Methods, Evidence, and Impact, McMaster University, Hamilton, ON L8S 4L8, Canada; ziesmana@mcmaster.ca (A.Z.); ruhikiflen27@gmail.com (R.K.); derubevg@mcmaster.ca (V.D.R.)
[2] Department of Health Promotion, Chronic Disease and Injury Prevention, Public Health Ontario, Toronto, ON M5G 1V2, Canada; Brendan.Smith@oahpp.ca
[3] Dalla Lana School of Public Health, University of Toronto, Toronto, ON M5T 3M7, Canada; jonathon.maguire@utoronto.ca (J.L.M.); catherine.birken@sickkids.ca (C.S.B.)
[4] Applied Health Research Centre of the Li Ka Shing Knowledge Institute of St. Michael's Hospital, University of Toronto, Toronto, ON M5G 1B1, Canada
[5] Department of Nutritional Sciences, Faculty of Medicine, University of Toronto, Toronto, Ontario M5S 1A8, Canada
[6] Department of Pediatrics, St. Michael's Hospital, Toronto, ON M5C 2T2, Canada
[7] Division of Child Health Evaluative Sciences (CHES), Sick Kids Research Institute, Toronto, ON M5G 0A4, Canada
* Correspondence: ln.anderson@mcmaster.ca; Tel.: +19-05-525-9140 (ext. 21725)

Received: 30 July 2019; Accepted: 27 September 2019; Published: 1 October 2019

Abstract: Sugar-containing beverages (SCBs) are a major source of sugar intake in children. Early life intake of SCBs may be a strong predictor of SCB intake later in life. The primary objective of this study was to evaluate if SCB intake (defined as 100% fruit juice, soda, and sweetened drinks) in early childhood (≤2.5 years of age) was associated with SCB intake in later childhood (5–9 years of age). A prospective cohort study was conducted using data from the TARGet Kids! primary care practice network ($n = 999$). Typical daily SCB intake was measured by parent-completed questionnaires. Odds ratios (OR) and 95% confidence intervals (CI) were estimated using logistic regression. A total of 43% of children consumed ≥0.5 cups/day of SCBs at ≤2.5 years and this increased to 64% by 5–9 years. Daily SCB intake, compared to no daily intake, at ≤2.5 years was significantly associated with SCB intake at 5–9 years (adjusted OR: 4.03; 95% CI: 2.92–5.55) and this association was much stronger for soda/sweetened drinks (adjusted OR: 12.83; 95% CI: 4.98, 33.0) than 100% fruit juice (OR: 3.61; 95% CI: 2.63–4.95). Other early life risk factors for SCB intake at 5–9 years were presence of older siblings, low household income, and shorter breastfeeding duration. Daily intake of SCBs in early childhood was strongly associated with greater SCB intake in later childhood. Early life may be an important period to target for population prevention strategies.

Keywords: sugars; fruit juices; life-course epidemiology; infant; child

1. Introduction

Sugar intake in children has been associated with increased cardiometabolic risk factors, including obesity [1,2], high blood pressure, and dyslipidemia [3–8] and dental caries [9,10]. These adverse conditions in childhood are associated with disease outcomes in later adulthood, such as type 2 diabetes and cardiovascular disease [11–13]. Further, sugar intake across the life-course is associated with similar adult disease outcomes [2,14]. In 2015, fruit juice and sweetened fruit drinks were among the top ten most common sources of sugar intake for children ages two to eight in Canada, amounting to approximately 11.6% and 3.5% of their total sugar intake, respectively [15]. In the US, it was found that children and adolescents derive 10%–15% of total calories from sugar-containing beverages [16]. Recent estimates suggest that the percentage of U.S. children 19–23 months of age consuming any 100% fruit juice is about 50%, and 38% consumed any sugar-sweetened beverages [17].

Sugar-sweetened beverages are frequently defined as liquids that are sweetened with various forms of added sugars (e.g., sodas, fruit drinks) [18]. Fruit juices which contain 100% juice are often excluded from the definition of sugar-sweetened beverages, as they are sometimes perceived as a healthier beverage option. However, 100% fruit juice also contains a high sugar content and lacks dietary fibre. The World Health Organization defines free sugars as "all monosaccharides and disaccharides added to foods and beverages by the manufacturer, cook or consumer, as well as sugars that are naturally present in honey, syrups, fruit juices and fruit juice concentrates", and recommends that both adults and children reduce their intake of free sugars to less than 10% of total energy intake [19]. The American Academy of Pediatrics recommends limiting children's fruit juice intake [20], which is consistent with the newly released 2019 Canada Food Guide, which also suggests limiting, if not replacing sugar-containing drinks with water [21]. The term sugar-containing beverages (SCBs) is used here to include the following beverages containing free sugars: Soda or pop, sweetened drinks, and 100% fruit juice.

Dietary patterns in childhood have been shown to track into adulthood [22,23], thus the early introduction of sugar-sweetened beverages may increase intake across the life-course [24–26]. Studies conducted in the United States [26] and Norway [25] reported that intake of soda and sweetened beverages before two years of age is strongly associated with increased intake in later childhood at age six and seven, respectively. However, neither study included 100% fruit juice, and only one study controlled for potential confounders [26]. Further, a recent systematic review of risk factors associated with sugar beverage intake in children <6 years of age also did not include 100% fruit juice intake in its definition of SCB [27]. This systematic review identified 17 determinants of sugar beverages that support an association with intake of SCB in early childhood, including child age, parental knowledge, household rules, screen time, snack consumption, formula fed, lower parental socioeconomic status, and early introduction of solids [10,27]. Few studies have prospectively evaluated SCB intake during infancy and risk factors in early life that may predict SCB intake in later childhood.

The primary objective of this study was to evaluate if SCB intake in early childhood (≤2.5 years of age) was associated with increased SCB intake in later childhood (5–9 years of age). The secondary objective was to evaluate if the association between early life SCB intake and later childhood intake was independent of other established SCB risk factors.

2. Materials and Methods

2.1. Study Design

A prospective cohort study was conducted among children in The Applied Research Group for Kids (TARGet Kids!) primary care practice-based research network. The TARGet Kids! network recruits children <6 years of age at scheduled well-child visits and follows them prospectively through childhood [28]. Children were recruited between 2008 and 2017 from primary care pediatricians or family practice clinics in the Greater Toronto Area, Canada. Children were excluded at enrollment if they had associated health conditions affecting growth (i.e., failure to thrive, cystic fibrosis), severe developmental delay, or other chronic conditions except for asthma or if the families were not English speaking. Informed written consent was obtained from parents and TARGet Kids! is approved by the research ethics boards at the Hospital for Sick Children (#1000012436) and St. Michael's Hospital, Toronto, Canada.

2.2. Study Population and Sample Size

For this study, children were included if they had at least one visit before or at 2.5 years of age (defined as early childhood) and a follow-up at 5–9 years of age (defined as later childhood) from 2008 to 2017. A total of 5478 children were identified in the TARGet Kids! database who had a visit at ≤2.5 years of age, 3574 children were excluded because they were recruited recently and would not have a minimum of 2.5 year follow-up time (i.e., these children had not yet reached 5 years of age). Of the remaining 1904 children who had a visit at ≤2.5 years of age there were 1267 with a follow-up visit between 5 and 9 years of age. However, 268 children were missing data on SCB intake at both time points (≤2.5 years of age, and 5–9 years of age) and therefore were removed, resulting in a total sample size of 999 children (Figure 1).

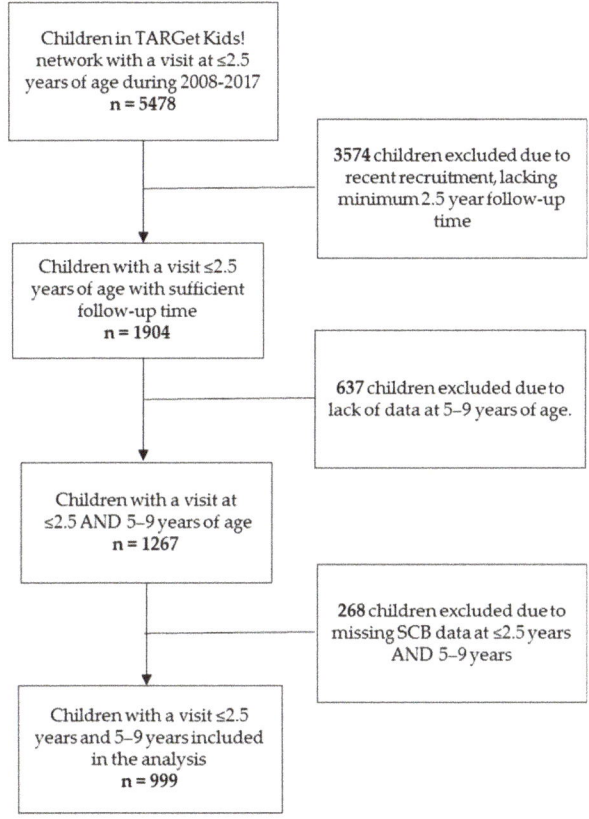

Figure 1. Participant flow diagram.

2.3. Measurement of Sugar-Containing Beverage Intake

At each TARGet Kids! visit, parents completed an age-specific nutrition and health questionnaire [28]. To determine a child's sugar beverage intake, parents were asked to "circle how many cups of each drink your child has currently in a typical day, if none then circle 0 (1 cup = 8 ounces = 250 mL)" and the response options were: 0, 1/2, 1, 2, 3, 4, and 5+. The list of beverages included "100% juice (apple, orange etc.)", "sweetened drinks (Kool Aid, Sunny D, etc.)" and "soda or pop". In this analysis, three primary exposure variables were investigated: (1) 100% fruit juice, (2) soda and sweetened drinks (combined sweetened drinks with soda/pop), and (3) total sugar-containing beverages (SCBs), which includes all three beverages

(100% fruit juice, soda, and sweetened drinks). Similar to previous studies in this age group [26] all variables were dichotomized into 0 cups/day and ≥0.5 cups/day since a high proportion of respondents indicated no intake.

2.4. Other Variables

Based on a recent systematic review of risk factors for sugar beverage intake in children <6 years of age [27], we identified variables of interest associated with SCB consumption, including age, sex, mother's education, mother's ethnicity, family income, number of siblings, zBMI, weekday free time play, parental BMI, and length of breastfeeding. Child and parent BMI were measured through the measurement of height and weight by trained research assistants using standard methods [29]. Child overweight and obesity categories were defined using BMI z-scores calculated using the WHO growth reference standard which is recommended for child growth monitoring in Canada [30]. All other variables were captured through questionnaires completed by parents.

2.5. Statistical Analysis

All statistical analyses were conducted using SAS software 9.4 (Raleigh, NC, USA). Descriptive analyses for all variables were conducted, including the frequency and percent, and when applicable, the mean and standard deviation. To assess whether the intake of SCBs in early childhood (ages ≤2.5) was associated with intake in later childhood (ages 5–9), multivariable logistic regression was used to estimate odds ratios (OR) and corresponding 95% confidence intervals (CI). Multivariable models included all potential risk factors identified a priori based on a previous systematic review. Results are presented for three dependent variables at 5–9 years of age: (1) total SCB intake; (2) sugar and sweetened drink; and (3) 100% fruit juice. All variables were adjusted for as presented in Table 1. Multicollinearity was assessed by inspection of the variance of inflation factors (VIFs). None of the VIFs were greater than 2.5, suggesting no concerns with multicollinearity. To account for missing data in the covariates, multiple imputation was performed using PROC MI and PROC MIANALYZE on SAS. Participants who were missing data on both the primary exposures and outcome variables were excluded from the study thus these variables were not imputed, only data on covariates was imputed. A minimum of 20 datasets were imputed for each analysis. All variables listed in Table 2 were included in the imputation model, including the exposure (≤2.5 years of age) and outcome (5–9 years of age). All statistical tests were 2-sided and a significance level of less than 0.05 was specified.

Table 1. Baseline characteristics of children from the TARGet Kids cohort with a visit at both ≤2.5 years old and 5–9 years old (*n* = 999).

Characteristic	n	%
Sex		
Male	533	53.4
Female	466	46.7
Maternal Education		
College/University	917	91.8
High school or Less	56	5.6
Missing	26	2.6
Number of Siblings		
0	442	44.2
1	378	37.8
≥2	107	10.7
Missing	72	7.2
Family Income [1]		
<$49,999	65	6.5
$50,000-99,999	161	16.1
$100,000-149,999	173	17.3
≥$150,000	555	55.6

Table 1. Cont.

Characteristic	n	%
Missing	45	4.5
Child zBMI [2]		
Normal and underweight (zBMI ≤ 1.0)	790	79.1
Risk of overweight (1 > zBMI ≤ 2.0)	141	14.1
Overweight and obesity (zBMI > 2)	34	3.4
Missing	34	3.4
Maternal Ethnicity		
European	729	73.0
East, South, or South-East Asian	135	13.5
Other	102	10.2
Missing	33	3.3
	Mean	SD
Age (months)	18.1	4.8
Parent BMI (kg/m^2) [3]	24.6	4.6
Typical Weekday Free Play (minutes)	58.7	56.5
Child zBMI	0.09	1.07
Breastfeeding duration (months)	11.3	5.7

[1] Canadian dollars. [2] Based on WHO Growth Standards. [3] Responses were from either mothers (84%) or fathers (16%).

Table 2. Distribution of sugar-containing beverage consumption in children at ≤2.5 years of age and 5–9 years of age among children in TARGet Kids! in Ontario, Canada (n = 999).

Sugar-Containing Beverages	≤2.5 Years of Age N (%)	5–9 Years of Age N (%)
100% Fruit Juice [1]		
0 cups/day	564 (57.1)	368 (37.1)
≥0.5 cups/day	424 (42.9)	623 (62.9)
Soda and Sweetened Drinks [2]		
0 cups/day	934 (96.8)	829 (93.9)
≥0.5 cups/day	31 (3.2)	54 (6.1)
Total SCB consumption		
0 cups/day	570 (57.1)	361 (36.1)
≥0.5 cups/day	429 (42.9)	638 (63.9)

[1] 11 missing 100% fruit juice intake at ≤2.5 years of age and 8 missing 100% fruit juice intake at 5–9 years of age; [2] 34 missing soda and sweetened drink intake at ≤2.5 years of age and 116 missing soda and sweetened drink intake at 5–9 years of age.

3. Results

A total of 999 children had measures of sugar-containing beverage intake between 0 and 2.5 years of age and 5 and 9 years of age (Figure 1). Baseline demographic characteristics are described in Table 1. Approximately 46.7% of children were female, 73.0% were of European ethnicity, and the mean age at the first visit was 18.1 months. There were 141 children (14.1%) who met the WHO defined cut-points for 'risk of overweight' and 34 (3.4%) who were 'overweight or obese' based on their zBMI at ≤2.5 years of age.

3.1. Description of SCB Consumption at ≤2.5 Years and 5–9 Years

The proportion of children that typically drank ≥0.5 cups/day of 100% fruit juice was 42.9% at ≤2.5 years of age, which increased to 62.9% at ages 5–9 (Table 2). At ≤2.5 years of age, only 3.2% of children typically consumed ≥0.5 cups/day of soda and sweetened drinks, which increased to 6.1% of children at 5–9 years of age. At ≤2.5 years, 57.1% of children consumed 0 cups/day of SCB, which decreased to 36.1% at ages 5–9. This represents a 21% decrease in those who consumed 0 cups/day of SCB. At ≤2.5 years of age, the majority of children (51.6%) who consumed soda and sweetened drinks, drank 0 to 1 cups (Figure 2a), whereas at 5–9 years of age, almost half of children (48.1%) drank 1 to 2 cups of soda and sweetened drinks (Figure 2b). Among children who consumed any SCB, 49.2% consumed 0 to 1 cups at ≤2.5 years of age (Figure 2a), and 42.5% consumed this amount at 5–9 years of age (Figure 2b).

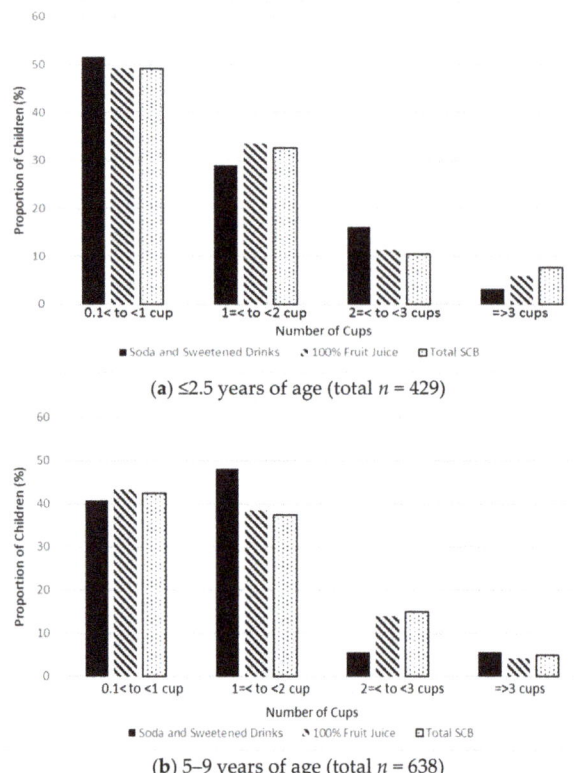

Figure 2. Distribution of sugar-containing beverage (SCB) intake among children in the TARGet Kids! Research Network among children who "ever consumed" SCBs.

3.2. SCB Intake at ≤2.5 Years Association with SCB Intake at 5–9 Years

A strong and significant association was observed between SCB intake at ≤2.5 years of age and later in childhood (5–9 years) and this persisted in the fully adjusted model after controlling for potential SCB risk factors (Table 3). For total SCB intake, children who consumed ≥0.5 cups/day of SCBs at ≤2.5 years old had 4.03 times greater odds of consuming SCBs in later childhood (fully adjusted OR: 4.03; 95% CI: 2.92–5.55) (Table 3). Children who had consumed soda and sweetened drinks at ≤2.5 years of age had 16.91 times greater odds of consuming soda and sweetened drinks at ages 5–9 (fully adjusted OR:16.91; 95% CI: 3.68, 77,69) (Table 3). Those who consumed ≥0.5 cups/day of 100% fruit juice at ≤2.5 years of age had 3.6 times greater odds of consuming fruit juice at ages 5–9 (fully adjusted OR: 3.61; 95% CI: 2.63–4.95) (Table 3). The associations for 100% fruit juice and total SCB changed minimally in the models adjusted only for age and sex versus the fully adjusted models (Table 3). However, the association for soda and sweetened drinks changed substantially with an OR of 23.48 (95% CI: 9.55, 57.76) for the model adjusted only for age and sex, and this decreased to 16.91 (95% CI: 3.68, 77.69) for the fully adjusted model (Table 3).

Table 3. Odds ratio estimates for total sugar-containing beverages (SCB), soda and sweetened drinks, and fruit juice intake at 5–9 years of age, for children with a visit at ≤2.5 years of age and 5–9 years in the TARGet Kids! Research Network (n = 999).

	Minimally Adjusted Model Total SCB Intake at 5–9 Years of Age OR [1]	95% CI	Fully Adjusted Model Total SCB Intake at 5–9 Years of Age OR [2]	95% CI	Minimally Adjusted Model Soda and Sweetened Drink at 5–9 Years of Age OR [1]	95% CI	Fully Adjusted Model Soda and Sweetened drink at 5–9 Years of Age OR [2]	95% CI	Minimally Adjusted Model Fruit Juice Intake at 5–9 Years of Age OR [1]	95% CI	Fully Adjusted Model Fruit Juice Intake at 5–9 Years of Age OR [2]	95% CI
Total SCB at ≤2.5 years												
0 cups/day	1.00		1.00									
≥0.5 cups/day	4.33	3.18, 5.91	4.03	2.92, 5.55								
Soda & sweetened drinks at ≤2.5 years												
0 cups/day					1.00		1.00					
≥0.5 cups/day					23.48	9.55, 57.76	16.91	3.68, 77.69				
Fruit Juice at age ≤2.5 years												
0 cups/day									1.00		1.00	
≥0.5 cups/day									3.85	2.83, 5.23	3.61	2.63, 4.95
Age (months)	1.02	0.99, 1.05	0.98	0.95, 1.01	0.99	0.94, 1.06	0.98	0.87, 1.09	0.98	0.95, 1.00	0.98	0.95, 1.01
Sex												
Male	1.00		1.00		1.00		1.00		1.00		1.00	
Female	1.33	1.01, 1.75	0.77	0.58, 1.02	0.67	0.36, 1.23	0.43	0.65, 8.21	0.78	0.59, 1.02	0.80	0.60, 1.05
Education (mother)												
College/University			1.00				1.00				1.00	
Highschool or less			1.90	0.89, 4.05			2.31	0.69, 4.92			1.26	0.63, 2.52
Sibling												
0			1.00				1.00				1.00	
1			1.10	0.80, 1.50			3.72	1.20, 11.47			1.15	0.85, 1.56
2+			1.14	0.71, 1.82			2.55	0.54, 11.89			1.08	0.67, 1.75
Ethnicity (mother)												
European			1.00				1.00				1.00	
East, South or South-East Asian			1.34	0.86, 2.08			7.39	2.47, 22.11			1.37	0.88, 2.12
Other			0.93	0.58, 1.50			1.12	0.43, 2.90			0.85	0.53, 1.35
Household income												
<49,999			1.64	0.79, 3.39			11.96	2.86, 50.09			1.67	0.82, 3.40
50,000–99,999			1.33	0.89, 1.80			1.16	0.38, 4.79			1.34	0.90, 2.00
100,000–149,999			1.22	0.82, 1.80			1.46	0.41, 5.20			1.19	0.82, 1.76
≥150,000			1.00				1.00				1.00	
Child zBMI (per 1-unit increase)			1.04	0.91, 1.19			0.79	0.49, 1.27			1.04	0.91, 1.20
Breastfeeding duration (per 2 months increase)			0.94	0.89, 0.99			0.89	0.82, 0.97			0.95	0.91, 1.01
Parental BMI [3] (per 5-unit increase)			1.12	0.93, 1.34			1.09	1.01, 1.18			1.12	0.94, 1.33
Typical weekday free play (per 30 min increase)			1.04	0.96, 1.28			1.00	1.00, 1.01			1.03	0.95, 1.13

[1] Adjusted for age and sex only. [2] Adjusted for all listed variables. [3] 84% of responses were collected from the mother and 16% were from fathers. Bolded text within the table represent statistically significant values ($p < 0.05$).

3.3. Other Early Childhood Risk Factors

Other possible risk factors for SCB consumption in later childhood (5–9 years of age) are also presented in Table 3. For total SCB intake, breastfeeding decreased odds of consuming SCBs in later childhood (fully adjusted OR: 0.94; 95% CI: 0.89, 0.99 per two months longer breast feeding) (Table 3). Having a household income of <$49,999 (fully adjusted OR: 1.64; 95% CI: 0.79, 3.39), in comparison to a household income ≥$150,000, was suggestive of increased odds of consuming SCBs at 5–9 years, although not statistically significant. Significant risk factors for increased odds of soda and sweetened drink intake included children with one sibling in comparison to having no siblings (fully adjusted OR: 3.72; 95% CI, 1.20, 11.47); a mother of East, South, or South-East Asian ethnicity in comparison to European (fully adjusted OR: 7.39; 95% CI: 2.47, 22.11); a household income of <$49,999 (fully adjusted OR: 11.96; 95% CI: 2.86, 50.09) in comparison to a household income ≥$150,000; parental BMI (per 5 kg/m^2) increase (OR: 1.09; 95% CI: 1.01, 1.18). For every two-month increase of breastfeeding duration there was a 0.89 reduction of odds of consuming soda and sweetened drink consumption when adjusting for all potential risk factors (95% CI: 0.82, 0.97) (Table 3). No variables were statistically significant risk factors for the consumption of 100% fruit juice at 5–9 years of age (Table 3), but, consistent with total SCB consumption, the fully adjusted ORs were greater than 1.0 for household income <$49,999 (fully adjusted OR: 1.67; 95% CI: 0.82, 3.40), and having a mother of East, South, or South-East Asian ethnicity (fully adjusted OR: 1.37; 95% CI: 0.88, 2.12).

4. Discussion

In this study 100% fruit juice, soda, and sweetened drinks, and total SCB intake at ≤2.5 years of age, were each strongly associated with higher SCB intake at 5–9 years of age. These associations were observed after adjustment for other potential early life risk factors of SCB intake. The findings from this study contribute evidence to the current literature supporting the association of SCB intake in later childhood when SCBs are introduced in early childhood. These results are consistent with two previous longitudinal cohort studies from the United States [26] and Norway [25]. The U.S. study included 1333 children with data at <1 and at 6 years of age. They found that children with intake of sugar-sweetened beverages at any time during infancy, compared to no intake during infancy, had 2.22 times higher odds for consuming sugar-sweetened beverages at least once per day at 6 years of age (OR: 2.22; 95% CI: 1.59-3.10) [26]. The Norwegian study included 9025 children participating at three time points (18 months, 36 months, and 7 years). They observed that children with low, medium, and high frequency of sugar-sweetened beverage intake at 18 months continued to drink sugar-sweetened beverage just as often at 36 months and 7 years of age [25]. However, these studies only examined the intake of soda and sweetened drinks and excluded 100% fruit juice. To the best of our knowledge, no previous studies have investigated 100% fruit juice in early childhood as a risk factor for later childhood SCB consumption. Understanding the impact of early childhood consumption of SCBs is imperative, as dietary patterns and behaviors have been found to influence behaviors throughout the life-course [31–33].

Following a recent systematic review that identified risk factors for SCB intake in children [27], we prospectively evaluated the associations between potential risk factors during early childhood (≤2.5 years of age) and later childhood (5–9 years of age) SCB intake, including 100% fruit juice, soda, and sweetened drinks, and total SCB intake (which included all three beverages). No statistically significant risk factors were identified for fruit juice intake and all associations evaluated were close to null, except for early life fruit juice intake. However, children were at greater odds of consuming soda and sweetened drinks if they had one sibling (compared to none), if maternal ethnicity was East, South, or South-East Asian (compared to European), if the household income was <$49,999 (compared to ≥$150,000), and increasing parental BMI, with ORs suggesting 1.09- to 11.96-fold increased odds for each of these previously identified risk factors. Child BMI was not associated with any SCB intake, which may reflect the young age of our study population. However, parent BMI was significantly associated with increased odds of soda and sweetened drink intake among children, and although not statistically significant, the ORs were in the same direction for total SCB and fruit juice intake. Results of the analyses for total SCBs (combining fruit juice, soda, and sweetened drinks) only identified one significant risk factor, breastfeeding duration, and this is likely explained by the fact that 100% fruit juice was the major contributor to total SCBs.

The proportion of children who consumed SCBs in this study population may appear low, but is somewhat comparable to national data from the 2015 Canadian Community Health Survey (CCHS). In the 2015 CCHS for children 1–8 years of age [34], 46.1% consumed fruit juice, 6.3% of children consumed regular soft drinks, and 14.4% consumed fruit drinks. The 2017 American Academy of Pediatric recommendations for children's juice intake indicate that infants <12 months of age should not be introduced to juice, children aged 1–3 should be limited to 4 oz. (0.5 cup) daily, children aged 4–6 should only consume 4–6 oz. daily, and children aged 7–18 should be limited to 8 oz. daily [20]. At ≤2.5 years of age, 21.9% of children from this sample were drinking more than the recommended limit of 4 oz. daily, and at 5–9 years of age 35.3% were drinking more than the recommended 4–6 oz. per day. The new 2019 Canada Food Guide suggests limiting, if not replacing sugar-containing drinks (e.g., fruit juice, soft drinks, sports drinks, fruit flavored drinks, and punches) with water, as there are many benefits to this practice [21]. These guidelines are general and not specific to age groups, perhaps to encourage the same healthy lifestyles throughout the life-course.

A potential limitation of this study is measurement error because of parent-reported questionnaires used to measure child SCB intake. However, studies have found that parent reported SCB is a valid measure of true SCB intake in children [35,36]. It is possible that parents misclassified 100% fruit juices as sweetened drinks or vice versa, however, regulations state that juices that are not 100% fruit juice must be identified as a juice "drink", "beverage", or "cocktail", and thus should have been classified as a sweetened drink, not as 100% fruit juice [37,38]. Further, we did not collect data on all possible SCBs. For example, data were not available on sweetened milks or energy drinks. Infant formula was also not included within the analysis. Since a high proportion of respondents indicated no intake, all SCB variables were dichotomized into 0 cups/day and ≥0.5 cups/day for all analyses, which may contribute to a loss of power. Another potential limitation of this study is the missing data for SCB intake at ≤2.5 years and 5–9 years, and all covariates. To address this limitation, we applied a multiple imputation approach to account for the missing data and limit potential biases associated with missing data. In addition, the study had a follow-up response rate of 52%. The low follow-up rates contributed to a reduction in the sample size leading to lower certainty in the ability to make strong conclusions. Lastly, the sample of the current study may not be representative of all Canadian children, since TARGet Kids! is an urban primary care network with relatively high family income and education levels. Although it may be assumed that SCB consumption during early childhood is mainly influenced by parents, it is possible that other factors, such as school environment, may also play a role in consumption patterns, especially in older children. We were unable to evaluate these factors, but future studies may be able to further understand the role these factors play in defining the consumption patterns of SCBs in children. Strengths of this study include the prospective nature of the study design with comprehensive measurements of potential SCB risk factors in early life, including 100% fruit juice. Few studies have investigated SCB intake and risk factors in early childhood. However, this may be a sensitive period in development from both a biologic sugar exposure and behavioural development point of view. Literature shows that taste preferences are developed in the prenatal, neonatal, infancy, and early childhood stages of life [39,40]. Early introduction to sugary beverages is only one of the many biological, social, and environmental factors that continue to be influence taste preferences throughout the lifespan.

5. Conclusions

The results of this study suggest that SCB intake in early childhood is strongly associated with SCB intake in later childhood and this association persists even after adjustment for many other possible predictors of SCB intake. Future studies are needed to understand longer-term associations and to evaluate if interventions targeting SCB reduction in early life reduce both SCB intake and adverse health outcomes in later childhood, adolescence and adulthood.

Author Contributions: Conceptualization, A.Z., R.K., B.T.S., J.L.M., C.S.B. and L.N.A.; Formal analysis, A.Z., V.D.R. and L.N.A.; Writing—Original draft, A.Z. and R.K.; Writing—Review & editing, A.Z., R.K., V.D.R., B.T.S., J.L.M., C.S.B. and L.N.A.

Funding: This research was funded by the Canadian Institutes of Health Research.

Acknowledgments: We thank all the participating families for their time and involvement in TARGet Kids! and are grateful to all practitioners who are currently involved in the TARGet Kids! practice-based research network. *TARGet Kids!* Collaborators—Co-Leads: Catherine S. Birken, Jonathon L. Maguire; Advisory Committee: Ronald Cohn, Eddy Lau, Andreas Laupacis, Patricia C. Parkin, Michael Salter, Peter Szatmari, Shannon Weir; Science Review and Management Committees: Laura N. Anderson, Cornelia M. Borkhoff, Charles Keown-Stoneman, Christine Kowal, Dalah Mason; Site Investigators: Murtala Abdurrahman, Kelly Anderson, Gordon Arbess, Jillian Baker, Tony Barozzino, Sylvie Bergeron, Dimple Bhagat, Gary Bloch, Joey Bonifacio, Ashna Bowry, Caroline Calpin, Douglas Campbell, Sohail Cheema, Elaine Cheng, Brian Chisamore, Evelyn Constantin, Karoon Danayan, Paul Das, Mary Beth Derocher, Anh Do, Kathleen Doukas, Anne Egger, Allison Farber, Amy Freedman, Sloane Freeman, Sharon Gazeley, Charlie Guiang, Dan Ha, Curtis Handford, Laura Hanson, Leah Harrington, Sheila Jacobson, Lukasz Jagiello, Gwen Jansz, Paul Kadar, Florence Kim, Tara Kiran, Holly Knowles, Bruce Kwok, Sheila Lakhoo, Margarita Lam-Antoniades, Eddy Lau, Denis Leduc, Fok-Han Leung, Alan Li, Patricia Li, Jessica Malach, Roy Male, Vashti Mascoll, Aleks Meret, Elise Mok, Rosemary Moodie, Maya Nader, Katherine Nash, Sharon Naymark, James Owen, Michael Peer, Kifi Pena, Marty Perlmutar, Navindra Persaud, Andrew Pinto, Michelle Porepa, Vikky Qi, Nasreen Ramji, Noor Ramji, Danyaal Raza, Alana Rosenthal, Katherine Rouleau, Caroline Ruderman, Janet Saunderson, Vanna Schiralli, Michael Sgro, Hafiz Shuja, Susan Shepherd, Barbara Smiltnieks, Cinntha Srikanthan, Carolyn Taylor, Stephen Treherne, Suzanne Turner, Fatima Uddin, Meta van den Heuvel, Joanne Vaughan, Thea Weisdorf, Sheila Wijayasinghe, Peter Wong, John Yaremko, Ethel Ying, Elizabeth Young, Michael Zajdman; Research Team: Farnaz Bazeghi, Vincent Bouchard, Marivic Bustos, Charmaine Camacho, Dharma Dalwadi, Christine Koroshegyi, Tarandeep Malhi, Sharon Thadani, Julia Thompson, Laurie Thompson; Project Team: Mary Aglipay, Imaan Bayoumi, Sarah Carsley, Katherine Cost, Karen Eny, Theresa Kim, Laura Kinlin, Jessica Omand, Shelley Vanderhout, Leigh Vanderloo; Applied Health Research Centre: Christopher Allen, Bryan Boodhoo, Olivia Chan, David W.H. Dai, Judith Hall, Peter Juni, Gerald Lebovic, Karen Pope, Kevin Thorpe; Mount Sinai Services Laboratory: Rita Kandel, Michelle Rodrigues, Hilde Vandenberghe.

Conflicts of Interest: J.L.M. received an unrestricted research grant for a completed investigator-initiated study from the Dairy Farmers of Canada (2011–2012) and Ddrops provided non-financial support (vitamin D supplements) for an investigator-initiated study on vitamin D and respiratory tract infections (2011–2015). The other authors declare no conflict of interest.

References

1. Frantsve-Hawley, J.; Bader, J.D.; Welsh, J.A.; Wright, J.T. A systematic review of the association between consumption of sugar-containing beverages and excess weight gain among children under age 12. *J. Public Health Dent.* **2017**, *77* (Suppl. S1), S43–S66. [CrossRef] [PubMed]
2. Hu, F.B. Resolved: There is sufficient scientific evidence that decreasing sugar-sweetened beverage consumption will reduce the prevalence of obesity and obesity-related diseases. *Obes. Rev.* **2013**, *14*, 606–619. [CrossRef]
3. Vos, M.B.; Kaar, J.L.; Welsh, J.A.; Van Horn, L.V.; Feig, D.I.; Anderson, C.A.M.; Patel, M.J.; Cruz Munos, J.; Krebs, N.F.; Xanthakos, S.A.; et al. Added Sugars and Cardiovascular Disease Risk in Children: A Scientific Statement From the American Heart Association. *Circulation* **2017**, *135*, e1017–e1034. [CrossRef] [PubMed]
4. Fidler Mis, N.; Braegger, C.; Bronsky, J.; Campoy, C.; Domellöf, M.; Embleton, N.D.; Hojsak, I.; Hulst, J.; Indrio, F.; Lapillonne, A.; et al. Sugar in Infants, Children and Adolescents: A Position Paper of the European Society for Paediatric Gastroenterology, Hepatology and Nutrition Committee on Nutrition. *J. Pediatr. Gastroenterol. Nutr.* **2017**, *65*, 681–696. [CrossRef] [PubMed]
5. Ambrosini, G.L.; Oddy, W.H.; Huang, R.C.; Mori, T.A.; Beilin, L.J.; Jebb, S.A. Prospective associations between sugar-sweetened beverage intakes and cardiometabolic risk factors in adolescents. *Am. J. Clin. Nutr.* **2013**, *98*, 327–334. [CrossRef]
6. Kell, K.P.; Cardel, M.I.; Bohan Brown, M.M.; Fernández, J.R. Added sugars in the diet are positively associated with diastolic blood pressure and triglycerides in children. *Am. J. Clin. Nutr.* **2014**, *100*, 46–52. [CrossRef] [PubMed]
7. Kosova, E.C.; Auinger, P.; Bremer, A.A. The relationships between sugar-sweetened beverage intake and cardiometabolic markers in young children. *J. Acad. Nutr. Diet.* **2013**, *113*, 219–227. [CrossRef]

8. Wang, J.W.; Mark, S.; Henderson, M.; O'Loughlin, J.; Tremblay, A.; Wortman, J.; Paradis, G.; Gray-Donald, K. Adiposity and glucose intolerance exacerbate components of metabolic syndrome in children consuming sugar-sweetened beverages: QUALITY cohort study. *Pediatr. Obes.* **2013**, *8*, 284–293. [CrossRef]
9. Moynihan, P.J.; Kelly, S.A. Effect on caries of restricting sugars intake: Systematic review to inform WHO guidelines. *J. Dent. Res.* **2014**, *93*, 8–18. [CrossRef]
10. Scientific Advisory Committee On Nutrition. *Carbohydrates and Health*; The Stationary Office: London, UK, 2015; pp. 1–384.
11. Malik, V.S.; Popkin, B.M.; Bray, G.A.; Després, J.P.; Willett, W.C.; Hu, F.B. Sugar-sweetened beverages and risk of metabolic syndrome and type 2 diabetes: A meta-analysis. *Diabetes Care* **2010**, *33*, 2477–2483. [CrossRef]
12. Sahoo, K.; Sahoo, B.; Choudhury, A.K.; Sofi, N.Y.; Kumar, R.; Bhadoria, A.S. Childhood obesity: Causes and consequences. *J. Fam. Med. Prim. Care* **2015**, *4*, 187–192. [CrossRef]
13. Raitakari, O.T.; Juonala, M.; Kahonen, M.; Taittonen, L.; Laitinen, T.; Maki-Torkko, N.; Jarvisalo, M.J.; Uhari, M.; Jokinen, E.; Ronnemaa, T.; et al. Cardiovascular risk factors in childhood and carotid artery intima-media thickness in adulthood: The Cardiovascular Risk in Young Finns Study. *JAMA* **2003**, *290*, 2277–2283. [CrossRef] [PubMed]
14. Nissinen, K.; Mikkilä, V.; Männistö, S.; Lahti-Koski, M.; Räsänen, L.; Viikari, J.; Raitakari, O.T. Sweets and sugar-sweetened soft drink intake in childhood in relation to adult BMI and overweight. The Cardiovascular Risk in Young Finns Study. *Public Health Nutr.* **2009**, *12*, 2018–2026. [CrossRef] [PubMed]
15. Langlois, K.; Garriguet, D.; Gonzalez, A.; Sinclair, S.; Colapinto, C.K. Change in total sugars consumption among Canadian children and adults. *Health Rep.* **2019**, *30*, 10–19.
16. Wang, Y.C.; Bleich, S.N.; Gortmaker, S.L. Increasing caloric contribution from sugar-sweetened beverages and 100% fruit juices among US children and adolescents, 1988-2004. *Pediatrics* **2008**, *121*, e1604–e1614. [CrossRef]
17. Hamner, H.C.; Perrine, C.G.; Gupta, P.M.; Herrick, K.A.; Cogswell, M.E. Food Consumption Patterns among U.S. Children from Birth to 23 Months of Age, 2009-2014. *Nutrients* **2017**, *9*, 942. [CrossRef] [PubMed]
18. U.S. Department of Health and Human Services. 2015–2020 Dietary Guidelines for Americans. Available online: https://health.gov/dietaryguidelines/2015/guidelines/ (accessed on 10 June 2019).
19. World Health Organization. *Guideline: Sugar Intake for Adults and Children*; World Health Organization: Geneva, Switzerland, 2014.
20. Heyman, M.B.; Abrams, S.A.; SECTION ON GASTROENTEROLOGY, HEPATOLOGY, AND NUTRITION, COMMITTEE ON NUTRITION. Fruit Juice in Infants, Children, and Adolescents: Current Recommendations. *Pediatrics* **2017**, *139*. [CrossRef]
21. Health Canada. *Canada's Dietary Guidelines 2019*; Health Canada: Ottawa, ON, Canada, 2019.
22. Mikkilä, V.; Räsänen, L.; Raitakari, O.T.; Pietinen, P.; Viikari, J. Consistent dietary patterns identified from childhood to adulthood: The cardiovascular risk in Young Finns Study. *Br. J. Nutr.* **2005**, *93*, 923–931. [CrossRef]
23. Movassagh, E.Z.; Baxter-Jones, A.D.G.; Kontulainen, S.; Whiting, S.J.; Vatanparast, H. Tracking Dietary Patterns over 20 Years from Childhood through Adolescence into Young Adulthood: The Saskatchewan Pediatric Bone Mineral Accrual Study. *Nutrients* **2017**, *9*, 990. [CrossRef]
24. Kvaavik, E.; Andersen, L.F.; Klepp, K.I. The stability of soft drinks intake from adolescence to adult age and the association between long-term consumption of soft drinks and lifestyle factors and body weight. *Public Health Nutr.* **2005**, *8*, 149–157. [CrossRef]
25. Bjelland, M.; Brantsæter, A.L.; Haugen, M.; Meltzer, H.M.; Nystad, W.; Andersen, L.F. Changes and tracking of fruit, vegetables and sugar-sweetened beverages intake from 18 months to 7 years in the Norwegian Mother and Child Cohort Study. *BMC Public Health* **2013**, *13*, 793. [CrossRef] [PubMed]
26. Park, S.; Pan, L.; Sherry, B.; Li, R. The association of sugar-sweetened beverage intake during infancy with sugar-sweetened beverage intake at 6 years of age. *Pediatrics* **2014**, *134*, S56–S62. [CrossRef] [PubMed]
27. Mazarello, P.; Hesketh, K.; O'Malley, C.; Moore, H.; Summerbell, C.; Griffin, S.; Ong, K.; Lakshman, R. Determinants of sugar-sweetened beverage consumption in young children: A systematic review. *Obes. Rev.* **2015**, *16*, 903–913. [CrossRef] [PubMed]
28. Carsley, S.; Borkhoff, C.M.; Maguire, J.L.; Birken, C.S.; Khovratovich, M.; McCrindle, B.; Macarthur, C.; Parkin, P.C.; Collaboration, T.K. Cohort Profile: The Applied Research Group for Kids (TARGet Kids!). *Int. J. Epidemiol.* **2015**, *44*, 776–788. [CrossRef] [PubMed]

29. Carsley, S.E.; Anderson, L.N.; Plumptre, L.; Parkin, P.C.; Maguire, J.L.; Birken, C.S. Severe Obesity, Obesity, and Cardiometabolic Risk in Children 0 to 6 Years of Age. *Child. Obes.* **2017**, *13*, 415–424. [CrossRef]
30. De Onis, M. WHO Child groth standards based on length/height, weight, and age. *Int. J. Paediatr.* **2006**, *450*, 76–85.
31. Skinner, J.D.; Carruth, B.R.; Bounds, W.; Ziegler, P.; Reidy, K. Do food-related experiences in the first 2 years of life predict dietary variety in school-aged children? *J. Nutr. Educ. Behav.* **2002**, *34*, 310–315. [CrossRef]
32. Fiorito, L.M.; Marini, M.; Mitchell, D.C.; Smiciklas-Wright, H.; Birch, L.L. Girls' early sweetened carbonated beverage intake predicts different patterns of beverage and nutrient intake across childhood and adolescence. *J. Am. Diet. Assoc.* **2010**, *110*, 543–550. [CrossRef]
33. Teegarden, D.; Lyle, R.M.; Proulx, W.R.; Johnston, C.C.; Weaver, C.M. Previous milk consumption is associated with greater bone density in young women. *Am. J. Clin. Nutr.* **1999**, *69*, 1014–1017. [CrossRef]
34. Garriguet, D. Changes in beverage consumption in Canada. *Health Rep.* **2019**, *30*, 20–30. [CrossRef]
35. Lora, K.R.; Davy, B.; Hedrick, V.; Ferris, A.M.; Anderson, M.P.; Wakefield, D. Assessing Initial Validity and Reliability of a Beverage Intake Questionnaire in Hispanic Preschool-Aged Children. *J. Acad. Nutr. Diet.* **2016**, *116*, 1951–1960. [CrossRef] [PubMed]
36. Lewis, K.H.; Skelton, J.A.; Hsu, F.C.; Ezouah, P.; Taveras, E.M.; Block, J.P. Implementing a novel electronic health record approach to track child sugar-sweetened beverage consumption. *Prev. Med. Rep.* **2018**, *11*, 169–175. [CrossRef] [PubMed]
37. Clemens, R.; Drewnowski, A.; Ferruzzi, M.G.; Toner, C.D.; Welland, D. Squeezing fact from fiction about 100% fruit juice. *Adv. Nutr.* **2015**, *6*, 236S–243S. [CrossRef] [PubMed]
38. Government of Canada. Juice and Juice Products. Available online: http://www.inspection.gc.ca/food/requirements-and-guidance/labelling/industry/processed-fruit-or-vegetable-products/juice-and-juice-products/eng/1348153076023/1348153220895 (accessed on 19 August 2019).
39. Ventura, A.K.; Mennella, J.A. Innate and learned preferences for sweet taste during childhood. *Curr. Opin. Clin. Nutr. Metab. Care* **2011**, *14*, 379–384. [CrossRef] [PubMed]
40. Ventura, A.K.; Worobey, J. Early influences on the development of food preferences. *Curr. Biol.* **2013**, *23*, R401–R408. [CrossRef]

© 2019 by the authors. Licensee MDPI, Basel, Switzerland. This article is an open access article distributed under the terms and conditions of the Creative Commons Attribution (CC BY) license (http://creativecommons.org/licenses/by/4.0/).

Article

Tracking of Dietary Intake and Diet Quality from Late Pregnancy to the Postpartum Period

Audrée Lebrun [1,2,3], Anne-Sophie Plante [2], Claudia Savard [1,2,3], Camille Dugas [1,2,3], Bénédicte Fontaine-Bisson [4,5], Simone Lemieux [1,3], Julie Robitaille [1,2,3] and Anne-Sophie Morisset [1,2,3,*]

1. School of Nutrition, Laval University, Québec City, QC G1V 0A6, Canada
2. Endocrinology and Nephrology Unit, CHU de Québec-Université Laval Research Center, Québec City, QC G1V 4G2, Canada
3. Institute of Nutrition and Functional Foods (INAF), Laval University, Québec City, QC G1V 0A6, Canada
4. School of Nutrition Sciences, University of Ottawa, Ottawa, ON K1N 6N5, Canada
5. Institut du Savoir Montfort, Montfort Hospital, Ottawa, ON K1K 0T2, Canada
* Correspondence: anne-sophie.morisset@fsaa.ulaval.ca; Tel.: +1-418-656-2131 (ext. 413982)

Received: 19 July 2019; Accepted: 29 August 2019; Published: 3 September 2019

Abstract: The present study aimed to characterize dietary intake and diet quality from late pregnancy to six months postpartum. Participants ($n = 28$) completed 2–3 Web-based 24 h recalls at three distinct periods: (1) during the third trimester of pregnancy; (2) three months and (3) six months after delivery. Energy, macro-and micronutrient intakes (from foods and supplements), as well as the Canadian healthy eating index (C-HEI) were derived from the dietary recalls. No significant variation in energy and macronutrient intakes was observed between time points. The proportion of women taking at least one supplement decreased over time ($p = 0.003$). The total intake of several micronutrients (vitamins A, C, D, group B vitamins, iron, magnesium, zinc, calcium, phosphorus, manganese, and copper) decreased significantly over time ($p < 0.05$ for all micronutrients). The total C-HEI score and its components did not change, except for the total vegetables and fruit subscore, which decreased over time (8.2 ± 2.0 in the 3rd trimester, 7.1 ± 2.2 at three months postpartum, 6.9 ± 2.4 at 6 months postpartum, $p = 0.04$). In conclusion, we observed a general stability in diet quality, energy, and macronutrient intakes from the third trimester of pregnancy to six months postpartum. However, several micronutrient intakes decreased over time, mostly due to changes in supplement use.

Keywords: pregnancy; postpartum; dietary intake; energy intake; supplements; dietary reference intakes (DRIs); diet quality; Healthy Eating Index

1. Introduction

Adopting healthy eating behaviors is crucial during pregnancy in order to positively influence both the mother's and the child's health [1–3]. Maintaining a balanced diet after childbirth is also important to ensure optimal maternal health, both in the short and long term [4–7]. In the short term, a woman's diet after delivery can influence weight retention since it is associated with the total energy intake [4]. In the long term, postpartum weight retention has been identified as a contributor to obesity, the latter being associated with an increase in morbidity and mortality risk [5,6]. Furthermore, adequate diet quality and dietary intake are essential to support the energy demand associated with lactation and ensure optimal early life nutrition for the newborn. In fact, greater maternal diet quality during pregnancy and lactation has been inversely associated with infant weight and adiposity in the early postpartum period, which could prevent obesity later in life [7]. Hence, it is important to maintain healthy eating behaviors both during and after pregnancy.

However, few studies have investigated this continuum and the changes that can occur from pregnancy to the postpartum period. One study showed that Swedish women's diet quality tended to decrease after delivery, mostly due to an increased intake of discretionary food (e.g., sweets, cakes, cookies, crisps, and ice cream), and a decreased intake of vegetables and fruit [8]. Although a different study found a significant increase in the proportion of women engaging in more positive behaviors (drinking two or more cups of milk per day, consuming three or more servings of vegetables and fruit per day, and eating breakfast every day) from pre-pregnancy to pregnancy, that proportion decreased dramatically at six months postpartum [9]. It was also reported, in a cohort of low-income women with diverse ethnicity, that following childbirth, mean daily servings of grains, vegetables and fruit declined, while the percentage of energy from fat and added sugar increased in comparison with pregnancy [10]. Overall, although an improvement in diet quality has been reported during pregnancy [9,11], healthy eating habits adopted in the prenatal period are not often maintained after childbirth [8–12].

Nevertheless, studies examining maternal diet from late pregnancy to the postpartum period rarely detail women's adherence to dietary recommendations [8–12], despite the significant impact of nutrition on maternal and child health. Thus, it appears relevant to examine women's diet during their transition to maternity as well as look at their adherence to nutritional recommendations. The aims of this study are to characterize dietary intake and diet quality from late pregnancy to the postpartum period and to investigate women's adherence to current Canadian nutritional recommendations at each time point. Firstly, we hypothesize that diet quality decreases from the third trimester to six months postpartum. Secondly, we hypothesize that adherence to micronutrient intake recommendations will be low in the postpartum period, especially for lactating women in whom nutritional needs are increased.

2. Materials and Methods

2.1. Study Population

The ANGE (Apports Nutritionnels durant la GrossessE) project included eighty-six (86) pregnant women recruited in Quebec City, QC, Canada. At the recruitment of the initial cohort, the exclusion criteria were to be under 18 years of age or to present a severe medical condition (i.e., type 1 or 2 diabetes, renal disease, inflammatory and autoimmune disorders). Previously published analyses included 79 women with complete nutritional data at each trimester of pregnancy [13,14]. In the present paper, the final sample includes 28 women for whom we have complete nutritional information for the 3rd trimester of pregnancy, as well as for the 3- and 6-months postpartum time points (see Figure 1). The Institutional Ethics Committee approved the project (Reference number: 2016–2866) and participants gave their informed written consent.

2.2. The Automated Web-Based 24 h Recall (R24W)

In the 3rd trimester of pregnancy (range: 31.9–36.1 gestational weeks, gestational age confirmed by ultrasound in the 1st trimester), at 3 months postpartum (range: 9.4–13.7 weeks after delivery) and 6 months postpartum (range: 23.0–26.4 weeks after delivery), participants used the R24W (Rappel de 24 h Web) platform. They completed, at each period, 2–3 Web-based 24 h dietary recalls. The average time between recalls was 1.0 ± 0.5, 1.4 ± 0.9 and 1.3 ± 0.7 weeks in the 3rd trimester, at 3 months postpartum, and at 6 months postpartum, respectively. The R24W has been previously described and was validated in pregnant women as well as in the general adult population [15–17]. Data regarding intake of total energy and 22 nutrients were analyzed.

2.3. Healthy Eating Index

The R24W platform automatically calculates the 2007 version of the Canadian Healthy Eating Index (C-HEI), an adaptation of the HEI developed by Kennedy et al. [18,19] that was used to assess diet quality at each time point (3rd trimester, 3rd and 6th month of postpartum). The HEI has been validated in the general population and used by various authors to assess diet quality among pregnant women [18,20–22]. The assessment of the C-HEI relies on the number of servings reported by an individual, according to age and sex, as specified in the 2007 version of Canada's Food Guide [18,23]. In brief, the C-HEI is divided into eight adequacy components and three moderation components [18]. Based on the scoring criteria, participants were allocated a score between 0 and the potential maximum (5, 10, or 20). In accordance with the method described earlier [18,23], scores are then added up for a maximum of 100 points, representing a perfect adherence to the 2007 version of Canada's Food Guide (see detailed method in Supplementary Table S1).

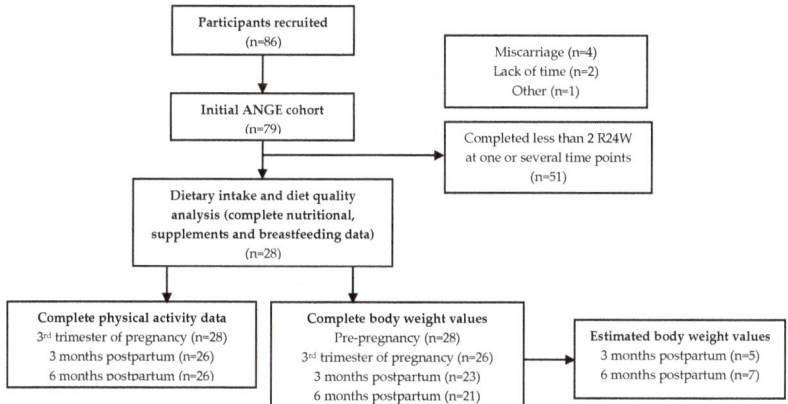

Figure 1. Flowchart of the study sample.

2.4. Supplement Use

All participants completed, at each time point, a Web questionnaire collecting information on supplement use. This questionnaire was previously described elsewhere [14]. In brief, for each supplement they reported taking, participants had to provide its name, its drug identification number, its measurement unit (e.g., tablet, drop, gram, milliliter, etc.) and its dose, as well as the frequency at which the reported dose was taken (e.g., once a day, twice a week, etc.). Nutritional data for all supplements were obtained by using the Health Canada Licensed Natural Health Product Database or the companies' product labels [24]. A research assistant reached out to participants when any supplement's characteristic was missing or incomplete. The use of supplements was compiled according to the supplement's type (multivitamin or one-nutrient supplement) and the number of women that reported taking each supplement.

2.5. Other Web Questionnaires

Women completed questionnaires on socioeconomic status and general health early in pregnancy to obtain data on sample descriptive characteristics. A Web-based self-administered questionnaire was completed by 27 women at 3 and 6 months postpartum to report the infant feeding methods and a research assistant contacted women with missing information.

2.6. Estimated Energy and Protein Requirements

Pre-pregnancy BMI was calculated using self-reported pre-pregnancy weight and measured height. The validated version of the pregnancy physical activity questionnaire (PPAQ) was completed by women as well as the International physical activity questionnaire (IPAQ) at 3 and 6 months postpartum [25–27]. Physical activity levels (PALs) were calculated with the total number of minutes per day participants engaged in moderate and high-intensity activities. Based on the dietary reference intakes (DRIs), we characterized participants as either sedentary, low-active, active or very active [28]. Estimated energy requirements (EERs) in the 3rd trimester were calculated with the participant's age, height, pre-pregnancy weight, physical activity level, to which 452 kcal was added [28]. The weight of 26 women was measured during an on-site visit at the 3rd trimester of pregnancy. In the postpartum periods, a Web-based self-administered questionnaire regarding maternal weight was completed by 23 women at 3 months and by 21 women at 6 months. Mean weight difference between each period and pre-pregnancy was calculated (see Table 1). The difference between postpartum periods and pre-pregnancy weights can then be defined as weight retention. Since body weight values were missing for 5 women at 3 months postpartum and for 7 women at 6 months postpartum (see Figure 1), an estimation of their weight difference (in kg) was made at those time points, in order to calculate their estimated energy and protein requirements. The estimated weight difference was calculated by using the mean weight difference (vs. pre-pregnancy weight) for the 23 women at 3 months (4.3 kg) and the mean weight difference for the 21 women at 6 months (2.8 kg). We were therefore able to calculate EERs at each postpartum period by using age, height, weight, physical activity levels and by adding 500 kcal only for women who were breastfeeding at the time they completed the questionnaires [28]. In the 3rd trimester, daily estimated protein requirements (EPRs) were determined with the following calculation: 1.1 g/kg of pre-pregnancy weight + 25 g of protein [28]. For postpartum periods, daily EPRs were estimated as 1.3 g/kg of reported or estimated weight in breastfeeding mothers and 0.8 g/kg for others [28].

Table 1. Participants' characteristics ($n = 28$).

Variables	Mean ± SD or n (%)
Age at study enrollment (years)	32.7 ± 3.6
Primiparous	12 (43)
Pre-pregnancy BMI (kg/m^2)	24.9 ± 5.1
Normal weight	18 (64)
Overweight	6 (21)
Obese	4 (14)
Weight (kg)	
Pre-pregnancy	69.3 ± 16.6
3rd trimester of pregnancy (difference with pre-pregnancy) [a]	12.8 ± 4.5
3 months postpartum (difference with pre-pregnancy) [b]	4.3 ± 6.3
6 months postpartum (difference with pre-pregnancy) [c]	2.8 ± 5.7
Ethnicity-Caucasian	28 (100)
Education	
College	2 (7)
University	26 (93)
Household income	
<60,000$	4 (14)
60,000–79,999$	5 (18)
80,000–99,999$	10 (36)
>100,000$	9 (32)
Breastfeeding (exclusively or not)	
At 3 months postpartum	27 (96)
At 6 months postpartum	25 (89)
Duration of exclusive breastfeeding postpartum (months) [c]	5.2 ± 1.0

[a] $n = 26$; [b] $n = 23$; [c] $n = 21$.

2.7. Statistical Analyses

Based on the automatically generated data from the completed dietary recalls, means and standard deviations were determined for energy, macro- and micronutrient intakes at each time point. Micronutrient intake from supplements and from food sources were compiled to estimate the total micronutrient intake. The intake of energy and nutrients were compared to their respective dietary reference intakes (DRIs). We then obtained a number of women that had intakes below the estimated average intakes (EARs) or above the upper intake limit (UL), when relevant [29]. A comparison was made between folate intake expressed as dietary folate equivalents (DFE) and the folate EAR. Comparison with the folic acid UL was only made with fortified foods and supplements because this limit is only applicable to synthetic forms of folic acid [28]. Similarly, only intake from supplements were compared to the UL for niacin and magnesium intakes [28]. The average energy intake (EI) and percentages of energy provided by each macronutrient (protein, carbohydrate, fats) were compared respectively with EERs and the acceptable ranges for each macronutrient [28]. The acceptable macronutrient distribution ranges (AMDR) are recommended intervals for energy proportion provided from each macronutrient, in order to ensure an adequate intake of essential nutrients [28]. Percentages of women with data below or above their EERs or AMDR were determined. Comparison of protein intake (g/day) with the EPRs was performed [28]. To assess changes in energy, macro- and micronutrient intakes and HEI scores from the 3rd trimester of pregnancy to 6 months postpartum, repeated measure analyses of variance (ANOVA) were computed. Tukey's honest significant difference post-hoc tests were then performed to identify specific differences between time points. Chi-square tests were conducted to compare supplement use (categorical variables) over time. All statistical analyses were performed with JMP version 14 (SAS Institute Inc., Cary, NC, USA).

3. Results

Participants' characteristics are presented in Table 1. All 28 included women were Caucasians and on average in their early thirties. The sample covered a wide range of pre-pregnancy BMI, but most of the participants had a normal weight. The majority was also multiparous and had a high level of education as well as a substantial household income. Most women breastfed, exclusively or not, up to six months after delivery. This sample's characteristics ($n = 28$) do not differ significantly from the ANGE project's original sample ($n = 79$) [13].

3.1. Energy and Macronutrients

Table 2 shows that no significant variation was observed for energy and macronutrient intakes across time points. Energy intake was below EER for 61%, 78% and 74% of the participants in the 3rd trimester of pregnancy, at three months and at six months postpartum, respectively. When macronutrient intake was examined in grams (Supplementary Table S2), protein intake of at least 64% of the participants exceeded the EPR at each period. However, at each time points, all women had protein intake as a percentage of energy within the AMDR. At each time point, more than 57% of women reported fat intake as percentages of energy intake that was above the AMDR. Inversely, carbohydrate intake as percentages of energy was below the AMDR range for up to 50% of participants. Dietary fiber intake decreased over time ($p = 0.01$) and more than 89% of participants had intake below the DRI at each period (Supplementary Table S2). No significant difference in energy and nutrient intakes was observed between normal weight, overweight or obese participants (data not shown). Secondary analyses were conducted to test associations between energy intake and weight retention (data not shown). No significant difference in weight retention was observed between participants who were below or over their EERs. However, Spearman's correlation showed that energy intake at six months after delivery was positively associated with postpartum weight retention at the same period ($r = 0.52$, $p = 0.02$).

Table 2. Period-specific energy and macronutrient intakes as percentages of energy intake in comparison with dietary reference intakes.

	3rd Trimester of Pregnancy			3 Months Postpartum			6 Months Postpartum			Overall p-Value [b]
	Mean ± SD	%Below AMDR, EER or UL	%Above AMDR, EER or UL	Mean ± SD	%Below AMDR, EER or UL	%Above AMDR, EER or UL	Mean ± SD	%Below AMDR, EER or UL	%Above AMDR, EER or UL	
EER (kcal/day) [a]	2497 ± 203	-	-	2578 ± 201	-	-	2678 ± 327	-	-	-
Energy intake (kcal/day)	2321 ± 429	61	39	2305 ± 506	78	22	2227 ± 474	74	26	0.48
AMDR protein, E%	10–35			10–35			10–35			-
Protein, E%	17.5 ± 2.8	0	0	17.1 ± 2.8	0	0	16.9 ± 3.2	0	0	0.70
AMDR carbohydrate, E%	45–65			45–65			45–65			-
Carbohydrate, E%	47.0 ± 5.9	29	0	44.8 ± 6.1	46	0	45.3 ± 5.9	50	0	0.08
UL added sugar, E%	25			25			25			-
Added sugar, E%	9.8 ± 3.8	100	0	9.4 ± 4.0	96	0	9.1 ± 3.2	100	0	0.71
AMDR, total fat, E%	20–35			20–35			20–35			-
Total fat, E%	35.5 ± 5.0	0	57	36.6 ± 4.6	0	64	36.3 ± 4.7	0	64	0.57
SFA, E%	13.6 ± 2.1	-	-	12.9 ± 1.6	-	-	13.0 ± 2.7	-	-	0.32
MUFA%	12.4 ± 2.2	-	-	13.0 ± 2.1	-	-	13.0 ± 2.1	-	-	0.30
PUFA%	6.7 ± 2.1	-	-	7.8 ± 2.8	-	-	7.4 ± 1.8	-	-	0.08

[a] $n = 27$ for 3 and 6 months post-natal; [b] p-value for repeated measures ANOVA performed to assess variations in energy and macronutrient intakes across periods. When no dietary reference intake is established for a nutrient, the "-" is used instead of a 0. AMDR: acceptable macronutrient distribution range; EER: estimated energy requirements, calculated with the following formula: 354 − (6.91 × age) + physical activity coefficient × [(9.36 × weight) + (726 × height)], to which an additional 452 kcal were added in the 3rd trimester or an additional 500 kcal were added in the postpartum period if the participant was breastfeeding; UL: upper limit of 25% of the energy provided by added sugar according to the dietary reference intakes. SFA: saturated fatty acids; MUFA: monounsaturated fatty acids; PUFA: polyunsaturated fatty acids.

3.2. Vitamins and Minerals

Micronutrient intake derived from the R24W (food sources only) and proportions of women that reported intake above or below the corresponding DRIs are presented in Table 3. Micronutrient intake from food alone decreased from late pregnancy to six months postpartum for vitamins A and C as well as thiamin, riboflavin, calcium, and phosphorus ($p \leq 0.03$ for all micronutrients). A downward trend was observed for vitamin D, vitamin B_{12}, and iron intakes. As presented in Table 4, fewer women reported the use of supplements after delivery (89%, 68% and 46% at the 3rd trimester of pregnancy, at three and six months postpartum, respectively, $p = 0.003$), and multivitamins remained the most commonly used supplement. A secondary analysis showed that 67% and 52% of participants who breastfed were taking at least one supplement at three and six months after delivery, respectively. As shown in Table 5, when food sources and dietary supplements were combined, total micronutrient intake decreased from late pregnancy to the postpartum period for vitamins A, B_6, B_{12}, C and D as well as for thiamin, riboflavin, niacin, pantothenic acid, iron, magnesium, zinc, calcium, phosphorus, manganese and copper ($p \leq 0.03$ for all micronutrients). At each period, a higher proportion of women had a total micronutrient intake below the EAR, particularly for vitamins A (4%, 54%, 54%) and D (11%, 36%, 57%). Total intake of sodium was above the UL for more than 79% of women at each time point. Similar results regarding the total micronutrient intake as well as EAR and UL adherence were observed, both at three and six months postpartum, when looking only at lactating women, which represents 89%–96% of our sample.

3.3. Diet Quality

Mean HEI scores are presented in Table 6. Total and sub-scores did not significantly vary from late pregnancy to six months postpartum, except for the total vegetables and fruit sub-score, which decreased over time ($p = 0.04$) and more specifically between the third trimester of pregnancy and six months postpartum ($p = 0.03$).

Table 3. Period-specific micronutrient intake from food alone in comparison with dietary reference intakes.

	3rd Trimester of Pregnancy			3 Months Postpartum			6 Months Postpartum			Overall p-Value [a]
	Mean ± SD	%Below EAR	%Above UL	Mean ± SD	%Below EAR	%Above UL	Mean ± SD	%Below EAR	%Above UL	
Vitamin D IU/day	292 ± 128	75	0	252 ± 116	86	0	236 ± 105	93	0	0.06
Iron, mg/day	14.8 ± 3.7	96	0	14.9 ± 3.3	0	0	14.0 ± 3.5	0	0	0.09
Folate, µg DFE/day	518 ± 151	57	-	519 ± 130	36	-	474 ± 125	43	-	0.11
Folic acid, µg/day	136 ± 75	-	0	151 ± 67	-	0	134 ± 66	-	0	0.28
Vitamin B6, mg/day	1.8 ± 0.4	32	0	1.8 ± 0.1	36	0	1.7 ± 0.4	50	0	0.13
Magnesium, mg/day	424 ± 79	4	-	399 ± 99	7	-	400 ± 100	11	-	0.12
Vitamin A, µg RAE/day	975 ± 371	7	0	817 ± 317	68	0	752 ± 334	71	0	0.005
Zinc, mg/day	13.8 ± 3.9	7	0	12.2 ± 3.6	32	0	12.4 ± 3.9	36	0	0.06
Calcium, mg/day	1560 ± 453	0	0	1252 ± 386	18	0	1217 ± 377	18	0	0.002
Vitamin C, mg/day	148 ± 73	11	0	109 ± 58	25	0	108 ± 53	43	0	0.03
Thiamin, mg/day	1.8 ± 0.5	4	-	1.8 ± 0.6	7	-	1.7 ± 0.5	14	-	0.01
Vitamin B12, µg/day	5.8 ± 2.6	0	-	4.8 ± 1.7	7	-	4.7 ± 1.7	11	-	0.09
Riboflavin, mg/day	2.7 ± 0.7	0	-	2.4 ± 0.5	0	-	2.3 ± 0.6	7	-	0.004
Niacin, mg NE/day	46.1 ± 9.0	0	-	47.0 ± 9.4	0	-	45.9 ± 11.3	0	-	0.70
Pantothenic acid, mg/day	6.9 ± 1.6	-	-	6.4 ± 1.5	-	-	6.4 ± 1.6	-	-	0.18
Phosphorus, mg/day	1768 ± 362	0	0	1614 ± 346	0	0	1549 ± 354	0	0	0.01
Sodium, mg/day	3303 ± 893	-	86	3229 ± 921	-	86	3236 ± 1107	-	79	0.93
Manganese, mg/day	4.3 ± 1.3	-	0	4.1 ± 1.2	-	0	3.9 ± 1.2	-	0	0.17
Selenium, µg/day	136 ± 27	0	0	137 ± 34	0	0	136 ± 49	0	0	0.99
Copper, mg/day	1.6 ± 0.3	0	0	1.5 ± 0.3	4	0	1.5 ± 0.8	4	0	0.50

[a] p-value for repeated measures ANOVA performed to assess variations in micronutrient intake across periods. When no EAR or UL was established for a nutrient, the "-" is used instead of a 0. EAR: estimated average requirements; UL: upper intake limit; DFE: dietary folate equivalent; RAE: retinol activity equivalents; NE: niacin equivalent.

Table 4. Proportions of vitamin- and mineral-supplement users among participants.

	3rd Trimester of Pregnancy	3 Months Postpartum	6 Months Postpartum	Overall p-Value [a]
		n (%)		
≥1 supplement (all types)	25 (89)	19 (68)	13 (46)	0.003
Number of supplements (all types) taken during each period				
0	3 (11)	9 (32)	15 (54)	0.003
1	21 (75)	14 (50)	11 (39)	0.03
≥2	4 (14)	5 (18)	2 (7)	0.69
Types of supplements most commonly taken				
Multivitamins	23 (92)	16 (84)	12 (92)	0.01
Folic acid supplement	2 (8)	1 (5)	1 (8)	0.77
Vitamin D supplement	2 (8)	2 (11)	1 (8)	0.81
Iron supplement	1 (4)	1 (5)	0 (0)	0.60
Omega-3 supplement (mostly EPA-DHA)	1 (4)	1 (5)	1 (8)	1.00

[a] p-value for chi-square tests performed to assess variations in supplement use across periods.

Table 5. Period-specific total micronutrient intake (including food sources and supplements) in comparison with dietary reference intakes.

	3rd Trimester of Pregnancy			3 Months Postpartum			6 Months Postpartum			Overall p-Value [a]
	Mean ± SD	%Below EAR	%Above UL	Mean ± SD	%Below EAR	%Above UL	Mean ± SD	%Below EAR	%Above UL	
Vitamin D IU/day	841 ± 776	11	4	684 ± 848	36	4	586 ± 846	57	4	0.001
Iron, mg/day	45.5 ± 32.7	18	39	33.8 ± 18.5	0	25	26.7 ± 15.7	0	14	0.03
Folate, μg DFE/day	1175 ± 993	18	-	1608 ± 2039	18	-	1095 ± 1045	21	-	0.30
Folic acid, μg/day	793 ± 999	-	46	1240 ± 1991	-	61	755 ± 1023	-	46	0.29
Vitamin B_6, mg/day	6.1 ± 4.2	4	0	14.1 ± 46.9	14	4	3.8 ± 3.7	32	0	0.01
Magnesium, mg/day	474 ± 81	0	0	429 ± 106	7	0	420 ± 106	11	0	0.01
Vitamin A, μg RAE/day	1473 ± 607	4	0	924 ± 356	54	0	838 ± 360	54	0	<0.0001
Zinc, mg/day	23.1 ± 7.3	0	0	19.1 ± 8.6	18	4	17.1 ± 8.1	25	0	0.001
Calcium, mg/day	1779 ± 471	0	4	1415 ± 387	11	0	1330 ± 433	14	0	<0.0001
Vitamin C, mg/day	229 ± 70	4	0	171 ± 86	25	0	151 ± 78	25	0	0.001
Thiamin, mg/day	3.5 ± 1.1	0	-	3.2 ± 1.6	0	-	2.5 ± 1.2	7	-	0.001
Vitamin B_{12}, μg/day	11.2 ± 5.7	0	-	8.8 ± 6.1	4	-	7.2 ± 4.6	7	-	0.002
Riboflavin, mg/day	4.5 ± 1.5	0	-	3.8 ± 1.6	0	-	3.2 ± 1.4	4	-	0.0002
Niacin, mg NE/day	61.7 ± 12.2	0	0	58.1 ± 15.2	0	4	53.4 ± 15.8	0	0	0.01
Pantothenic acid, mg/day	11.6 ± 2.9	-	-	9.9 ± 3.4	-	-	8.9 ± 3.6	-	-	0.001
Phosphorus, mg/day	1177 ± 349	0	0	1614 ± 346	0	0	1549 ± 354	0	0	0.01
Sodium, mg/day	3303 ± 893	-	86	3229 ± 921	-	86	3236 ± 1107	-	79	0.93
Manganese, mg/day	5.4 ± 1.2	-	0	4.8 ± 1.4	-	0	4.5 ± 1.1	-	0	0.002
Selenium, μg/day	150 ± 31	0	0	148 ± 38	0	0	145 ± 55	0	0	0.91
Copper, mg/day	2.8 ± 0.9	0	0	2.3 ± 1.1	4	0	2.1 ± 1.1	4	0	0.02

[a] p-value for repeated measures ANOVA performed to assess variations in micronutrient intake across periods. When no EAR or UL was established for a nutrient, the "-" is used instead of a 0. EAR: estimated average requirements; UL: upper intake limit; DFE: dietary folate equivalent; RAE: retinol activity equivalents; NE: niacin equivalent.

Table 6. Healthy Eating Index total and subscores from late pregnancy to postpartum.

HEI	Score Range	3rd Trimester	3 Months Postpartum	6 Months Postpartum	Overall p-Value [a]
Total	0–100	64.1 ± 12.2	61.2 ± 11.4	60.5 ± 8.3	0.13
Adequacy	0–60	46.4 ± 7.6	44.5 ± 8.3	43.3 ± 7.5	0.08
Total vegetables and fruit	0–10	8.2 ± 2.0	7.1 ± 2.2	6.9 ± 2.4	0.04
Whole fruit	0–5	4.4 ± 1.5	3.6 ± 1.9	3.4 ± 1.7	0.09
Dark green and orange vegetables	0–5	3.3 ± 1.6	3.4 ± 1.5	2.9 ± 1.6	0.25
Total grains products	0–5	4.4 ± 0.7	4.5 ± 0.8	4.2 ± 1.0	0.29
Whole grains	0–5	2.6 ± 1.8	2.7 ± 1.7	2.3 ± 1.9	0.62
Milk and alternatives	0–10	9.5 ± 1.4	8.8 ± 1.9	8.8 ± 2.4	0.08
Meat and alternatives	0–10	8.9 ± 1.6	8.6 ± 2.0	8.5 ± 1.9	0.16
Unsaturated fat	0–10	5.2 ± 3.7	5.8 ± 3.7	6.4 ± 3.3	0.29
Moderation	0–40	17.7 ± 7.2	16.7 ± 6.7	17.2 ± 7.6	0.65
Saturated fats	0–10	2.8 ± 2.6	3.5 ± 2.4	3.7 ± 3.1	0.33
Sodium	0–10	4.6 ± 2.8	4.8 ± 2.9	4.8 ± 3.2	0.93
Other foods	0–20	10.3 ± 5.1	8.4 ± 5.2	8.7 ± 4.8	0.16

[a] p-value for repeated measures ANOVA performed to assess variations in HEI-scores across periods.

4. Discussion

Our prospective evaluation of women's dietary intake revealed stability in energy and macronutrient intakes from late pregnancy to six months postpartum. Most women were below their energy estimated requirements and above their protein estimated requirements. Total micronutrient intake decreased from late pregnancy to six months after delivery for many vitamins and minerals. We also observed a decrease in diet quality regarding the total consumption of vegetables and fruit.

Stability in energy intake was found from late pregnancy to six months after delivery. Likewise, Talai Rad et al., as well as Moran et al. found no significant variation in women's energy intake from pregnancy to the postpartum period [20,30]. In contrast, George et al. found that the transition from pregnancy to the postpartum period was associated with a decrease in the mean energy intake in the overall sample, and in both lactating and nonlactating low-income women [10]. However, our small study sample consisted mostly of lactating women, for whom the energy estimated requirements were similar from late pregnancy to the postpartum period, which may explain the stability in energy intake. We also found that most women were under their respective EERs in the third trimester of pregnancy as well as in the postpartum period. In contrast, Moran et al. found that most of 301 overweight or obese women met the Australian Nutrient Reference Values in energy from pregnancy to four months postpartum [20]. In comparison with our participants, women in this study were all overweight or obese and came from an area of greater social deprivation [20]. Also, at four months after delivery, 57% of Australian women were breastfeeding versus 96% of our participants at three months postpartum [20]. Considering that breastfeeding requires additional energy intake, we suggest that lactating mothers encounter more difficulties in meeting these caloric recommendations. The underreporting of energy intake may also have influenced mean caloric intake of our study sample considering that other studies have reported divergent percentages (between 13% and 49%) of under-reporters during pregnancy [31–34]. However, there is a lack of consistency in the methods and thresholds used to evaluate the misreporting of energy intake in pregnancy, indicating a need to further investigate which method would be the most appropriate to use.

Regarding macronutrient intake, most of our participants had protein intake (as percentages of energy intake) within their respective acceptable distribution ranges, at all time points. However, the majority of our participants had protein intake that exceeded their EPRs at all periods, similarly to Moran et al. [20]. Further, up to 50% of women had carbohydrate consumption (as a percentage of energy intake) below the AMDR, with a higher proportion in the postpartum period compared to what has been observed in overall pregnancy [14]. Furthermore, many participants had fat intake as a percentage of energy intake that was above the acceptable distribution range from late pregnancy to six months after delivery. Similarly, Talai Rad et al. found that in 32 healthy women, the proportion of fat intake within the total caloric intake (36%) slightly exceeded the German Nutrition Society recommendation from early pregnancy to six weeks after delivery [30]. Additionally, a previous analysis in this cohort found that macronutrient intake, as percentages of energy intake, was stable throughout pregnancy [14]. Hence, the mean intake of macronutrients, in comparison with acceptable distribution ranges, does not seem to change from early pregnancy to six months after delivery in our cohort of high-income women. Finally, we observed that almost all women did not meet the adequate intake recommended for dietary fiber, at each period, similarly to the results obtained during pregnancy in the initial ANGE cohort [14] and in the general population [35]. Also, fiber intake decreased from late pregnancy to six months after delivery, in contrast to Moran et al. who found a stability from pregnancy to four months postpartum [20].

Despite our small sample size, we found that total intake of 16 of the 20 vitamins and minerals had significantly decreased from the third trimester of pregnancy to six months postpartum, which is concordant with the decrease in vegetables and fruit intakes observed with the HEI-score. In addition, most participants did not meet the recommendation for vitamin A in the postpartum period since the EAR for this vitamin almost doubles in the context of lactation compared to pregnancy [29]. Also, a significant proportion of women failed to meet the recommendation for vitamin D after delivery. A decreasing trend observed for the milk and alternatives HEI subscore may partially explain the observed decrease in vitamin D since most of these foods are fortified with vitamin D. Nevertheless, and more importantly, the decrease in micronutrient intake might be explained by the decline in supplement use following the delivery, despite the Society of Obstetricians and Gynaecologists of Canada's recommendation for women to keep taking a prenatal multivitamin as long as breastfeeding continues [36]. Similar results regarding total micronutrient intake and adherence to recommendations were observed in the postpartum period when looking only at lactating women, which represent most of our sample. A decrease in total micronutrient intake of breastfeeding participants persisted even if more than half of them continued taking at least one supplement after delivery. Moran et al. also found lower total intake of iron, zinc and calcium as well as vitamins A, B_6 and C from the third trimester to four months postpartum, with a significant decrease in supplement use over this period [20]. It would therefore appear that women reduced their supplement use after childbirth, which may put lactating women at risk of non-adherence to micronutrient recommendations, since their nutritional needs are increased compared to non-lactating women.

Diet quality remained stable from late pregnancy to the postpartum period in this limited sample size, except for the total vegetables and fruit sub-score. Total C-HEI score and its components are based on the number of servings consumed, therefore participants seemed to decrease their vegetable and fruit intakes after delivery in comparison to late pregnancy. This might have had an impact on the observed decrease in dietary intake of fiber, vitamin A, and vitamin C. As previously published by our research team (initial ANGE cohort), total C-HEI scores did not significantly vary throughout pregnancy [13]. However, the adequacy sub-score decreased significantly from early to late pregnancy, mostly due to a decreased intake in vegetables and fruit [13]. We can hypothesize that vegetables and fruit intakes decrease throughout pregnancy, which continues in the postpartum period, as reported in other studies [8–10]. This is concordant with the supposition that motivation for healthy eating might decrease as pregnancy progresses and after delivery. Interestingly, a study found that multiparous women, who make up the majority of this sample, have lower intentions to eat in a healthier manner compared to new and non-parents [37]. The same authors hypothesized that mothers may find it difficult to put time and energy in preparing healthy meals for multiple children, thus leading to the subsequent decrease in motivation for their own dietary behaviors [37]. As a possible strategy to counter the decrease in micronutrient intake and diet quality, healthcare providers should reinforce recommendations regarding multivitamin supplementation and address the importance of vegetables and fruit consumption during postpartum follow-ups.

To our knowledge, this is the first study to prospectively assess whether women adhere or not to current Canadian nutritional recommendations up to six months after delivery. A major strength of this study is the use of detailed information collected on dietary intake with the completion of 2–3 validated Web-based 24 h recalls at each period combined with a Web questionnaire on supplements use. However, some limitations need to be acknowledged, mainly regarding the small sample size and lack of representativeness of our sample. The small sample size could have attenuated the statistical significance, however, the results we observed were similar to those from other studies in larger cohorts. Our sample can also include a potential proportion of under-reporters and a possible overestimation of the energy requirement of overweight or obese breastfeeding women [38], which would have inflated the proportion of women not meeting their EERs. Since all women were Caucasian and most of them were of higher socioeconomic status, our results may not be representative of mothers from a more ethnically and socioeconomically diverse population. Diet quality and dietary intake should be further investigated in a larger and more representative cohort, from pregnancy to the postpartum period.

5. Conclusions

In summary, we observed relative stability in diet quality, energy and macronutrient intakes from late pregnancy to the postpartum period, but intake of several micronutrients decreased from the third trimester of pregnancy to six months after delivery. Decreased use of supplements, energy intake below the estimated requirements, lower intake of vegetables and fruit as well as increased micronutrient needs may put lactating women at risk of nutritional deficiencies. However, further research is needed, first, to confirm in a larger and more diverse cohort the decrease we observed in micronutrient intake, as well as in vegetables and fruit consumption from pregnancy to the postpartum period, and second, to identify women at a higher risk of inadequate postpartum diet that could be targeted in future interventions.

Supplementary Materials: The following are available online at http://www.mdpi.com/2072-6643/11/9/2080/s1, Table S1: Components of Canadian adaptation of Healthy Eating Index, range of scores and scoring criteria, Table S2: Period-specific macronutrient intake in comparison with dietary reference intakes.

Author Contributions: All listed authors contributed substantially to the conception and design process that led to the first draft of this manuscript. They all revised a first draft of the manuscript and provided their input. A.L. collected the data under the supervision of A.-S.M. and conducted primary statistical analyses of the data with the help of A.-S.P., C.S., C.D., B.F.-B., S.L., J.R. and A.-S.M. All authors participated in the secondary analyses and interpretation of data. All authors gave their approval of the manuscript's final version to be published and therefore take public responsibility for the content of the manuscript. Finally, all authors agreed to be accountable for all aspects of the work.

Funding: The Danone Institute of Canada and startup founds (Fonds de recherche du Québec-Santé et Fondation du CHU de Québec) funded this project. All funding permitted the collection, analysis, and interpretation of data, but played no role in the writing of this manuscript.

Acknowledgments: We would like to acknowledge the valuable collaboration of the research team that developed the R24W in collaboration with S.L. and J.R.: Benoît Lamarche, Louise Corneau, Catherine Laramée, and Simon Jacques. We would also like to acknowledge Elizabeth Izedonmven, undergraduate student, for the revision of this manuscript.

Conflicts of Interest: The authors declare no conflict of interest.

References

1. Kaiser, L.; Allen, L.H. Position of the American Dietetic Association: Nutrition and lifestyle for a healthy pregnancy outcome. *J. Am. Diet. Assoc.* **2008**, *108*, 553–561. [PubMed]
2. Symonds, M.E.R. *Maternal-Fetal Nutrition during Pregnancy and Lactation*; Cambridge University Press: Cambridge, UK, 2010.
3. Rolfes, S.R.; Pinna, K.; Whitney, E. *Understanding Normal and Clinical Nutrition*; Wadsworth Publishing Company, Inc.: Belmont, CA, USA, 2002.

4. Boghossian, N.S.; Yeung, E.H.; Lipsky, L.M.; Poon, A.K.; Albert, P.S. Dietary patterns in association with postpartum weight retention. *Am. J. Clin. Nutr.* **2013**, *97*, 1338–1345. [CrossRef] [PubMed]
5. Flegal, K.M.; Graubard, B.I.; Williamson, D.F.; Gail, M.H. Cause-specific excess deaths associated with underweight, overweight, and obesity. *JAMA* **2007**, *298*, 2028–2037. [CrossRef] [PubMed]
6. Rong, K.; Yu, K.; Han, X.; Szeto, I.M.; Qin, X.; Wang, J.; Ning, Y.; Wang, P.; Ma, D. Pre-pregnancy BMI, gestational weight gain and postpartum weight retention: A meta-analysis of observational studies. *Public Health Nutr.* **2015**, *18*, 2172–2182. [CrossRef] [PubMed]
7. Tahir, M.J.; Haapala, J.L.; Foster, L.P.; Duncan, K.M.; Teague, A.M.; Kharbanda, E.O.; McGovern, P.M.; Whitaker, K.M.; Rasmussen, K.M.; Fields, D.A.; et al. Higher Maternal Diet Quality during Pregnancy and Lactation Is Associated with Lower Infant Weight-For-Length, Body Fat Percent, and Fat Mass in Early Postnatal Life. *Nutrients* **2019**, *11*. [CrossRef] [PubMed]
8. Wennberg, A.L.; Isaksson, U.; Sandstrom, H.; Lundqvist, A.; Hornell, A.; Hamberg, K. Swedish women's food habits during pregnancy up to six months post-partum: A longitudinal study. *Sex. Reprod. Healthc.* **2016**, *8*, 31–36. [CrossRef] [PubMed]
9. Olson, C.M. Tracking of food choices across the transition to motherhood. *J. Nutr. Educ. Behav.* **2005**, *37*, 129–136. [CrossRef]
10. George, G.C.; Hanss-Nuss, H.; Milani, T.J.; Freeland-Graves, J.H. Food choices of low-income women during pregnancy and postpartum. *J. Am. Diet. Assoc.* **2005**, *105*, 899–907. [CrossRef]
11. Wiltheiss, G.A.; Lovelady, C.A.; West, D.G.; Brouwer, R.J.; Krause, K.M.; Ostbye, T. Diet quality and weight change among overweight and obese postpartum women enrolled in a behavioral intervention program. *J. Acad. Nutr. Diet.* **2013**, *113*, 54–62. [CrossRef]
12. George, G.C.; Milani, T.J.; Hanss-Nuss, H.; Freeland-Graves, J.H. Compliance with dietary guidelines and relationship to psychosocial factors in low-income women in late postpartum. *J. Am. Diet. Assoc.* **2005**, *105*, 916–926. [CrossRef]
13. Savard, C.; Lemieux, S.; Carbonneau, E.; Provencher, V.; Gagnon, C.; Robitaille, J.; Morisset, A.S. Trimester-Specific Assessment of Diet Quality in a Sample of Canadian Pregnant Women. *Int. J. Environ. Res. Public Health* **2019**, *16*. [CrossRef] [PubMed]
14. Savard, C.; Lemieux, S.; Weisnagel, S.J.; Fontaine-Bisson, B.; Gagnon, C.; Robitaille, J.; Morisset, A.S. Trimester-Specific Dietary Intakes in a Sample of French-Canadian Pregnant Women in Comparison with National Nutritional Guidelines. *Nutrients* **2018**, *10*. [CrossRef] [PubMed]
15. Jacques, S.; Lemieux, S.; Lamarche, B.; Laramee, C.; Corneau, L.; Lapointe, A.; Tessier-Grenier, M.; Robitaille, J. Development of a Web-Based 24-h Dietary Recall for a French-Canadian Population. *Nutrients* **2016**, *8*. [CrossRef] [PubMed]
16. Lafreniere, J.; Laramee, C.; Robitaille, J.; Lamarche, B.; Lemieux, S. Assessing the relative validity of a new, web-based, self-administered 24 h dietary recall in a French-Canadian population. *Public Health Nutr.* **2018**, *21*, 2744–2752. [CrossRef] [PubMed]
17. Savard, C.; Lemieux, S.; Lafreniere, J.; Laramee, C.; Robitaille, J.; Morisset, A.S. Validation of a self-administered web-based 24-hour dietary recall among pregnant women. *BMC Pregnancy Childbirth* **2018**, *18*, 112. [CrossRef] [PubMed]
18. Garriguet, D. Diet quality in Canada. *Health Rep.* **2009**, *20*, 41–52. [PubMed]
19. Kennedy, E.T.; Ohls, J.; Carlson, S.; Fleming, K. The Healthy Eating Index: Design and applications. *J. Am. Diet. Assoc.* **1995**, *95*, 1103–1108. [CrossRef]
20. Moran, L.J.; Sui, Z.; Cramp, C.S.; Dodd, J.M. A decrease in diet quality occurs during pregnancy in overweight and obese women which is maintained post-partum. *Int. J. Obes. (Lond.)* **2013**, *37*, 704–711. [CrossRef] [PubMed]
21. Shin, D.; Bianchi, L.; Chung, H.; Weatherspoon, L.; Song, W.O. Is gestational weight gain associated with diet quality during pregnancy? *Matern. Child Health J.* **2014**, *18*, 1433–1443. [CrossRef] [PubMed]
22. Tsigga, M.; Filis, V.; Hatzopoulou, K.; Kotzamanidis, C.; Grammatikopoulou, M.G. Healthy Eating Index during pregnancy according to pre-gravid and gravid weight status. *Public Health Nutr.* **2011**, *14*, 290–296. [CrossRef] [PubMed]
23. *Eating Well with Canada's Food Guide*; Health Canada: Ottawa, ON, Canada, 2007.
24. Health Canada. Licensed Natural Health Product Database. Available online: https://health-products.canada.ca/lnhpd-bdpsnh/index-eng.jsp (accessed on 1 May 2019).

25. Chandonnet, N.; Saey, D.; Almeras, N.; Marc, I. French Pregnancy Physical Activity Questionnaire compared with an accelerometer cut point to classify physical activity among pregnant obese women. *PLoS ONE* **2012**, *7*, e38818. [CrossRef] [PubMed]
26. Chasan-Taber, L.; Schmidt, M.D.; Roberts, D.E.; Hosmer, D.; Markenson, G.; Freedson, P.S. Development and validation of a Pregnancy Physical Activity Questionnaire. *Med. Sci. Sports Exerc.* **2004**, *36*, 1750–1760. [CrossRef] [PubMed]
27. Craig, C.L.; Marshall, A.L.; Sjostrom, M.; Bauman, A.E.; Booth, M.L.; Ainsworth, B.E.; Pratt, M.; Ekelund, U.; Yngve, A.; Sallis, J.F.; et al. International physical activity questionnaire: 12-country reliability and validity. *Med. Sci. Sports Exerc.* **2003**, *35*, 1381–1395. [CrossRef] [PubMed]
28. Otten, J.J.; Hellwig, J.P.; Meyers, L.D. (Eds.) *Dietary Reference Intakes: The Essential Guide to Nutrient Requirements*; National Academies Press: Washington, DC, USA, 2006.
29. Institute of Medicine (US) Subcommittee on Interpretation and Uses of Dietary Reference Intakes; Institute of Medicine (US) Standing Committee on the Scientific Evaluation of Dietary Reference Intakes. Application of DRIs for Group Diet Assessment. In *DRI Dietary Reference Intakes: Applications in Dietary Assessment*; National Academies Press: Washington, DC, USA, 2000.
30. Talai Rad, N.; Ritterath, C.; Siegmund, T.; Wascher, C.; Siebert, G.; Henrich, W.; Buhling, K.J. Longitudinal analysis of changes in energy intake and macronutrient composition during pregnancy and 6 weeks post-partum. *Arch. Gynecol. Obstet.* **2011**, *283*, 185–190. [CrossRef] [PubMed]
31. Derbyshire, E.; Davies, G.J.; Costarelli, V.; Dettmar, P.W. Habitual micronutrient intake during and after pregnancy in Caucasian Londoners. *Matern. Child Nutr.* **2009**, *5*, 1–9. [CrossRef] [PubMed]
32. Horan, M.K.; McGowan, C.A.; Gibney, E.R.; Byrne, J.; Donnelly, J.M.; McAuliffe, F.M. Maternal Nutrition and Glycaemic Index during Pregnancy Impacts on Offspring Adiposity at 6 Months of Age—Analysis from the ROLO Randomised Controlled Trial. *Nutrients* **2016**, *8*. [CrossRef] [PubMed]
33. Lindsay, K.L.; Heneghan, C.; McNulty, B.; Brennan, L.; McAuliffe, F.M. Lifestyle and dietary habits of an obese pregnant cohort. *Matern. Child Health J.* **2015**, *19*, 25–32. [CrossRef] [PubMed]
34. Moran, L.J.; McNaughton, S.A.; Sui, Z.; Cramp, C.; Deussen, A.R.; Grivell, R.M.; Dodd, J.M. The characterisation of overweight and obese women who are under reporting energy intake during pregnancy. *BMC Pregnancy Childbirth* **2018**, *18*, 204. [CrossRef] [PubMed]
35. Brassard, D.; Laramee, C.; Corneau, L.; Begin, C.; Belanger, M.; Bouchard, L.; Couillard, C.; Desroches, S.; Houle, J.; Langlois, M.F.; et al. Poor Adherence to Dietary Guidelines among French-Speaking Adults in the Province of Quebec, Canada: The PREDISE Study. *Can. J. Cardiol.* **2018**, *34*, 1665–1673. [CrossRef]
36. Wilson, R.D.; Wilson, R.D.; Audibert, F.; Brock, J.A.; Carroll, J.; Cartier, L.; Gagnon, A.; Johnson, J.A.; Langlois, S.; Murphy-Kaulbeck, L.; et al. Pre-Conception Folic Acid and Multivitamin Supplementation for the Primary and Secondary Prevention of Neural Tube Defects and Other Folic Acid-Sensitive Congenital Anomalies. *J. Obstet. Gynaecol. Can.* **2015**, *37*, 534–552. [CrossRef]
37. Bassett-Gunter, R.L.; Levy-Milne, R.; Naylor, P.J.; Symons Downs, D.; Benoit, C.; Warburton, D.E.; Blanchard, C.M.; Rhodes, R.E. Oh baby! Motivation for healthy eating during parenthood transitions: A longitudinal examination with a theory of planned behavior perspective. *Int. J. Behav. Nutr. Phys. Act.* **2013**, *10*, 88. [CrossRef] [PubMed]
38. Hanson, M.A.; Bardsley, A.; De-Regil, L.M.; Moore, S.E.; Oken, E.; Poston, L.; Ma, R.C.; McAuliffe, F.M.; Maleta, K.; Purandare, C.N.; et al. The International Federation of Gynecology and Obstetrics (FIGO) recommendations on adolescent, preconception, and maternal nutrition: "Think Nutrition First". *Int. J. Gynaecol. Obstet.* **2015**, *131* (Suppl. 4), S213–S253. [CrossRef]

© 2019 by the authors. Licensee MDPI, Basel, Switzerland. This article is an open access article distributed under the terms and conditions of the Creative Commons Attribution (CC BY) license (http://creativecommons.org/licenses/by/4.0/).

Article

Neonatal Consumption of Oligosaccharides Greatly Increases L-Cell Density without Significant Consequence for Adult Eating Behavior

Gwenola Le Dréan [1,2,3,*], Anne-Lise Pocheron [1,2,3], Hélène Billard [1,2,3], Isabelle Grit [1,2,3], Anthony Pagniez [1,2,3], Patricia Parnet [1,2,3], Eric Chappuis [4], Malvyne Rolli-Derkinderen [2,3,5] and Catherine Michel [1,2,3]

1. Nantes Université, INRA, UMR1280, PhAN, F-44000 Nantes, France
2. IMAD, F-44000 Nantes, France
3. CRNH-Ouest, F-44000 Nantes, France
4. Olygose, parc Technologique des Rives de l'Oise, F 60280 Venette, France
5. Nantes Université, INSERM, UMR 1235, TENS, F-44000 Nantes, France
* Correspondence: gwenola.ledrean@univ-nantes.fr; Tel.: +33-244-768-076

Received: 30 May 2019; Accepted: 14 August 2019; Published: 21 August 2019

Abstract: Oligosaccharides (OS) are commonly added to infant formulas, however, their physiological impact, particularly on adult health programming, is poorly described. In adult animals, OS modify microbiota and stimulate colonic fermentation and enteroendocrine cell (EEC) activity. Since neonatal changes in microbiota and/or EEC density could be long-lasting and EEC-derived peptides do regulate short-term food intake, we hypothesized that neonatal OS consumption could modulate early EECs, with possible consequences for adult eating behavior. Suckling rats were supplemented with fructo-oligosaccharides (FOS), beta-galacto-oligosaccharides/inulin (GOS/In) mix, alpha-galacto-oligosaccharides (αGOS) at 3.2 g/kg, or a control solution (CTL) between postnatal day (PND) 5 and 14/15. Pups were either sacrificed at PND14/15 or weaned at PND21 onto standard chow. The effects on both microbiota and EEC were characterized at PND14/15, and eating behavior at adulthood. Very early OS supplementation drastically impacted the intestinal environment, endocrine lineage proliferation/differentiation particularly in the ileum, and the density of GLP-1 cells and production of satiety-related peptides (GLP-1 and PYY) in the neonatal period. However, it failed to induce any significant lasting changes on intestinal microbiota, enteropeptide secretion or eating behavior later in life. Overall, the results did not demonstrate any OS programming effect on satiety peptides secreted by L-cells or on food consumption, an observation which is a reassuring outlook from a human perspective.

Keywords: prebiotic; gut-brain; programming; microbiota; L-cell; eating behavior

1. Introduction

Preventing unhealthy feeding behavior is highly desirable since deleterious eating habits are associated with health problems, including a higher risk of overweight and obesity [1]. Since eating behavior is the result of integrated central and peripheral biological systems that are influenced by genetic, psychological, and environmental factors [2], its optimization is highly complex and requires the full elucidation of the mechanisms that control eating behavior. Central regulation of appetite is mediated by peripheral inputs generated by stomach distension, through signals from the gut epithelium when it senses the availability of nutrients, such as satiety-regulating peptides synthetized and released by enteroendocrine cells (EECs), as well as by long-term energy signals released by adipose tissue and cerebral inputs generated by hedonics and rewards circuits [2,3].

In addition to the evident progress in understanding these interconnections, recent advances include two major findings: first, eating behavior may be programmed very early in life, and second, it could be regulated by intestinal microbiota.

According to the developmental origin of health and disease (DOHaD) theory, adverse early-life conditions may predispose a person to disordered eating [4]. Among the environmental stressors that may have an effect, it is suggested in both animal and human studies that perinatal nutrition could program the appetite (see [5,6] for reviews). In rodents, experiments based on restricting maternal nutrition and/or manipulating litter size have demonstrated that both pre- and post-natal nutrition may alter food intake [7–9] and/or food preference [10] in offspring, with subsequent repercussions in adulthood. In humans, although controversial results have been observed concerning the influence of prenatal nutrition on later eating behavior (see [11] for review), some observational evidence suggests that early nutrition/growth affects appetite regulation [12–14] and food preference programming, as demonstrated after repeated exposure to new flavors [15].

With regard to the involvement of intestinal microbiota in feeding behavior, although it has been known for several years that fermentation catalyzed by intestinal microbiota stimulates the expression of satiety peptides by EECs [16,17], it is only recently, in connection with the growing appreciation of the role that intestinal microbiota play in regulating host physiology, that this topic has generated renewed interest [18,19]. As pointed out in these reviews, some observations objectively support the involvement of intestinal microbiota in the regulation of feeding behavior. Thus, in ascending order of convincing power, we can quote: (i) the differences observed in microbiota composition or diversity in patients with anorexia nervosa (see [18] for review, [20]); (ii) the fact that feeding behavior differs between germ-free and conventional animals (see [21] for an example), (iii) the ability of certain microbiota modulating agents—e.g., certain prebiotic oligosaccharides [22,23]—to affect feeding behavior, and (iv) the delineation of mechanistic pathways that link microbiota with central and peripheral neuroendocrine systems responsible for feeding behavior, a finding which supports the existence of a causative link. For example, EECs that secrete appetite-regulating peptides can be mentioned since they have a large diversity of receptors enabling them to sense microbial inputs such as fermentation-derived short chain fatty acids (SCFA), secondary biliary salts or pathogen-associated molecular patterns [see 18 for review].

Reconciling these two emerging issues related to the regulation of feeding behavior, i.e., its possible programming in early life and its control by intestinal microbiota, we hypothesized that early modifications to microbiota may program adult feeding behavior. This programming could stem from either the programming of intestinal microbiota (e.g., [24]) or the early impacts of microbiotal changes with long-lasting consequences for the peripheral neuroendocrine systems that control adult feeding behavior and/or the central sensing of it. In this respect, it is worth mentioning the ability of microbiota-modulating agents to affect the hypothalamic expression of neurogenic factor (BDNF) during the neonatal stage [25], and the potential programmable character of both the EECs [26] and the vagal sensitivity [9]. In addition, the putative ability of the gut microbiota to act through epigenetic mechanisms (see [27] for review) as well as the ability of the microbiota presence [28] and certain microbiota modulating agents [e.g., in the case of prebiotics [29,30]) to modulate some behaviors in adults mice can be cited, assuming that they are transposable in the neonatal period.

Using the rat as a model, we therefore evaluated whether neonatal modulation of the microbiota induced by prebiotics could program eating behavior and the secretion of gastrointestinal peptides in adulthood. We first verified that the presuppositions underlying our hypothesis were present in our case, by investigating the immediate impact of the neonatal prebiotic supplementation on both the intestinal microbiota and the maturation and functioning of EEC in suckled rats. We decided to use indigestible oligosaccharides (OS) to modify the intestinal microbiota of the neonatal rats for two reasons: first, OS are recognized as intensively-fermented prebiotics [31], which are also operant in neonatal rats [24] and infants (see [32] for review) and have been shown to stimulate EEC proliferation and activity in adult animals [33,34]; second, they represent relevant nutrients in neonatal nutrition since they are commonly added to infant formula to better mimic maternal milk [35].

2. Materials and Methods

2.1. Ethics Statement

All experiments were conducted in accordance with the European Union Directive on the protection of animals used for scientific purposes (2010/63/EU). The protocols were approved by the Ethics Committee for Animal Experiments for the Pays de la Loire region (France) and the French Ministry of Research (APAFIS#3652-20 160 1 1910192893 v3). The animal facility is registered by the French Veterinary Department as A44276.

2.2. Animal Experiment

Primiparous female Sprague-Dawley rats ($n = 16$) were obtained on day one of gestation (G1) from Janvier-labs (le Genest Saint Isle, France) and housed individually (22 ± 2 °C, 12:12-h light/dark cycle) with free access to water and chow (A03, Safe Diet, Augy, France). At birth, 8 litters were culled to 8 male pups per mother with systematic cross fostering as previously described [24]. From day 5 to day 14/15 of life (PND5 to PND14/15), the pups were given various solutions of FOS, GOS/In mix (9:1), αGOS or a mix of the monomers present in the OS solutions (Table 1) by oral gavage. These OS were selected either because they are already used in infant formula (GOS/In, FOS [35]) or because they constitute a new source of OS, the physiological properties of which are to be characterized (αGOS). Two pups from each litter were given one of the 4 solutions daily.

Table 1. Composition of solutions administered by gavage to pups from PND5 to PND14/15 (g·mL^{-1}).

	CTL	FOS	GOS/In	αGOS
GOS syrup (VivinalGOS, FrieslandCampina Domo, LE Amersfoort, The Netherlands)			0.65	
Inulin powder (Raftiline HP, BENEO-Orafti S.A., Tienen, Belgium)			0.03	
FOS powder (Beneo P95, BENEO-Orafti S.A., Tienen, Belgium)		0.34		
αGOS powder (Olygose, Venette, France)				0.32
α-Lactose monohydrate (L3625, Sigma-Aldrich, St. Quentin Fallavier, France)	0.096	0.096		0.096
D(+)-glucose monohydrate (108342, Merck Santé SAS, Fontenay sous Bois, France)	0.087	0.082		0.087
D(+)-galactose monohydrate (104058, Merck Santé SAS, Fontenay sous Bois, France)	0.005	0.005		0.004
D(−)-fructose (F0127, Sigma-Aldrich, St. Quentin Fallavier, France)	0.015		0.015	0.015
Saccharose (S9378, Sigma-Aldrich, St. Quentin Fallavier, France)	0.002		0.002	0.002
Total oligosaccharides §		0.30	0.30	0.30
Total digestible sugars § !!	0.20	0.20	0.20	0.20

CTL, control; FOS, fructo-oligosaccharides; 93.2% dry matter composed of 90.4% oligomers and 6.6% monomers, providing 0.015 g·mL^{-1} of fructose, 0.005 g·mL^{-1} of glucose and 0.002 g·mL^{-1} of saccharose; GOS/In, mix (9:1) of galacto-oligosaccharides and long chain fructo-oligosaccharides (In, inuline). For GOS: 75% dry matter composed of 59% oligomers and 41% monomers; for inulin: 97% dry matter composed of 99.5% oligomers, the mix was providing 0.095 g·mL^{-1} of lactose, 0.086 g·mL^{-1} of glucose and 0.005 g·mL^{-1} of galactose; αGOS: alpha galacto-oligosaccharides (95.9% dry matter composed of 99.4% oligomers, providing 0.001 g·mL^{-1} of galactose. !!, § These calculations take into account the dry matter of the components, their purity, and the amount of digestible sugars they contain.

The pups were weighed daily and the administered volume was adapted to body weight to reach 3.2 g/kg in order to approximate the dosage actually consumed by babies fed with prebiotic enriched formula, taking into account both the difference in metabolic rate between rats and humans and the true prebiotic content of infant formula [35].

Four of the 8 litters were used for our main objective, i.e., to assess eating behavior programming: rats from these 4 litters ($n = 8$ per group) were weaned at PND21 onto standard chow (A03, Safe Diet, Augy, France) in individual cages until PND124/126, when they were sacrificed by decapitation after induction of deep anesthesia (isoflurane/O_2, 5 L·min^{-1}). During the follow-up, food consumption was measured 3 times a week. Rats from the 4 remaining litters ($n = 8$ per group except for FOS where

$n = 7$ as explained below) were sacrificed at PND14/15 by the method described above to investigate the immediate impact of the neonatal prebiotic supplementation on both intestinal microbiota and the maturation and functioning of EECs.

This experimental set-up was designed to form 8 supplemented males, originating from 4 different litters, per group at each studied age. Due to the death of one of the pups during the supplementation period (this pup was then replaced by an untreated one to equilibrate the litter size), the number of pups in the FOS group at PND14/15 was reduced to 7. These values are maximum numbers that are not always found in each of the analyses (see the illustration legends). This stemed from either physiological reasons (e.g., 2 animals did not eat at all during the fasting-refeeding test), or because of quality requirements (e.g., reliable data from in physiological cages could only be obtained for $n = 7$ in CTL and GOS/In groups; $n = 6$ in FOS group and $n = 5$ in αGOS group), or statistical inconsistency (e.g., outliers identified by the Dixon's Q test were excluded in RT-qPCR analysis as well as food/beverage consumption follow-ups), or technical problems (e.g., accidental spillage of supernatant before analysis of bacterial end-products or sequencing failure during 16S rDNA analysis or poor quality of some tissue sections in the case of immunochemical analysis). Nevertheless, in all analyses, the 4 different litters were always represented.

2.3. Tissue Collection

Under anesthesia, intracardiac blood was collected in a tube containing EDTA (Microtubes 1.3 mL K3E, Sarstedt MG & Co, Marnay, France) and plasma collected after centrifugation 2000× g, 15 min, 4 °C) was frozen at −20 °C for further analysis. The contents of the most distal 15 cm of the ileum were harvested by flushing, using 1 mL of Hanks' Balanced Salt Solution (HBSS, Thermo Fisher Scientific, St-Herblain, France), and the cecocolonic (PND14/15) or cecal (PND124/126) content was collected, weighed, mixed with 5-fold or 2-fold their volume of sterile water (PND14/15 and PND124/126, respectively). After complete homogenization, these cecocolonic/cecal suspensions were centrifuged 7800× g, 20 min, 4 °C) then both supernatants and pellets were frozen at −20 °C for analysis of the fermentation end-products (SCFA and lactate) and microbiota, respectively. Intestinal tissues (ileum and proximal colon) were rapidly collected and frozen in liquid nitrogen for RNA analysis. Additional tissue samples were fixed in 4% paraformaldehyde for immunofluorescence analysis.

2.4. Eating Behavior

2.4.1. Meal Pattern

Between PND74 and PND99, eating behavior was analyzed in physiological cages (Phecomb cages, Bioseb, Vitrol, France) as previously described [8]. Briefly, the rats were housed individually and following 24 h of acclimatization to the cage and refilling with fresh food between 9.00 a.m. and 11.00 a.m., data were recorded every 5 s over a 20-h period. Due to the intervention during the diurnal phase, the analysis was reduced to 8 h whereas the nocturnal phase was 12 h. The exact feeding pattern was defined with a minimal size of 0.1 g, a minimum duration of 10 s and a minimum inter-meal interval of 10 min. Events such as large vibrations (contact with the feed tray without eating) were filtered by the Phecomb system monitoring software (Compulse v1.1.01). The reliability percentage of the quality signal was calculated by the software and only experiments with a percentage >80% were used. Meal parameters extracted from Compulse software included number of meals, meal size and duration, inter-meal intervals and satiety ratio.

2.4.2. Taste Preference

Preference for sweet taste was measured at PND110 using the bottle test experiment [36]. After a two-day habituation to the presence of two bottles in their own cages, the animals had the choice of the two bottles, one containing tap water and the other 0.05% saccharin (Sigma-Aldrich, St. Quentin Fallavier, France). Drink intake was measured daily for three days. The position of the two bottles was reversed each day to prevent position preference bias. The sweet preference score was calculated as the ratio between the volume of saccharin solution consumed and the total drink intake in 24 h, then multiplied by 100. Preference was defined as a percentage greater than 50.

2.5. Fasting-Refeeding Test, Kinetics of GLP-1 and PPY Release and Response to Glucose

At PND105, a 4 h kinetic of GLP-1 and PYY release in plasma was carried out. Rats were not fed for 16 h to induce hunger and then fed for 20 min with a calibrated quantity of chow (A03, safe Diet). Food intake was weighed at the end of the 20 min period. Any crumbs that fell in the cage were weighed and deducted from the food intake. Blood samples were collected from the tail vein in tubes containing EDTA (Microvette CB300 EDTA 3K, Sarstedt, Marnay, France) at 0 (15 min before refeeding), 30, 60, 120, and 180 min after the beginning of the meal.

At PND124/126, the rats were not fed for 16 h, and 2 h before being sacrificed they were given an oral bolus of glucose (2 kg/kg BW) in order to challenge glucose sensing in GLP-1/PYY-producing EECs.

2.6. Plasma Gastrointestinal Peptides

The plasma concentration of total GLP-1 and total PYY was assayed by the ELISA technique using kits from Millipore (Merck- Millipore, Molsheim, France) and Phoenix Pharmaceutical (Phoenix France S.A.S, Strasbourg, France), respectively.

2.7. Fermentation End-Products

Ileal and cecal supernatants were centrifuged 8000× g, 20 min, 4 °C, diluted (1/10) with 0.5 M oxalic acid and SCFA (acetate, propionate, butyrate, isobutyrate, valerate and isovalerate) were analyzed by gas chromatography as previously described [37]. The D-and L-lactates were measured in the supernatants after heating to 80 °C for 20 min with a D/L-lactic acid enzymatic kit following the manufacturer's instructions (Biosentec, Toulouse, France).

2.8. Immunochemistry

Tissue sections (4–5 µm) of fixed ileum and proximal colon were double-stained with a goat polyclonal antibody raised against GLP-1 diluted at 1/200 (Santa Cruz Biotechnology Inc, Santa Cruz, USA) and a rabbit anti-chromograninA (chrgA, diluted at 1/1000 (ImmunoStar Inc, Hudson, USA), followed by incubation with anti-goat and anti-rabbit fluorescent secondary antibodies (1/1000). Nuclei were counterstained with 4′,6-Diamidine-2′-phenylindole dihydrochloride (DAPI, Sigma-Aldrich, St. Quentin Fallavier, France). Tissues sections were mounted in Prolong Gold anti-fading medium (Molecular Probes, Thermo Scientific, Courtaboeuf, France). Three sections per sample were analyzed with a Nanozoomer (×20) (Hamamatsu Photonics France, Massy, France). The number of fluorescent cells along the crypt-villus axis unit was counted twice by a blind operator, using the NDP view software (Hamamatsu, Photonics France, Massy, France). A total of 40 to 60 crypt-villus units per section were counted.

2.9. Quantitative Real-Time PCR

Total RNA extraction from the ileum and colon was carried out using a QIAamp RNA Blood Mini kit (Qiagen, Courtaboeuf, France) following the manufacturer's instructions. Two micrograms of RNA were reverse-transcribed using M-MLV reverse transcriptase (Promega, Charbonnières-les-Bains, France). Five microliters of 1/40 dilution of cDNA solution were subjected to RT-qPCR in a Bio-Rad iCycler iQ system (Biorad, Marnes-la-Coquette, France) using a qPCR SYBR Green Eurobiogreen®Mix (Eurobio, Les Ulis, France). The quantitative PCR consisted of 40 cycles, 15 s at 95 °C, 15 s at 60 °C and 15 s at 72 °C each. Primers sequences are shown in Table S1 of the Supplementary Material. For quantification of Neurog3, rat PrimePCR™ SYBR®GreenAssay Neurog3 (Biorad,

Marnes-la-Coquette, France) was used. Relative mRNA quantification was expressed using the $2^{-\Delta\Delta Cq}$ method with actin gene as a reference.

2.10. Bacterial 16S rDNA Sequencing of Cecal Contents

DNA was extracted from pellets of ceco-colonic content (max 250 mg) using the QIAamp Fast DNA Stool Mini kit (Qiagen, Courtaboeuf, France) after enzymatic and mechanical disruptions as described previously [37] except that homogenization was carried out at 7800 rpm for 3 × 20 s intervals with 20 s rest between each interval in a Precellys® "evolution" bead-beater (Bertin, Montigny-le-Bretonneux France). The V4 hyper-variable region of the 16S rDNA gene was amplified from the DNA extracts during the first PCR step using composite primers (5'-CTTTCCCTACACGACGCTCTTCCGATCTGTGY CAGCMGCCGCGGTAA-3' and 5'-GGAGTTCAGACGTGTGCTCTTCCGATCTGGACTACHVGGG TWTCTAAT-3') based on the primers adapted from Caporaso et al. (i.e., 515F and 806R) [38]. Amplicons were purified using a PP201 PCR Purification Kit (Jena Bioscience, Jena, Germany). Paired-end sequencing was performed on a HiSeq 2500 System (Illumina, San Diego, CA, USA) with v3 reagents, producing 250 bp reads per end, following the manufacturer's instructions by the GeT+-PlaGe platform (INRA, Toulouse, France). The 16S rDNA raw sequences were analyzed with FROGS v2 pipeline (http://frogs.toulouse.inra.fr/) [39]. After de-multiplexing, quality filtering and chimera removing, the taxonomic assignments were conducted for OTUs with abundance >0.005% with Blast using Silva 128 database containing sequences with a pintail score at 80 to determine the bacterial compositions. FROGSSTAT Phyloseq tools were used to normalize raw abundances by rarefaction and to calculate alpha and beta diversity indices.

2.11. Statistical Analysis

Statistical analyses were carried out using GraphPad Prism 6 software (GraphPad Software Inc., San Diego, USA) or R (librairies "stats v3.5.1" and "corrplot v0.84", [40]). Differences between treatments were searched using one-way ANOVA followed by Tukey's multiple comparison tests for most data, with the exception of growth and food consumption data which were subjected to multiple t-tests with correction for multiple comparison using the Holm-Sidak method. Sweet taste preference test was analyzed by the one sample t-test to compare to compare data against the 50% (no preference) value. A p value < 0.05 was considered statistically significant.

3. Results

3.1. Neonatal OS Supplementation Did Not Substantially Affect Rat Growth

Both FOS and αGOS supplementation was associated with a significant transitory reduction of pup growth in the first days of intervention (PND7 to PND10 and PND6 to PND8 respectively, Figure S1). When compared with body weights from the CTL group, the differences observed were only 9.1 to 11.5% and did not significantly affect the cumulative weight gains measured either from birth until the end of supplementation or for the whole lactation period (Table 2).

Table 2. Bodyweight gain (g) during lactation.

Treatment	BW Gain PND0-14	BW Gain PND0-20
CTL	30.4 ± 4.2 [1]	50.5 ± 6.0
FOS	28.0 ± 3.4	45.7 ± 4.6
GOS/In	29.4 ± 3.3	49.6 ± 5.8
αGOS	27.9 ± 2.7	46.8 ± 5.1

[1] Data are means ± SD collected from the total effective of rats (n = 15–16 per group during PND0-14 and n = 8 during PND0-20). BW, bodyweight.

No significant differences in bodyweight were observed between groups after weaning (Figure S2).

3.2. Neonatal OS Supplementation Exerted a Marked Immediate Impact on Intestinal Environment

3.2.1. OS Supplementation Modified Both Composition and Activity of Neonatal Intestinal Microbiota

Following 16S rDNA sequencing, no significant differences were noticed in raw sequence numbers between cecocolonic samples collected at PND14/15 (355,245 ± 10,367, 30,306 ± 13,817, 40,275 ± 18,343 and 31,808 ± 10,101 for CTL, FOS, GOS/In and αGOS, respectively) or in percentages of sequences kept after quality filtering (83.8 ± 4.0, 76.4 ± 18.4, 83.9 ± 4.2, and 81.7 ± 7.1). The cecocolonic contents of animals supplemented with OS exhibited similar reductions in richness ($p < 0.001$) compared with CTL animals (Chao1 values: 66.2 ± 21.0, 72.9 ± 28.1, and 73.9 ± 35.3 for FOS, GOS/In and αGOS, respectively *versus* 180.0 ± 35.7 for CTL). The cluster dendrogram generated using weighed UniFrac metric which illustrates beta or between-sample diversity, highlighted an obvious dissimilarity between the microbiotas of the OS-supplemented animals and those of animals from the CTL group (Figure 1) but did not reveal any effect of the nature of the OS.

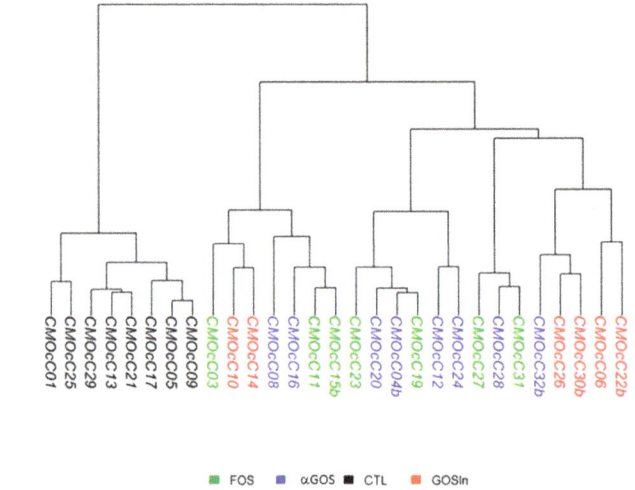

Figure 1. Hierarchical clustering based on the Ward's method of phylogenetically informed distance matrix computed using the weighted UniFrac metric for cecocolonic contents collected at postnatal day (PND) 14/15 (n = 6 to 8 per group).

When considering bacterial families occurring at more than 0.01% of the total sample abundances (Table 3), the OS impact was typified by significant decreases in Lactobacillaceae, Bacteroidales S24-7 group, Prevotellaceae, Streptococcaceae, Peptococcaceae, Coriobacteriaceae, Aerococcaceae, Family XIII, and Rikenellaceae. In addition, OS supplementation decreased Ruminococcaceae abundance but this impact only reached statistical significance for FOS and αGOS. These decreases in relative abundance were differently compensated according to the OS: increases in Bifidobacteriaceae reached statistical significance following FOS and αGOS supplementations, Enterobacteriaceae increased following αGOS supplementation and Lachnospiraceae increased following GOS/In supplementation.

Significant differences between OS were scarce and only occurred between GOS/In and αGOS in their impact on Lachnospiraceae (Table 3).

Concurring with these compositional changes, the 10-day supplementation greatly affected fermentation end-product concentrations in both ileal and colonic contents at PND14/15.

In the ileum, lactate concentration was below the detection limit (0.22 mM) in all animals, and the concentration of acetate—the sole SCFA present at this age in this intestinal segment—was significantly increased ($p < 0.005$) through FOS supplementation (6.9 ± 3.6 mM) compared to CTL (0.3 ± 0.4 mM), GOS/In (1.6 ± 2.0 mM) and αGOS (0.6 ± 0.8 mM).

Table 3. Relative abundances (%) for families with abundances > 0.01% at PND14/15 according to the postnatal OS supplementation.

Family	CTL	FOS	GOSIn	αGOS
Actinomycetaceae	0.095 ± 0.084 [1]	0.028 ± 0.032	0.076 ± 0.056	0.068 ± 0.058
Aerococcaceae	0.086 ± 0.029 [a,2]	0.011 ± 0.011 [b]	0.015 ± 0.012 [b]	0.020 ± 0.019 [b]
Alcaligenaceae	0.020 ± 0.041	0.030 ± 0.047	0.378 ± 0.576	0.036 ± 0.053
Bacteroidaceae	2.352 ± 0.991	6.510 ± 10.047	6.837 ± 5.729	3.719 ± 5.492
Bacteroidales.S24.7 group	6.812 ± 2.953 [a]	0.053 ± 0.080 [b]	0.098 ± 0.094 [b]	0.084 ± 0.103 [b]
Bifidobacteriaceae	0.624 ± 0.45 [a]	17.188 ± 12.735 [b]	7.894 ± 7.947 [ab]	13.577 ± 10.631 [b]
Campylobacteraceae	0.009 ± 0.024	0.093 ± 0.220	0.066 ± 0.151	0.294 ± 0.546
Clostridiaceae.1	0.273 ± 0.146	2.413 ± 3.231	5.509 ± 8.749	5.044 ± 4.581
Coriobacteriaceae	0.108 ± 0.039 [a]	0.039 ± 0.034 [b]	0.036 ± 0.041 [b]	0.023 ± 0.018 [b]
Corynebacteriaceae	0.032 ± 0.023	0.007 ± 0.011	0.020 ± 0.030	0.012 ± 0.023
Desulfovibrionaceae	0.098 ± 0.182	0.000 ± 0.000	0.003 ± 0.008	0.006 ± 0.014
Enterobacteriaceae	13.86 ± 5.97 [a]	23.48 ± 12.23 [ab]	19.51 ± 6.69 [b]	33.42 ± 11.99 [b]
Enterococcaceae	0.435 ± 0.707	0.145 ± 0.203	2.892 ± 6.293	0.542 ± 0.771
Erysipelotrichaceae	0.682 ± 0.387	4.080 ± 3.988	3.774 ± 4.950	2.766 ± 3.002
Family.XIII	0.062 ± 0.030 [a]	0.004 ± 0.008 [b]	0.000 ± 0.000 [b]	0.001 ± 0.003 [b]
Lachnospiraceae	6.327 ± 2.300 [a]	9.787 ± 6.180 [ab]	15.298 ± 9.544 [b]	4.962 ± 4.587 [b]
Lactobacillaceae	57.47 ± 8.72 [a]	28.74 ± 10.84 [b]	24.47 ± 5.71 [b]	31.13 ± 11.24 [b]
Micrococcaceae	0.140 ± 0.064	0.075 ± 0.074	0.071 ± 0.053	0.110 ± 0.105
Pasteurellaceae	0.582 ± 0.581	0.236 ± 0.235	0.456 ± 0.297	0.394 ± 0.446
Peptococcaceae	0.396 ± 0.182 [a]	0.006 ± 0.015 [b]	0.015 ± 0.019 [b]	0.007 ± 0.021 [b]
Peptostreptococcaceae	0.747 ± 0.485	0.471 ± 0.262	0.543 ± 0.108	0.640 ± 0.379
Porphyromonadaceae	1.242 ± 1.153	5.924 ± 9.747	9.826 ± 15.228	2.055 ± 5.475
Prevotellaceae	2.136 ± 1.540 [a]	0.014 ± 0.016 [b]	0.011 ± 0.018 [b]	0.028 ± 0.060 [b]
Rikenellaceae	0.034 ± 0.039 [a]	0.001 ± 0.004 [b]	0.000 ± 0.000 [b]	0.001 ± 0.003 [b]
Ruminococcaceae	3.242 ± 0.743 [a]	0.135 ± 0.147 [b]	1.610 ± 2.622 [b]	0.406 ± 0.665 [b]
Streptococcaceae	2.118 ± 0.620 [a]	0.510 ± 0.316 [b]	0.586 ± 0.156 [b]	0.643 ± 0.450 [b]

[1] Data are means ± SD (n = 6 to 8 per group). [2] Within a row, values followed by different letters (a,b,ab) differ significantly ($p < 0.05$).

In the cecum, the concentration of total end products increased in all OS groups compared to CTL (Figure 2). This was mainly due to an increase in SCFA concentration, which only reached statistical significance in the case of FOS and also an increase in lactate concentration in the case of αGOS.

Increases in total SCFA reflected acetate increases which were significant for both FOS and GOS/In groups, and paralleled significant decreases in pH values (Table 4). In addition, OS supplementation shifted microbiotal activity, as evidenced by significant changes in the relative proportions of acetate (93.8 ± 4.6, 93.1 ± 4.1, and 95.4 ± 2.9% for FOS, GOS/In and αGOS, respectively versus 86.3 ± 4.5% for CTL) and propionate (5.4 ± 4.6, 5.2 ± 3.4, and 3.6 ± 2.9% for FOS, GOS/In and αGOS, respectively versus 10.7 ± 3.0% for CTL). Concentration and relative proportions of butyrate—which is scarcely produced in the neonatal stage—were not affected significantly by supplementation.

Table 4. Concentration (mM) of major short chain fatty acids (SCFA) in cecocolonic contents at PND 14/15.

Treatment	Acetate	Propionate	Butyrate	pH
CTL	3.17 ± 1.05 [1,a,2]	0.39 ± 0.16	0.07 ± 0.04	6.9 ± 0.3 [a]
GOS/In	5.82 ± 1.32 [b]	0.33 ± 0.21	0.10 ± 0.09	6.3 ± 0.2 [b]
αGOS	5.69 ± 1.77 [ab]	0.28 ± 0.27	0.05 ± 0.00	6.1 ± 0.2 [b]
FOS	8.00 ± 2.94 [b]	0.47 ± 0.37	0.06 ± 0.04	6.2 ± 0.2 [b]

[1] Data are means ± SD (n = 7 to 8 per group). [2] Within columns, values followed by different letters (a,b,ab) differ significantly at $p < 0.05$.

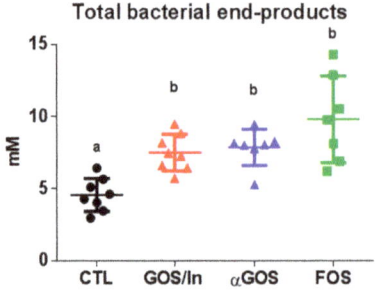

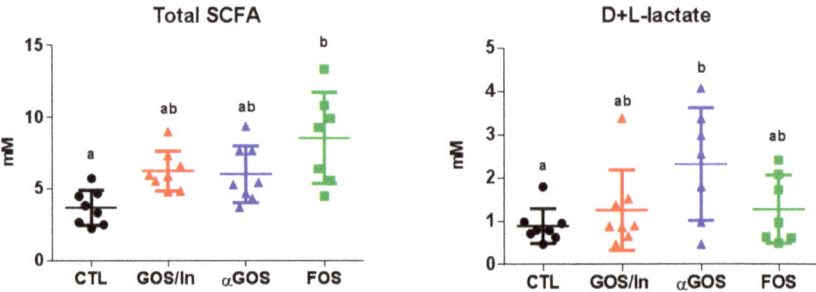

Figure 2. Cecocolonic concentrations of fermentation end-products. Individual, mean and SD values are plotted (n = 7 to 8 per group). Different letters indicated significant difference ($p < 0.05$) between groups.

3.2.2. OS Supplementation Modified both Differentiation and Activity of the Neonatal EEC

In the ileum, a profound effect on the enteroendocrine lineage was induced by neonatal OS supplementation, as revealed by a significant decrease in *Neurog3* expression in the OS groups compared to CTL, whereas, an early expressed marker in the commitment secretory lineage (*Atoh1*) was not affected significantly (Figure 3). The related expression of genes specifically implied in the differentiation of EECs (*Pax4* and *Pax6*) decreased significantly in OS supplemented groups compared to CTL, whereas expression of *Foxa1* did not vary between the groups. Similar to *Pax4* and *Pax6*, *Neurod1* expression decreased in OS groups compared with CTL, but this did not reach statistical significance for FOS. Regarding the expression of gene coding for peptides produced by mature L-cells, *Pyy* increased significantly in OS groups compared to CTL. At the same time, despite a 2-fold increase in *Gcg* expression in the OS groups compared to CTL, this effect was not statistically significant due to the widely varying expression between samples.

In the proximal colon, the impact of OS supplementation was much more moderate and their only significant effect was a decrease in the expression of *Pax4* (Figure S3).

Along with this profound remodeling in the expression of markers of L-cell differentiation, the number of GLP-1/ChgrA positive cells, i.e., mature EECs, was higher in the ileum of pups from OS groups compared to CTL but only reached statistical significance for villi (Figure 4A–C).

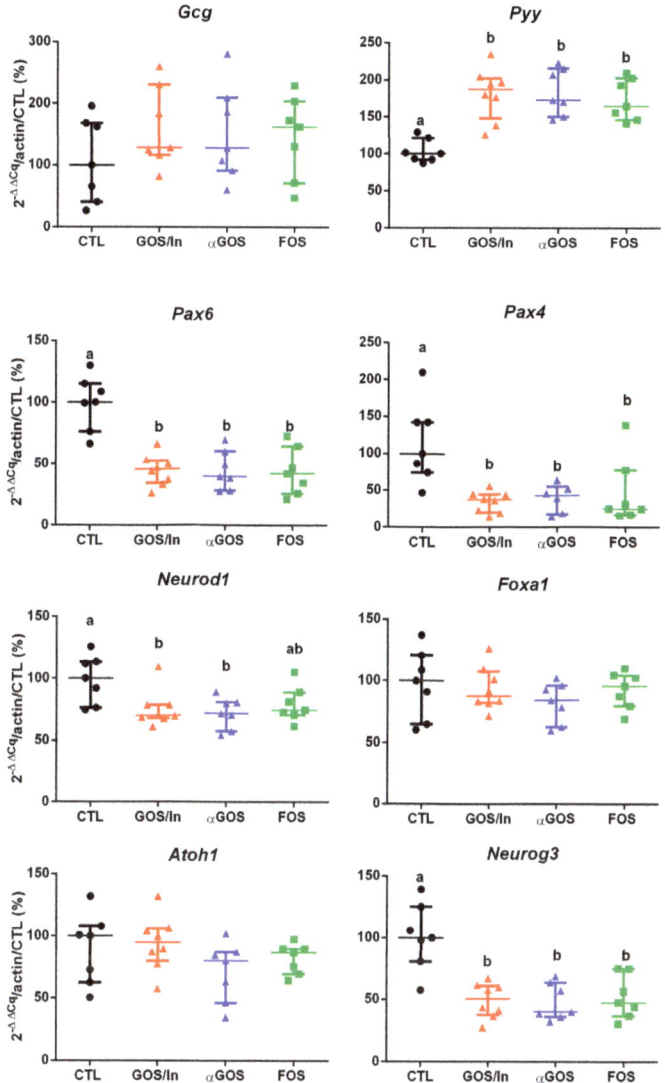

Figure 3. Relative expression of genes implied in the endocrine lineage and in L-cells differentiation in the ileum. Different letters indicate significant difference between groups ($p < 0.05$). Data are fold-change expressed in % of CTL group. Individual values, median with interquartile range are plotted (n = 7 to 8 per group).

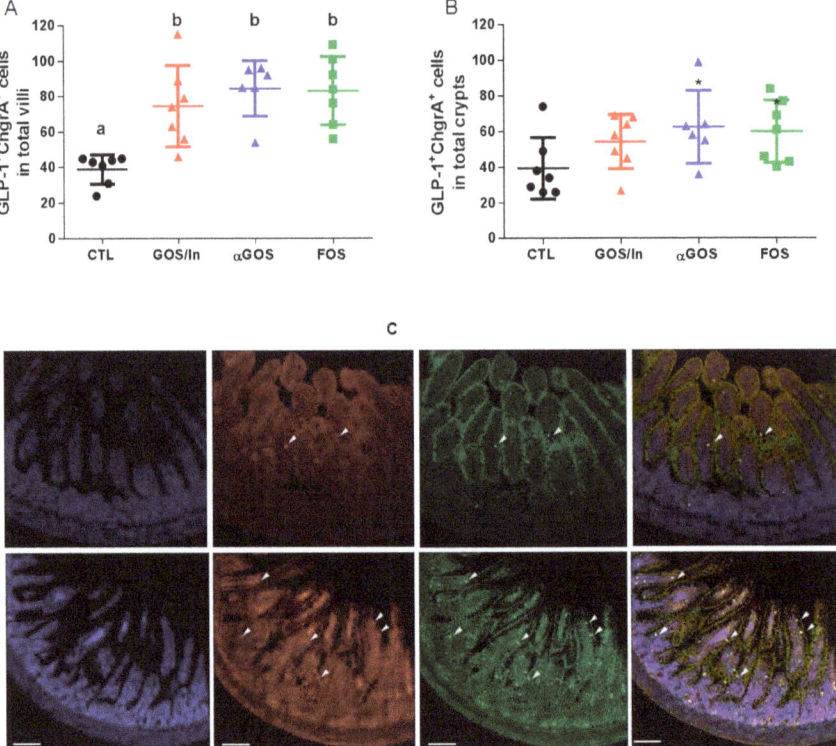

Figure 4. Effect of oligosaccharides (OS) supplementation on the density of GLP-1 cells in ileum: (**A**) in villi (**B**) in crypts. Different letters indicate significant differences among groups ($p < 0.05$); Individual, mean and SD values are plotted (n = 6 to 7 per group). (**C**) Representative images of immunofluorescence in ileal sections from a control solution (CTL) (top) and αGOS groups (down), arrows indicate positive fluorescence in cells: blue (DAPI, nuclei staining), red (GLP-1 cells), green (ChrgA cells) and merge (GLP-1/ChrgA cells). Bars indicate 100 µm.

In agreement with this rise in the number of mature enteroendocrine cell (EEC), plasma concentrations of GLP-1 (Figure 5A) and PYY (Figure 5B) were significantly increased by all the neonatal OS supplementations, as compared with CTL.

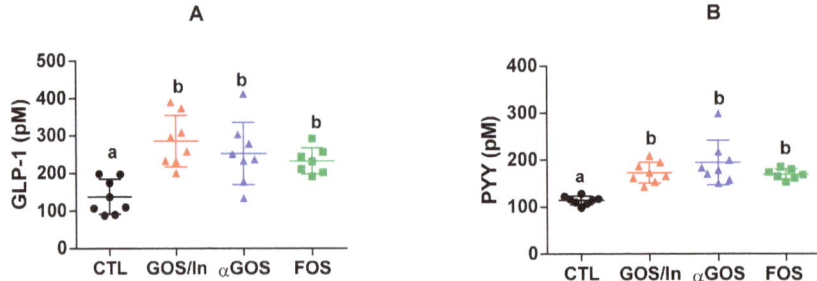

Figure 5. Plasma concentration of (**A**) Total GLP-1; (**B**) Total PYY at PND 14/15. Different letters indicate significant differences among groups ($p < 0.05$). Individual, mean and SD values are plotted (n = 7 to 8 per group).

Significant positive associations between plasma concentrations of GLP-1 and PYY and the ileal expression of their respective genes were evidenced (Figure 6A). Conversely, these plasma concentrations as well as the density of GLP-1 secreting cells, were inversely correlated with expressions of *Neurog3*, *Neurod1*, *Pax4*, and *Pax6*. With respect to associations between microbiota and EEC descriptors (Figure 6B), only some of the differentiating factors (*Pax4*, *Neurod1*, *Pax6* and *Neurog3*) exhibited significant positive correlations with the abundance of some bacterial families corresponding to those the abundance of which was significantly reduced by OS, except for Prevotellaceae. For these factors, the sole negative correlation was that between *Neurod1* and abundance of Clostridiaceae.1. Conversely, the PYY and GLP-1 plasmatic concentrations, EEC densities and *Pyy* expression, but not *Gcg* expression, were negatively correlated with the same families including Prevotellaceae.

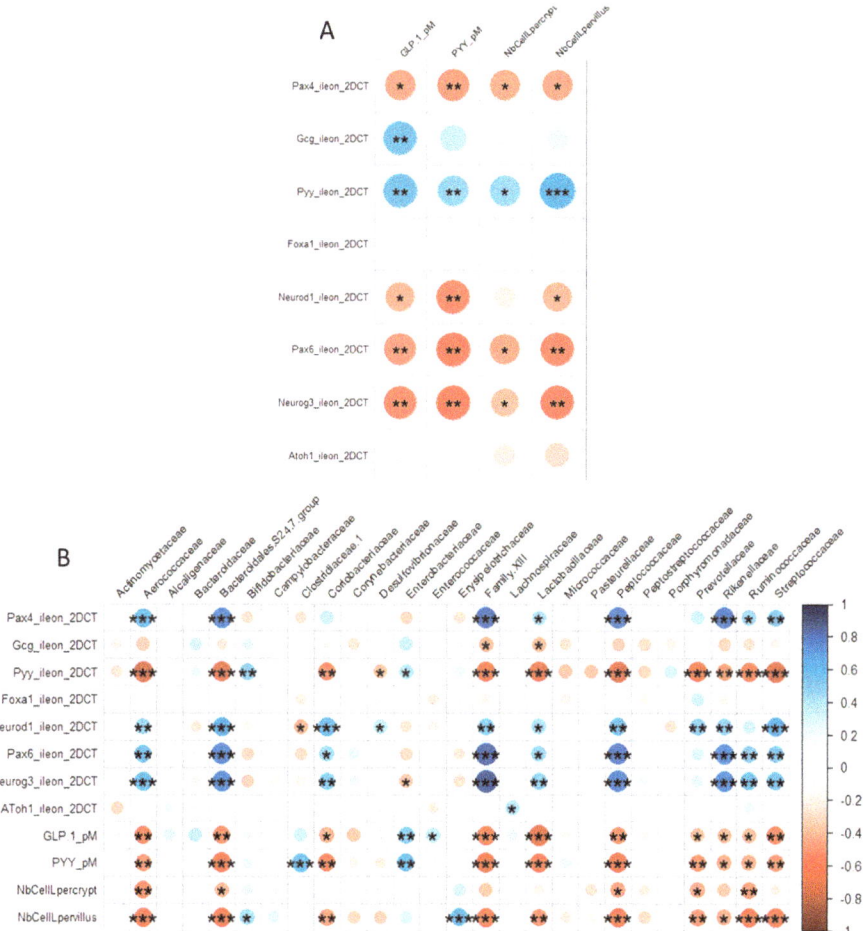

Figure 6. Correlograms within EEC descriptors (**A**) or between these descriptors and the relative abundances of main bacterial families (**B**). Positive correlations are displayed in blue and negative correlations in red. The intensity of the color and the size of the circles are proportional to the correlation coefficients. Asterisks indicate the level of significance (*, $p < 0.05$; **, $p < 0.01$; ***, $p < 0.001$). On the right of the correlogram, the color legend shows the correspondence between correlation coefficients and colors.

Overall, these results indicate that OS supplementation profoundly modulates neonatal microbiota in terms of both its composition and its fermentative activity, with repercussions not only in the cecocolon but also, as exemplified with FOS, in the ileum. An increased density of ileal L-EECs and their secreted anorectic hormones, GLP-1 and PYY, were observed and unexpectedly the expression of transcription factors beyond the stage of secretory cell engagement (Atoh1) was inhibited at the same time. Whether this strong impact of early OS supplementation on satiety peptide-related EECs could last into later life and affect eating behavior was investigated further.

3.3. Neonatal OS Supplementation Had No Significant Long-Term Consequences

3.3.1. Neonatal OS Supplementation Did Not Significantly Program Enteropeptide Production or Eating Behavior in Adulthood

To investigate the long-term effect of neonatal supplementation of OS on nutrient sensing in EECs, once pups reached adulthood, we studied the release of GLP-1 and PYY in response to both a 20-min test meal (PND 74/76) and an oral bolus of glucose (PND 124/126) after 16 h of fasting.

No significant differences were observed between groups in the amount of food consumed during the 20-min test meal (Figure 7A). In response to this meal, the plasma concentration of GLP-1 increased immediately after refeeding and returned to pre-prandial level 120 and 180 min later (Figure 7B). The total amount of GLP-1 secreted during this period, quantified by AUC, did not differ significantly between the groups (Figure 7C). PYY secretion did not show any postprandial peak or significant differences between the groups (data not shown).

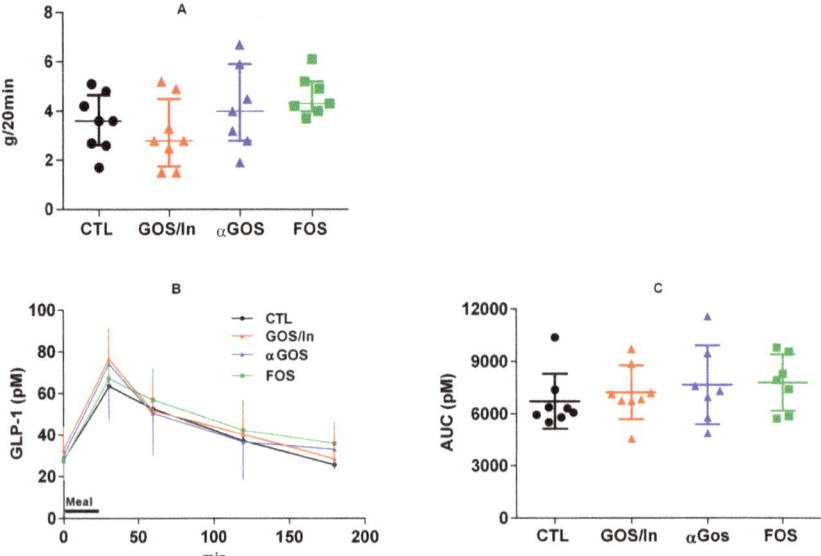

Figure 7. Fasting-refeeding test. (**A**) Food intake measured during refeeding (20 min-meal); (**B**) Plasma concentration of total GLP-1 measured during the 3h-kinetic follow-up (means ± SD); (**C**) Total amount of GLP-1 secreted during the 0-180min period expressed as AUC. Individuals, means and SD are plotted. (n = 7 to 8 per groups).

Similarly, at PND 124/126, plasma concentrations of GLP-1 (CTL: 34.4 ± 13.5; GOS/In: 38.6 ± 28.6; αGOS: 28.9 ± 10.0 and FOS: 37.9 ± 20.6 pM) and PYY (CTL: 84.7 ± 4.0; GOS/In: 88.7 ± 7.8; αGOS: 91.4 ± 7.3 and FOS: 91.5 ± 5.8 pM) measured 2h after an oral bolus of glucose did not show any significant difference between groups.

To investigate the long-term effect of a neonatal supplementation of OS on subsequent eating behavior, we followed up the food consumption from weaning to adulthood, performed a refined analysis of feeding pattern using physiological cages from PND75 to PND100 and assessed the preference for sugar taste between PND109 and PND111.

The analysis of food consumption during development, expressed per Kg of body weight to allow for strict comparison, only revealed a single significant difference which occurred at PND32 between animals from the FOS and CTL groups (Figure 8), an observation which indicates that neonatal supplementation with OS did not greatly influence the subsequent food intake in our experimental conditions.

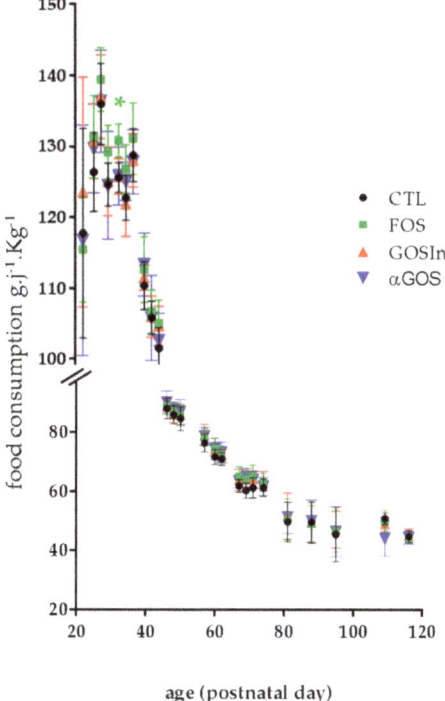

Figure 8. Daily consumption of food in the post-weaning stage, expressed as kilograms of bodyweight. The asterisk indicates a significant difference between FOS and CTL groups ($p < 0.05$). Data are means ± SD (n = 7 to 8 by group and day).

This absence of effect on daily food consumption was confirmed by a detailed analysis of food consumption: we observed no significant difference in meal patterns among the groups (food intake, food intake per meal, number and duration of meals, latency to eat the first nocturnal meal, satiety ratio and ingestion rate), whatever the period of measurement (total 20 h period of measurement, diurnal period (8 h) or nocturnal period (12 h) (Figure 9 and Figure S4).

In the sweet taste preference test, there was no significant difference between groups in terms of the consumption of saccharin solution expressed as a percentage of daily beverage intake, regardless of the day of testing (Figure 10). Strikingly, the preference for sweet taste for the GOS/In group did not reach statistical significance on the first day of the test, in contrast to the FOS and αGOS groups. However, this preference did not persist on day 2, contrary to what was observed for the CTL group. This suggests that neonatal supplementation with OS slightly reduced the persistence of sweet preference in adulthood.

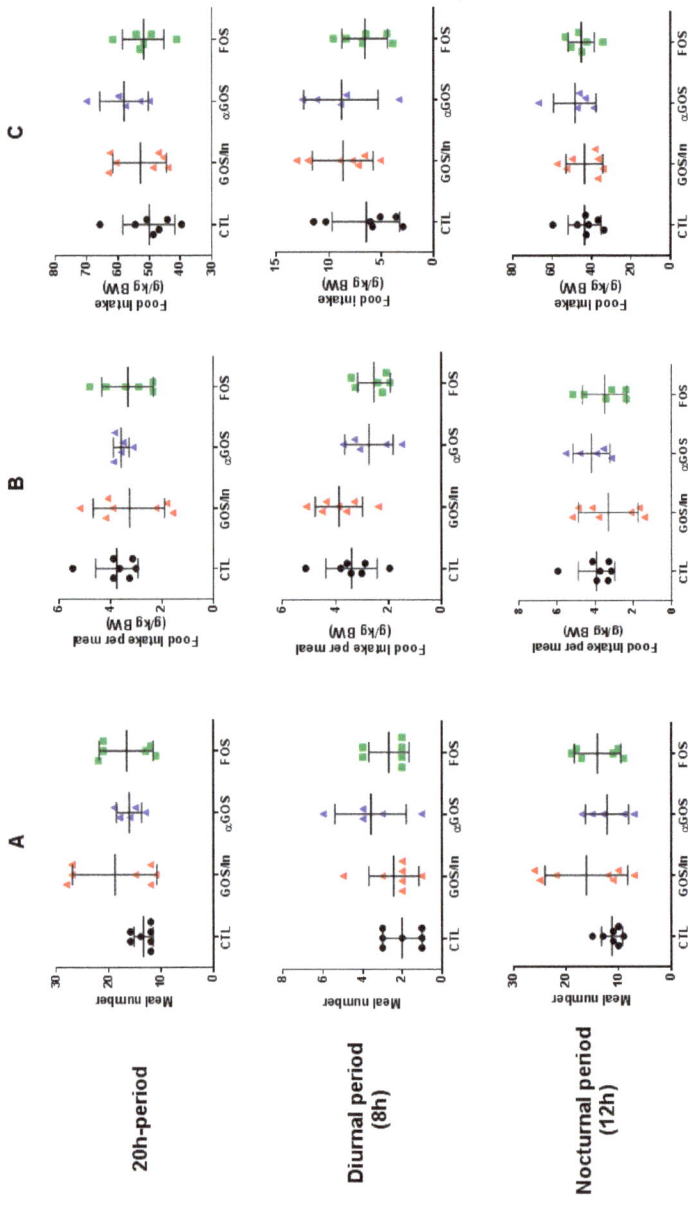

Figure 9. Feeding patterns illustrated by (**A**) Meal number; (**B**) Food intake per meal; (**C**) Food intake during the considered period analyzed in physiological cages at PND 75–100 (n = 5–7 per group). BW, bodyweight. Individuals, means and SD are plotted (n = 5 to 7 per groups).

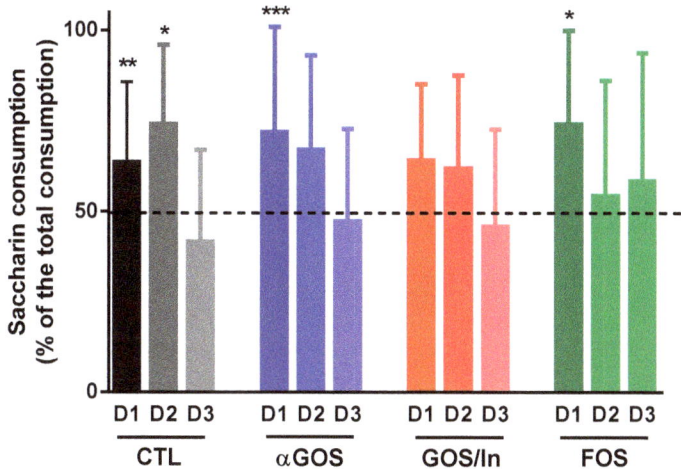

Figure 10. Preference for sweet taste. Data are means ± SD (n = 7 to 8). Asterisks represent significant preference as compared with no preference (i.e., 50%, dotted line): *, $p < 0.05$; **, $p < 0.01$; ***, $p < 0.001$.

3.3.2. Neonatal OS Supplementation Did Not Significantly Program Adult Intestinal Microbiota

At adult age (PND 124/126), no significant differences were observed between treatments with respect to the raw number of sequences obtained, percentages of sequences kept after quality filtering, or alpha-diversity indexes (data not shown). Similarly, β-diversity analysis (Figure S5), principal component analysis on OTU abundances (Data not shown) and comparisons of the cumulated relative abundances at family level (Figure 11) failed to show any significant difference between cecal samples with respect to neonatal supplementation. Finally, neither ileal nor cecal concentrations of SCFA showed significant differences between the groups (Table S2).

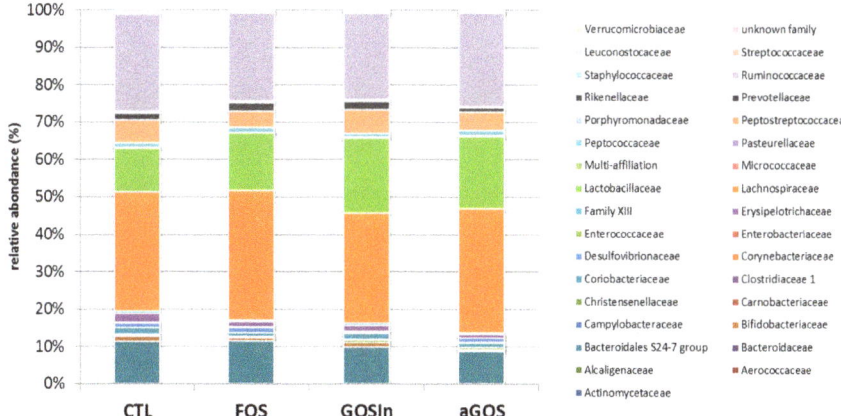

Figure 11. Impact of postnatal OS-supplementation on cecal microbiota composition at PND124/126: families distribution expressed as the average of cumulated relative abundances (n = 7 to 8 per group).

Overall, these data did not reveal that neonatal OS supplementation had any programming effect on adult microbiota.

4. Discussion

Considering that the regulation of feeding behavior could be programmed from the beginning of life and controlled by intestinal microbiota, we hypothesized that modifications to the neonatal microbiota could program adult feeding behavior. We therefore checked the ability of prebiotic-induced intestinal microbiota modulations to affect the maturation and functioning of L-EECs in suckled male rats, then assessed whether this resulted in delayed alterations in eating behavior and the secretion of GI peptides in adulthood. The observed effects are specifically attributable to OS since we adjusted the compositions of the administered solutions by taking into account the digestible sugar contents of commercial OS sources. In this study, we show that neonatal supplementation with 3 different OS strongly impacts cecocolonic microbiota, GLP-1 cell density in the ileum, and the production of satiety-related peptides during the neonatal period, but does not induce any significant enduring effect in adulthood on either eating behavior or gut peptide secretion.

The validity of this statement is obviously limited to our operating conditions which represent both strengths and limitations for our study.

Limitations include the fact that we only studied males in order to avoid the already described fluctuations in food intake throughout the estrous cycles [41], and did not characterize every components of eating behaviour such as motivation. However, we believe that the numerous components investigated allow for consideration of both its homeostatic and hedonic elements. We did not investigate immediate impact of OS supplementations on feeding behavior to avoid the recognized stress induced in pups by separation from the mother which would have been required to quantify milk intake either by gravimetric [42] or deuterated water turnover methods [43]. We did consider moreover whether neonatal prebiotic supplementation having an impact on the pups' eating behavior was beyond the scope of the programming of adult eating behavior. Nevertheless, we reported a transitory reduction in BW gain in FOS and GOS/In groups between PND6 and PND10 which suggests that the reducing impact of OS on food intake may also operate in the neonatal period.

Inversely, our study has three major advantages: the combination of hormonal, behavioral and microbiological analyzes; the minimizing of the influence of lactating mother influence by supplementing pups from the same litters with the different OS, and finally the use of OS doses comparable to those actually consumed by toddlers.

4.1. Neonatal OS Supplementation Affected Intestinal Microbiota Despite Its Immaturity

Corroborating our previous findings based on a non-exhaustive analysis of the microbiota [24], and in concurrence with several in vivo and in vitro studies investigating the impact in adulthood of OS (including those of the αGOS [44]) on intestinal microbiota, in humans and animals (e.g., [45,46]) and in human infants (see [32] for review), all the oligosaccharides used here dramatically affected neonatal microbiota in rat pups. This confirms that the prebiotic properties previously demonstrated in adult rodents (e.g., [30,45,46]), also operate in neonatal pups despite the immaturity of the microbiota at this stage of development [37].

In addition to these changes in composition and the reduction in microbiota richness, our neonatal OS supplementations also modified the activity of the microbiota by stimulating the production of acetate and lactate at the expense of that of propionate. This decrease in propionate concentration stands out from what is observed in adult rats, for which GOS and FOS are frequently reported as being particularly stimulating for propionate and/or butyrate production (e.g., [34,47]) and could be related to the known progressive maturation of the microbiotal capacity to synthesis the different SCFAs in neonates [37,48]. The production of butyrate is therefore barely detectable before the day 16 of life in rats [37]. In any case, the neonatal OS supplementation we performed resulted in microbiota that differed greatly from that of unsupplemented animals, an observation which was a prerequisite for investigating the ability of neonatal microbiota modulation to program adult eating behavior or gut peptide response.

4.2. OS Supplementation May Stimulate Ileal EECs to Produce GLP-1/PYY While Acting in Feedback on Endocrine Precursors

Our results showed that neonatal OS supplementation had immediate effects on mature ileal GLP-1-cells by increasing the density in villi and the mRNA expression of *Gcg* and *PYY* leading to enhanced plasma concentrations in these two anorectic peptides. These new observations in neonatal rats are consistent with those reported in adult rats for FOS and GOS/In [33,34,49–51] and are, to our knowledge, reported here for the first time for αGOS. In one of these previous studies, this increased production of GLP-1 was related to a higher differentiation of *Neurog3*-expressing EEC progenitors into L-cells in the colon [50]. Here, we demonstrate a drastic down-regulation of endocrine lineage-devoted genes during OS supplementation, mainly in the ileum. This unexpected result is difficult to reconcile as an effect of OS on early endocrine precursors leading to the production of more L-cell subtypes.

Neurog3 marks the endocrine progenitors and is essential for generating new EECs [52]. Post-neurog3 differentiation and maturation of EECs is controlled by dynamics in transcriptional factors such *Neurod1*, *Pax4* and *Pax6* and many others (*Arx*, *Pdx1*, *Foxa1* and *Foxa2*). The hierarchy of these events is still poorly understood [53] and the extrinsic factors that may interplay remain largely unknown. For this study, the well-known effect of OS prebiotics in stimulating L-cells cannot simply be explained by the impact on endocrine precursors, as suggested in the above-mentioned study [50]. Since we know that *Neurog3* expression is restricted to immature proliferative cells, the decreased Neurog3 expression we observed in the ileum may instead reflect a feedback regulation to limit new EEC generation in response to OS supplementation. A similar observation (decreased duodenal Neurog3 and increased EEC density) was reported in a model of maternal deprivation [26]. These data and our own suggest that the postnatal environment affects the differentiation of EEC precursors but not the proliferation of progenitors, leading to increased EEC density. High levels of circulating GLP-1 have been previously attributed to the increased number of ileal L-cells in *Gcgr*-deleted mice, and this effect involved up-regulation of post-*neurog3* transcription factors, affecting the proliferation of L-cells precursors [54]. Here, the expression of these factors, i.e., *Neurod1*, *Pax4* and *Pax6*, was reduced in OS-supplemented groups with high circulating levels of GLP-1, suggesting a different mechanism in the increased density of L-cells. In this respect, it should be noted that although EECs are still classified according to their major/unique hormone product (as for example GLP-1 for L-cells), it is now acknowledged that EECs are multihormonal [53,55]. In particular, more recent data has demonstrated that mature differentiated EECs display hormonal plasticity, allowing them to change their hormonal products in response to extrinsic factors, such as bone morphogenic proteins (BMP) during their migration along the crypt-villus axis [56,57]. Thus, the increased L-cell density observed here may be the result of the direct effect of OS on this plasticity to produce more GLP-1, independently of early markers of EEC proliferation and differentiation. Interestingly, in this study, the production of CCK—a key early-satiety peptide—was not affected by the OS supplementation at PND14/15 (data not shown) reinforcing the specificity of the effect of OS on EECs in producing GLP-1 and PYY in a segment of gut where CCK is not predominantly produced. How OS can modulate both the identity of EEC subtypes and/or the expression of GI peptides by acting on extrinsic factors (such as villus-produced BMP) needs further investigation.

4.3. What are the Putative Mediators of the Massive Effect of Neonatal OS Supplementation on Ileal L-Cells?

Identification of the small intestine rather than the colon as a privileged site for the action of OS on transcriptional activity has been previously reported in studies involving adult animals [58,59]. Conventionalization of germ-free mice led to similar observations (e.g., [60]). However, a nutritional modulation by OS supplementation may have a different impact on ileal epithelium compared to the absence or presence of microbiota. For example, in the Arora's study [60], conventionalization of germ-free mice led to the down-regulation of GLP-1 secreting vesicle process in L-cells, whereas we observed an increase in GLP-1 and PYY production. These contrasting results may stem from either inter-individual variability, or more likely the great differences in age between the animals studied.

Nevertheless, our data raise the question of how OS modulation of microbiota could act on ileal L-cells. The well-known capacity of SCFA (mainly butyrate but also propionate or even, non-consensually, acetate) to stimulate PYY and/or GLP-1 production [61–63] seems inconsistent with our observation of an OS-impact mainly localized in the ileum, within the context of no propionate/butyrate synthesis.

Others potential mechanisms include acidification of the luminal milieu or changes in the pathogen-associated molecular patterns (PAMPs). Zhou et al. [61] showed that changes in pH from 7.5 to 6.5 induce *per se* an increase in *Gcg* expression by STC-1 cells in vitro. Apart from this, it is known that EECs have receptors for PAMPs (i.e., Toll-like receptors) (see [18] for review). This is of particular interest since it has been demonstrated that some bacterial strains elicit GLP-1 secretion through signaling agents of the Toll-like receptor system, as illustrated by the fact that a MyD88 blockade triggers GLP-1 secretion induced by bacteria [64].

4.4. OS impact on Eating Behavior, Usually Observed Simultaneously with Their Consumption, Does Not Seem to Be Programmable

Despite a certain disparity in the literature, possibly related to the heterogeneity in dosage or methodology, several studies have reported the beneficial effect of OS prebiotics—mainly fructans but also αGOS, on the eating habits of healthy adults [65,66] or overweight adults [67,68], such as feelings of reduced hunger, increased satiety or reduced energy consumption. Note that the existing literature does not establish whether this is also true in infants, who are frequently given prebiotic supplements. In concurrence with human data, decreased food/energy intake has been evidenced in adult rodents supplemented with fructans [49,51] or βGOS [34]. In both models, these effects have been related to SCFA production by colonic bacteria during OS supplementation. For each of the 3 main SCFAs, i.e., acetate, propionate and butyrate, it has been demonstrated that they reduce energy intake, particularly in rodent models of diet-induced obesity [69–71], although conflicting results are reported [72], probably dependent on the mode (orogastric [71,72], intraperitoneal [70], intracerebroventricular [70], colonic delivery via fermentable fibers [69,71], etc.) and duration (acute [69,70] vs. chronic [69,72]) of SCFA or SCFA precursors administration. In humans, this hypothesis has been substantiated for both acetate and propionate by numerous studies focusing on appetite-related parameters (see [73] for review) as well as observations of reduced hedonic response to high-energy foods regulated in striatum [74] or reduced energy intake following the administration of propionate precursors in overweight adults [75]. How these SCFA regulate appetite directly at hypothalamic level [70] or via a vagal-dependent mechanism [71,72], whether or not implicating an enhanced intestinal satiety peptide (GLP-1 and PYY) secretion following SCFA interaction with FFAR receptors on L-cells is still a matter of experimental research in animal models and clinical trials in humans [18].

Since the perinatal environment [6,9,24,26] appears to have a long-lasting impact on each of the microbiota-EEC-brain axis actors, we had assumed that early modulation of the microbiota associated with changes in EECs could program eating behavior, a hypothesis which has remained unexplored until now. However, this hypothesis could not be corroborated in this study as adult feeding behavior did not seem to be significantly affected by early supplementation with OS, which nonetheless increased total SCFA, along with increased release of GLP-1 and PYY and L-cell density at the end of supplementation. This lack of eating behavior programming indicates that none of the presupposed events (i.e., programming of EEC or vagal sensitivity and/or microbiota programming) occurred under the test conditions. In fact, no difference in the expression of *c-Fos* was observed in the nucleus of the solitary tract in the rat's brainstem 2 h after administering a bolus of glucose in adult rats (data not shown). It therefore seems that depending on the nature and intensity of the perinatal stressor (maternal protein restriction [9], maternal deprivation [26] or postnatal modulation of microbiota by OS) the long-lasting impact is not systematic. For the microbiota, the lack of programming could be related to an inadequacy in the timing for applying the modulation, as discussed below.

4.5. Is Programming of the Microbiota Subject to Particular Timing?

In this study, we did not observe any programming effect of neonatal OS supplementation on adult microbiota. This result is in line with what we had previously observed for FOS [24] but contradicts the small-scale programming found after neonatal supplementation with GOS/In in this same study. This discrepancy may result from the difference in methods used to analyze the composition of the microbiota, even if it is counterintuitive, since the 16S rDNA sequencing used here is more exhaustive than the qPCR used previously. As this impact was minor, it may also not have been possible to reproduce under our new experimental conditions, i.e., a new batch of animals, a different room at our animal facility, or even a slight difference in the composition of the semi-purified diets we used, since all these parameters are known to affect the microbiota of laboratory animals (see [76] for review).

The disappearance of this nonetheless drastic effect in our animals at the end of supplementation raises the question of what is the most favorable period for sustainable modulation of the composition of the microbiota. In our experimental protocol, prebiotic supplementation was applied for a short postnatal period and ended before the onset of solid food consumption, whereas studies reporting programming effects for early supplementation with OS on the subsequent composition of the microbiota were based on longer-term supplementation, ranging from the prenatal period (i.e., supplementation of gestating mothers) to complete weaning and even beyond [77,78]. Whether the supplementation we applied was either not early enough, not late enough or not for a long enough time is difficult to establish on the sole basis of this comparison. However, in a study by Fugiwara et al. [77], a difference in adult microbiota composition was observed only in mice offspring that were supplemented with FOS beyond weaning. Whether this was also true in the Le Bourgot et al. study [78] cannot be evaluated since all piglets were supplemented with FOS for a few weeks after weaning. From this, it can be assumed that to be lastingly effective, prebiotics must be able to exert their microbiotal effect after full weaning, thereby controlling the impact of new bacterial sources and changes in dynamics of bacterial populations that result from the switch from maternal milk to solid food. Such a switch has been associated with dramatic changes in microbiota composition and activity both in humans [79] and rats [37]. This hypothesis would explain why the early-life events that are known to affect neonatal microbiota composition (i.e., birth mode, infant feeding etc.) are not associated with significant variations in adult microbiota composition [80], but strict comparisons between the time windows for supplementation are required for this to be validated.

4.6. All OS Studied Performed Similarly Despite Differences in Their Chemical Characteristics

In our study, the 3 OS studied led to comparable results in terms of both microbiotal impact and physiological repercussions. With regard to microbiotal changes, the observed modifications, in particular the acidification of the contents, the less diversified production of SCFA and the reduced richness of microbiota suggest that OS delays bacterial diversification. This is similar to what is supposed to happen in breast-fed babies compared with babies fed with unsupplemented formula [81]. The similarity is quite surprising in that the chemical nature of the constituent monomers and the pattern of glycoside linkages in different OS products are expected to influence the ability of individual bacteria to grow on them (see [31,82] for reviews). However, our results are consistent with Harris et al.'s findings [83] that the orientation of glycoside linkage is not a main driver of the SCFA production profile. When this chemical difference could act, it would primarily modulate the proportion of butyrate, an SCFA weakly produced in our immature animals. In addition, they also agree with the similarities of microbiotal impacts reported between βGOS and FOS on the one hand [30], and between αGOS and βGOS on the other [44].

Thus, our study confirms the prebiotic character of αGOS and, in addition, extends the well-known activity of FOS and GOS/In as secretagogues of satiety enteropeptides to this new prebiotic, a finding which is in accordance with the satietogenic effect described in humans [67].

5. Conclusions

In conclusion, our study depicts that the ability of the OS to modulate EECs as previously described in adults also operates in the neonatal period, despite the immaturity of the microbiota at this time. This observation therefore calls into question the nature of the mediators actually involved, as supposed so far. In addition, our in-depth study of the impacts of the OS on the genes regulating the differentiation of EEC precursors queries the current understanding of the ontogenesis of these cells.

Finally, our results do not demonstrate any programming impact of OS either on EECs and food consumption or on the constitution of the adult microbiota. If this holds true for humans, it is reassuring since this study concerns types and dosages of OS mimicking some of those commonly prescribed in formula for toddlers.

Supplementary Materials: The following are available online at http://www.mdpi.com/2072-6643/11/9/1967/s1, Figure S1: Postnatal growth of suckling rats in the different groups of OS supplementation, Table S1: Primer sequences, Figure S2: Growth of weaned rats until adulthood, Figure S3: Relative expression of gene implied in the endocrine lineage and in L-cells differentiation in the colon. Figure S4: Supplemental parameters of meal pattern. Figure S5: Hierarchical clustering of phylogenetically informed distance matrix computed using the unweighted UniFrac metric for cecal contents collected at PND 124/126. Table S2: Concentration of SCFA in ileal and cecal contents at PND 124/126.

Author Contributions: C.M., G.L.D. and M.R.-D. conceived and designed the experiments with the help of P.P.; C.M., M.R.-D., G.L.D. and A.-L.P. performed the experiments; A.-L.P., A.P. and I.G. contributed to biological analyses; C.M., G.L.D. and H.B. analyzed the data and prepared figures; G.L.D. and C.M. drafted the manuscript. E.C. and P.P. revised the manuscript.

Funding: The present study was partly funded by OLYGOSE.

Acknowledgments: Authors thank Vincent Paillé and Fanny Morel for their scientific advices; Edith Gouyon, Aurélie Reufflet, Agnès David-Sochard, Martine Rival, Guillaume Poupeau and Elise Beneteau for their technical assistance. They are also grateful to the Genotoul platforms (GetPlaGe, Genotoul-bioinfo and Sigenae, INRA, Toulouse-Midi-Pyrénées, France) for 16S rDNA sequencing and for providing help in computing and storage resources thanks to Galaxy instance.

Conflicts of Interest: The authors declare that they had a financial relationship with the organization that partly sponsored the research. E.C. is employee of Olygose.

References

1. Lindsay, A.C.; Sitthisongkram, S.; Greaney, M.L.; Wallington, S.F.; Ruengdej, P. Non-Responsive Feeding Practices, Unhealthy Eating Behaviors, and Risk of Child Overweight and Obesity in Southeast Asia: A Systematic Review. *Int. J. Env. Res. Public Health* **2017**, *14*, 436. [CrossRef] [PubMed]
2. MacLean, P.S.; Blundell, J.E.; Mennella, J.A.; Batterham, R.L. Biological control of appetite: A daunting complexity. *Obesity* **2017**, *25*, S8–S16. [CrossRef] [PubMed]
3. Berthoud, H.-R.; Sutton, G.M.; Townsend, R.L.; Patterson, L.M.; Zheng, H. Brainstem mechanisms integrating gut-derived satiety signals and descending forebrain information in the control of meal size. *Physiol. Behav.* **2006**, *89*, 517–524. [CrossRef] [PubMed]
4. Gaetani, S.; Romano, A.; Provensi, G.; Ricca, V.; Lutz, T.; Passani, M.B. Eating disorders: From bench to bedside and back. *J. Neurochem.* **2016**, *139*, 691–699. [CrossRef] [PubMed]
5. Cripps, R.L.; Martin-Gronert, M.S.; Ozanne, S.E. Fetal and perinatal programming of appetite. *Clin. Sci.* **2005**, *109*, 1–11. [CrossRef]
6. Ross, M.G.; Desai, M. Developmental programming of appetite/satiety. *Ann. Nutr. Metab.* **2014**, *64*, 36–44. [CrossRef]
7. Desai, M.; Gayle, D.; Han, G.; Ross, M.G. Programmed hyperphagia due to reduced anorexigenic mechanisms in intrauterine growth-restricted offspring. *Reprod. Sci.* **2007**, *14*, 329–337. [CrossRef]
8. Coupé, B.; Delamaire, E.; Hoebler, C.; Grit, I.; Even, P.; Fromentin, G.; Darmaun, D.; Parnet, P. Hypothalamus integrity and appetite regulation in low birth weight rats reared artificially on a high-protein milk formula. *J. Nutr. Biochem.* **2011**, *22*, 956–963. [CrossRef]

9. Ndjim, M.; Poinsignon, C.; Parnet, P.; Le Dréan, G. Loss of Vagal Sensitivity to Cholecystokinin in Rats Born with Intrauterine Growth Retardation and Consequence on Food Intake. *Front. Endocrinol.* **2017**, *8*, 65. [CrossRef]
10. Martin Agnoux, A.; Alexandre-Gouabau, M.-C.; Le Dréan, G.; Antignac, J.-P.; Parnet, P. Relative contribution of foetal and post-natal nutritional periods on feeding regulation in adult rats. *Acta Physiol.* **2014**, *210*, 188–201. [CrossRef]
11. Van Deutekom, A.W.; Chinapaw, M.J.M.; Jansma, E.P.; Vrijkotte, T.G.M.; Gemke, R.J.B.J. The Association of Birth Weight and Infant Growth with Energy Balance-Related Behavior—A Systematic Review and Best-Evidence Synthesis of Human Studies. *PLoS ONE* **2017**, *12*, e0168186. [CrossRef] [PubMed]
12. Lussana, F.; Painter, R.C.; Ocke, M.C.; Buller, H.R.; Bossuyt, P.M.; Roseboom, T.J. Prenatal exposure to the Dutch famine is associated with a preference for fatty foods and a more atherogenic lipid profile. *Am. J. Clin. Nutr.* **2008**, *88*, 1648–1652. [CrossRef] [PubMed]
13. Perälä, M.-M.; Männistö, S.; Kaartinen, N.E.; Kajantie, E.; Osmond, C.; Barker, D.J.P.; Valsta, L.M.; Eriksson, J.G. Body size at birth is associated with food and nutrient intake in adulthood. *PLoS ONE* **2012**, *7*, e46139. [CrossRef] [PubMed]
14. van Deutekom, A.W.; Chinapaw, M.J.M.; Vrijkotte, T.G.M.; Gemke, R.J.B.J. The association of birth weight and postnatal growth with energy intake and eating behavior at 5 years of age—A birth cohort study. *Int. J. Behav. Nutr. Phys. Act.* **2016**, *13*, 15. [CrossRef] [PubMed]
15. De Cosmi, V.; Scaglioni, S.; Agostoni, C. Early Taste Experiences and Later Food Choices. *Nutrients* **2017**, *9*, 107. [CrossRef] [PubMed]
16. Reimer, R.A.; McBurney, M.I. Dietary fiber modulates intestinal proglucagon messenger ribonucleic acid and postprandial secretion of glucagon-like peptide-1 and insulin in rats. *Endocrinology* **1996**, *137*, 3948–3956. [CrossRef] [PubMed]
17. Kok, N.N.; Morgan, L.M.; Williams, C.M.; Roberfroid, M.B.; Thissen, J.P.; Delzenne, N.M. Insulin, glucagon-like peptide 1, glucose-dependent insulinotropic polypeptide and insulin-like growth factor I as putative mediators of the hypolipidemic effect of oligofructose in rats. *J. Nutr.* **1998**, *128*, 1099–1103. [CrossRef] [PubMed]
18. Van de Wouw, M.; Schellekens, H.; Dinan, T.G.; Cryan, J.F. Microbiota-Gut-Brain Axis: Modulator of Host Metabolism and Appetite. *J. Nutr.* **2017**, *147*, 727–745. [CrossRef] [PubMed]
19. Glenny, E.M.; Bulik-Sullivan, E.C.; Tang, Q.; Bulik, C.M.; Carroll, I.M. Eating Disorders and the Intestinal Microbiota: Mechanisms of Energy Homeostasis and Behavioral Influence. *Curr. Psych. Rep.* **2017**, *19*, 51. [CrossRef]
20. Morita, C.; Tsuji, H.; Hata, T.; Gondo, M.; Takakura, S.; Kawai, K.; Yoshihara, K.; Ogata, K.; Nomoto, K.; Miyazaki, K.; et al. Gut Dysbiosis in Patients with Anorexia Nervosa. *PLoS ONE* **2015**, *10*, e0145274. [CrossRef]
21. Rabot, S.; Membrez, M.; Bruneau, A.; Gérard, P.; Harach, T.; Moser, M.; Raymond, F.; Mansourian, R.; Chou, C.J. Germ-free C57BL/6J mice are resistant to high-fat diet-induced insulin resistance and have altered cholesterol metabolism. *FASEB. J.* **2010**, *24*, 4948–4959. [CrossRef] [PubMed]
22. Cani, P.D.; Neyrinck, A.M.; Maton, N.; Delzenne, N.M. Oligofructose promotes satiety in rats fed a high-fat diet: Involvement of glucagon-like Peptide-1. *Obes. Res.* **2005**, *13*, 1000–1007. [CrossRef] [PubMed]
23. Maurer, A.D.; Chen, Q.; McPherson, C.; Reimer, R.A. Changes in satiety hormones and expression of genes involved in glucose and lipid metabolism in rats weaned onto diets high in fibre or protein reflect susceptibility to increased fat mass in adulthood. *J. Physiol.* **2009**, *587*, 679–691. [CrossRef] [PubMed]
24. Morel, F.B.; Oozeer, R.; Piloquet, H.; Moyon, T.; Pagniez, A.; Knol, J.; Darmaun, D.; Michel, C. Preweaning modulation of intestinal microbiota by oligosaccharides or amoxicillin can contribute to programming of adult microbiota in rats. *Nutrition* **2015**, *31*, 515–522. [CrossRef] [PubMed]
25. Williams, S.; Chen, L.; Savignac, H.M.; Tzortzis, G.; Anthony, D.C.; Burnet, P.W.J. Neonatal prebiotic (BGOS) supplementation increases the levels of synaptophysin, GluN2A-subunits and BDNF proteins in the adult rat hippocampus. *Synapse* **2016**, *70*, 121–124. [CrossRef] [PubMed]
26. Estienne, M.; Claustre, J.; Clain-Gardechaux, G.; Paquet, A.; Taché, Y.; Fioramonti, J.; Plaisancié, P. Maternal deprivation alters epithelial secretory cell lineages in rat duodenum: Role of CRF-related peptides. *Gut* **2010**, *59*, 744–751. [CrossRef] [PubMed]

27. Mischke, M.; Plösch, T. More than just a gut instinct-the potential interplay between a baby's nutrition, its gut microbiome, and the epigenome. *Am. J. Physiol. Regul. Integr. Comp. Physiol.* **2013**, *304*, R1065–R1069. [CrossRef]
28. Forssberg, H. Microbiome programming of brain development: Implications for neurodevelopmental disorders. *Dev. Med. Child. Neurol.* **2019**, *61*, 744–749. [CrossRef]
29. Savignac, H.M.; Couch, Y.; Stratford, M.; Bannerman, D.M.; Tzortzis, G.; Anthony, D.C.; Burnet, P.W.J. Prebiotic administration normalizes lipopolysaccharide (LPS)-induced anxiety and cortical 5-HT2A receptor and IL1-β levels in male mice. *Brain Behav. Immun.* **2016**, *52*, 120–131. [CrossRef]
30. Burokas, A.; Arboleya, S.; Moloney, R.D.; Peterson, V.L.; Murphy, K.; Clarke, G.; Stanton, C.; Dinan, T.G.; Cryan, J.F. Targeting the Microbiota-Gut-Brain Axis: Prebiotics Have Anxiolytic and Antidepressant-like Effects and Reverse the Impact of Chronic Stress in Mice. *Biol. Psych.* **2017**, *82*, 472–487. [CrossRef]
31. Macfarlane, G.T.; Steed, H.; Macfarlane, S. Bacterial metabolism and health-related effects of galacto-oligosaccharides and other prebiotics. *J. Appl. Microbiol.* **2008**, *104*, 305–344. [CrossRef]
32. Skórka, A.; Pieścik-Lech, M.; Kołodziej, M.; Szajewska, H. Infant formulae supplemented with prebiotics: Are they better than unsupplemented formulae? An updated systematic review. *Br. J. Nutr.* **2018**, *119*, 810–825. [CrossRef]
33. Delzenne, N.M.; Cani, P.D.; Neyrinck, A.M. Modulation of glucagon-like peptide 1 and energy metabolism by inulin and oligofructose: Experimental data. *J. Nutr.* **2007**, *137*, 2547S–2551S. [CrossRef]
34. Overduin, J.; Schoterman, M.H.C.; Calame, W.; Schonewille, A.J.; Ten Bruggencate, S.J.M. Dietary galacto-oligosaccharides and calcium: Effects on energy intake, fat-pad weight and satiety-related, gastrointestinal hormones in rats. *Br. J. Nutr.* **2013**, *109*, 1338–1348. [CrossRef]
35. Sabater, C.; Prodanov, M.; Olano, A.; Corzo, N.; Montilla, A. Quantification of prebiotics in commercial infant formulas. *Food Chem.* **2016**, *194*, 6–11. [CrossRef]
36. Silva, M.T. Saccharin aversion in the rat following adrenalectomy. *Physiol. Behav.* **1977**, *19*, 239–244. [CrossRef]
37. Fança-Berthon, P.; Hoebler, C.; Mouzet, E.; David, A.; Michel, C. Intrauterine growth restriction not only modifies the cecocolonic microbiota in neonatal rats but also affects its activity in young adult rats. *J. Pediatr. Gastroenterol. Nutr.* **2010**, *51*, 402–413. [CrossRef]
38. Caporaso, J.G.; Lauber, C.L.; Walters, W.A.; Berg-Lyons, D.; Lozupone, C.A.; Turnbaugh, P.J.; Fierer, N.; Knight, R. Global patterns of 16S rRNA diversity at a depth of millions of sequences per sample. *Proc. Natl. Acad. Sci. USA* **2011**, *108*, 4516–4522. [CrossRef]
39. Escudié, F.; Auer, L.; Bernard, M.; Mariadassou, M.; Cauquil, L.; Vidal, K.; Maman, S.; Hernandez-Raquet, G.; Combes, S.; Pascal, G. FROGS: Find, Rapidly, OTUs with Galaxy Solution. *Bioinformatics* **2018**, *34*, 1287–1294. [CrossRef]
40. R Core Team. *R: A Language and Environment for Statistical Computing*; R Foundation for Statistical Computing: Vienna, Austria, 2018; Available online: https://www.R-project.org/ (accessed on 10 May 2019).
41. Sample, C.H.; Davidson, T.L. Considering sex differences in the cognitive controls of feeding. *Physiol. Behav.* **2018**, *187*, 97–107. [CrossRef]
42. Pueta, M.; Abate, P.; Haymal, O.B.; Spear, N.E.; Molina, J.C. Ethanol exposure during late gestation and nursing in the rat: Effects upon maternal care, ethanol metabolism and infantile milk intake. *Pharmacol. Biochem. Behav.* **2008**, *91*, 21–31. [CrossRef]
43. Sevrin, T.; Alexandre-Gouabau, M.C.; Darmaun, D.; Palvadeau, A.; André, A.; Nguyen, P.; Ouguerram, K.; Boquien, C.Y. Use of water turnover method to measure mother's milk flow in a rat model: Application to dams receiving a low protein diet during gestation and lactation. *PLoS ONE* **2017**, *12*, e0180550. [CrossRef]
44. Fehlbaum, S.; Prudence, K.; Kieboom, J.; Heerikhuisen, M.; van den Broek, T.; Schuren, F.H.J.; Steinert, R.E.; Raederstorff, D. In Vitro Fermentation of Selected Prebiotics and Their Effects on the Composition and Activity of the Adult Gut Microbiota. *Int. J. Mol. Sci.* **2018**, *19*, 3097. [CrossRef]
45. Liu, F.; Li, P.; Chen, M.; Luo, Y.; Prabhakar, M.; Zheng, H.; He, Y.; Qi, Q.; Long, H.; Zhang, Y.; et al. Fructooligosaccharide (FOS) and Galactooligosaccharide (GOS) Increase Bifidobacterium but Reduce Butyrate Producing Bacteria with Adverse Glycemic Metabolism in healthy young population. *Sci. Rep.* **2017**, *7*, 11789. [CrossRef]

46. Wang, L.; Hu, L.; Yan, S.; Jiang, T.; Fang, S.; Wang, G.; Zhao, J.; Zhang, H.; Chen, W. Effects of different oligosaccharides at various dosages on the composition of gut microbiota and short-chain fatty acids in mice with constipation. *Food Funct.* **2017**, *8*, 1966–1978. [CrossRef]
47. Le Blay, G.; Michel, C.; Blottière, H.M.; Cherbut, C. Prolonged intake of fructo-oligosaccharides induces a short-term elevation of lactic acid-producing bacteria and a persistent increase in cecal butyrate in rats. *J. Nutr.* **1999**, *129*, 2231–2235. [CrossRef]
48. Midtvedt, A.C.; Midtvedt, T. Production of short chain fatty acids by the intestinal microflora during the first 2 years of human life. *J. Pediatr. Gastroenterol. Nutr.* **1992**, *15*, 395–403. [CrossRef]
49. Cani, P.D.; Dewever, C.; Delzenne, N.M. Inulin-type fructans modulate gastrointestinal peptides involved in appetite regulation (glucagon-like peptide-1 and ghrelin) in rats. *Br. J. Nutr.* **2004**, *92*, 521–526. [CrossRef]
50. Cani, P.D.; Hoste, S.; Guiot, Y.; Delzenne, N.M. Dietary non-digestible carbohydrates promote L-cell differentiation in the proximal colon of rats. *Br. J. Nutr.* **2007**, *98*, 32–37. [CrossRef]
51. Parnell, J.A.; Reimer, R.A. Prebiotic fibres dose-dependently increase satiety hormones and alter Bacteroidetes and Firmicutes in lean and obese JCR:LA-cp rats. *Br. J. Nutr.* **2012**, *107*, 601–613. [CrossRef]
52. Li, H.J.; Ray, S.K.; Singh, N.K.; Johnston, B.; Leiter, A.B. Basic helix-loop-helix transcription factors and enteroendocrine cell differentiation. *Diabetes Obes. Metab.* **2011**, *13*, 5–12. [CrossRef]
53. Engelstoft, M.S.; Egerod, K.L.; Lund, M.L.; Schwartz, T.W. Enteroendocrine cell types revisited. *Curr. Opin. Pharm.* **2013**, *13*, 912–921. [CrossRef]
54. Grigoryan, M.; Kedees, M.H.; Charron, M.J.; Guz, Y.; Teitelman, G. Regulation of mouse intestinal L cell progenitors proliferation by the glucagon family of peptides. *Endocrinology* **2012**, *153*, 3076–3088. [CrossRef]
55. Habib, A.M.; Richards, P.; Cairns, L.S.; Rogers, G.J.; Bannon, C.A.M.; Parker, H.E.; Morley, T.C.E.; Yeo, G.S.H.; Reimann, F.; Gribble, F.M. Overlap of endocrine hormone expression in the mouse intestine revealed by transcriptional profiling and flow cytometry. *Endocrinology* **2012**, *153*, 3054–3065. [CrossRef]
56. Beumer, J.; Artegiani, B.; Post, Y.; Reimann, F.; Gribble, F.; Nguyen, T.N.; Zeng, H.; Van den Born, M.; Van Es, J.H.; Clevers, H. Enteroendocrine cells switch hormone expression along the crypt-to-villus BMP signalling gradient. *Nat. Cell Biol.* **2018**, *20*, 909–916. [CrossRef]
57. Gehart, H.; van Es, J.H.; Hamer, K.; Beumer, J.; Kretzschmar, K.; Dekkers, J.F.; Rios, A.; Clevers, H. Identification of Enteroendocrine Regulators by Real-Time Single-Cell Differentiation Mapping. *Cell* **2019**, *176*, 1158–1173.e16. [CrossRef]
58. Everard, A.; Lazarevic, V.; Derrien, M.; Girard, M.; Muccioli, G.G.; Muccioli, G.M.; Neyrinck, A.M.; Possemiers, S.; Van Holle, A.; François, P.; et al. Responses of gut microbiota and glucose and lipid metabolism to prebiotics in genetic obese and diet-induced leptin-resistant mice. *Diabetes* **2011**, *60*, 2775–2786. [CrossRef]
59. Cani, P.D.; Possemiers, S.; Van de Wiele, T.; Guiot, Y.; Everard, A.; Rottier, O.; Geurts, L.; Naslain, D.; Neyrinck, A.; Lambert, D.M.; et al. Changes in gut microbiota control inflammation in obese mice through a mechanism involving GLP-2-driven improvement of gut permeability. *Gut* **2009**, *58*, 1091–1103. [CrossRef]
60. Arora, T.; Akrami, R.; Pais, R.; Bergqvist, L.; Johansson, B.R.; Schwartz, T.W.; Reimann, F.; Gribble, F.M.; Bäckhed, F. Microbial regulation of the L cell transcriptome. *Sci. Rep.* **2018**, *8*, 1207. [CrossRef]
61. Zhou, J.; Martin, R.J.; Tulley, R.T.; Raggio, A.M.; McCutcheon, K.L.; Shen, L.; Danna, S.C.; Tripathy, S.; Hegsted, M.; Keenan, M.J. Dietary resistant starch upregulates total GLP-1 and PYY in a sustained day-long manner through fermentation in rodents. *Am. J. Physiol. Endocrinol. Metab.* **2008**, *295*, E1160–E1166. [CrossRef]
62. Tolhurst, G.; Heffron, H.; Lam, Y.S.; Parker, H.E.; Habib, A.M.; Diakogiannaki, E.; Cameron, J.; Grosse, J.; Reimann, F.; Gribble, F.M. Short-chain fatty acids stimulate glucagon-like peptide-1 secretion via the G-protein-coupled receptor FFAR2. *Diabetes* **2012**, *61*, 364–371. [CrossRef] [PubMed]
63. Larraufie, P.; Martin-Gallausiaux, C.; Lapaque, N.; Dore, J.; Gribble, F.M.; Reimann, F.; Blottiere, H.M. SCFAs strongly stimulate PYY production in human enteroendocrine cells. *Sci. Rep.* **2018**, *8*, 74. [CrossRef] [PubMed]
64. Panwar, H.; Calderwood, D.; Gillespie, A.L.; Wylie, A.R.; Graham, S.F.; Grant, I.R.; Grover, S.; Green, B.D. Identification of lactic acid bacteria strains modulating incretin hormone secretion and gene expression in enteroendocrine cells. *J. Funct. Foods* **2016**, *23*, 348–358. [CrossRef]

65. Cani, P.D.; Lecourt, E.; Dewulf, E.M.; Sohet, F.M.; Pachikian, B.D.; Naslain, D.; De Backer, F.; Neyrinck, A.M.; Delzenne, N.M. Gut microbiota fermentation of prebiotics increases satietogenic and incretin gut peptide production with consequences for appetite sensation and glucose response after a meal. *Am. J. Clin. Nutr.* **2009**, *90*, 1236–1243. [CrossRef] [PubMed]
66. Pedersen, C.; Lefevre, S.; Peters, V.; Patterson, M.; Ghatei, M.A.; Morgan, L.M.; Frost, G.S. Gut hormone release and appetite regulation in healthy non-obese participants following oligofructose intake. A dose-escalation study. *Appetite* **2013**, *66*, 44–53. [CrossRef] [PubMed]
67. Morel, F.B.; Dai, Q.; Ni, J.; Thomas, D.; Parnet, P.; Fança-Berthon, P. α-Galacto-oligosaccharides Dose-Dependently Reduce Appetite and Decrease Inflammation in Overweight Adults. *J. Nutr.* **2015**, *145*, 2052–2059. [CrossRef] [PubMed]
68. Reimer, R.A.; Willis, H.J.; Tunnicliffe, J.M.; Park, H.; Madsen, K.L.; Soto-Vaca, A. Inulin-type fructans and whey protein both modulate appetite but only fructans alter gut microbiota in adults with overweight/obesity: A randomized controlled trial. *Mol. Nutr Food Res.* **2017**, *61*. [CrossRef]
69. Lin, H.V.; Frassetto, A.; Kowalik, E.J., Jr.; Nawrocki, A.R.; Lu, M.M.; Kosinski, J.R.; Hubert, J.A.; Szeto, D.; Yao, X.; Forrest, G.; et al. Butyrate and propionate protect against diet-induced obesity and regulate gut hormones via free fatty acid receptor 3-independent mechanisms. *PLoS ONE* **2012**, *7*, e35240. [CrossRef]
70. Frost, G.; Sleeth, M.L.; Sahuri-Arisoylu, M.; Lizarbe, B.; Cerdan, S.; Brody, L.; Anastasovska, J.; Ghourab, S.; Hankir, M.; Zhang, S.; et al. The short-chain fatty acid acetate reduces appetite via a central homeostatic mechanism. *Nat. Commun.* **2014**, *5*, 3611. [CrossRef]
71. Li, Z.; Yi, C.X.; Katiraei, S.; Kooijman, S.; Zhou, E.; Chung, C.K.; Gao, Y.; van den Heuvel, J.K.; Meijer, O.C.; Berbée, J.F.P.; et al. Butyrate reduces appetite and activates brown adipose tissue via the gut-brain neural circuit. *Gut* **2018**, *67*, 1269–1279. [CrossRef]
72. Perry, R.J.; Peng, L.; Barry, N.A.; Cline, G.W.; Zhang, D.; Cardone, R.L.; Petersen, K.F.; Kibbey, R.G.; Goodman, A.L.; Shulman, G.I. Acetate mediates a microbiome-brain-β-cell axis to promote metabolic syndrome. *Nature* **2016**, *534*, 213–217. [CrossRef] [PubMed]
73. Darzi, J.; Frost, G.S.; Robertson, M.D. Do SCFA have a role in appetite regulation? *Proc. Nutr. Soc.* **2011**, *70*, 119–128. [CrossRef] [PubMed]
74. Byrne, C.S.; Chambers, E.S.; Alhabeeb, H.; Chhina, N.; Morrison, D.J.; Preston, T.; Tedford, C.; Fitzpatrick, J.; Irani, C.; Busza, A.; et al. Increased colonic propionate reduces anticipatory reward responses in the human striatum to high-energy foods. *Am. J. Clin. Nutr.* **2016**, *104*, 5–14. [CrossRef] [PubMed]
75. Chambers, E.S.; Viardot, A.; Psichas, A.; Morrison, D.J.; Murphy, K.G.; Zac-Varghese, S.E.K.; MacDougall, K.; Preston, T.; Tedford, C.; Finlayson, G.S.; et al. Effects of targeted delivery of propionate to the human colon on appetite regulation, body weight maintenance and adiposity in overweight adults. *Gut* **2015**, *64*, 1744–1754. [CrossRef] [PubMed]
76. Tomas, J.; Langella, P.; Cherbuy, C. The intestinal microbiota in the rat model: Major breakthroughs from new technologies. *Anim. Health Res. Rev.* **2012**, *13*, 54–63. [CrossRef]
77. Fujiwara, R.; Takemura, N.; Watanabe, J.; Sonoyama, K. Maternal consumption of fructo-oligosaccharide diminishes the severity of skin inflammation in offspring of NC/Nga mice. *Br. J. Nutr.* **2010**, *103*, 530–538. [CrossRef]
78. Le Bourgot, C.; Ferret-Bernard, S.; Apper, E.; Taminiau, B.; Cahu, A.; Le Normand, L.; Respondek, F.; Le Huërou-Luron, I.; Blat, S. Perinatal short-chain fructooligosaccharides program intestinal microbiota and improve enteroinsular axis function and inflammatory status in high-fat diet-fed adult pigs. *FASEB J.* **2019**, *33*, 301–313. [CrossRef]
79. Koenig, J.E.; Spor, A.; Scalfone, N.; Fricker, A.D.; Stombaugh, J.; Knight, R.; Angenent, L.T.; Ley, R.E. Succession of microbial consortia in the developing infant gut microbiome. *Proc. Natl. Acad. Sci. USA* **2011**, *108*, 4578–4585. [CrossRef]
80. Falony, G.; Joossens, M.; Vieira-Silva, S.; Wang, J.; Darzi, Y.; Faust, K.; Kurilshikov, A.; Bonder, M.J.; Valles-Colomer, M.; Vandeputte, D.; et al. Population-level analysis of gut microbiome variation. *Science* **2016**, *352*, 560–564. [CrossRef]
81. Mackie, R.I.; Sghir, A.; Gaskins, H.R. Developmental microbial ecology of the neonatal gastrointestinal tract. *Am. J. Clin. Nutr.* **1999**, *69*, 1035S–1045S. [CrossRef]

82. Louis, P.; Flint, H.J.; Michel, C. How to Manipulate the Microbiota: Prebiotics. *Adv. Exp. Med. Biol.* **2016**, *902*, 119–142. [CrossRef] [PubMed]
83. Harris, H.C.; Edwards, C.A.; Morrison, D.J. Impact of Glycosidic Bond Configuration on Short Chain Fatty Acid Production from Model Fermentable Carbohydrates by the Human Gut Microbiota. *Nutrients* **2017**, *9*, 26. [CrossRef]

© 2019 by the authors. Licensee MDPI, Basel, Switzerland. This article is an open access article distributed under the terms and conditions of the Creative Commons Attribution (CC BY) license (http://creativecommons.org/licenses/by/4.0/).

Article

Impact of Maternal Malnutrition on Gut Barrier Defense: Implications for Pregnancy Health and Fetal Development

Sebastian A. Srugo [1], Enrrico Bloise [2], Tina Tu-Thu Ngoc Nguyen [3] and Kristin L. Connor [1,3,*]

1. Department of Health Sciences, Carleton University, Ottawa, ON K1S 5B6, Canada; sebastian.srugo@carleton.ca
2. Department of Morphology, Federal University of Minas Gerais, Belo Horizonte 31270-901, Brazil; enrricobloise@biof.ufrj.br
3. Lunenfeld-Tanenbaum Research Institute, Mount Sinai Hospital, Toronto, ON M5G 1X5, Canada; nguyen.tinattn@gmail.com
* Correspondence: kristin.connor@carleton.ca; Tel.: +1-613-520-2600 (ext. 4202)

Received: 11 May 2019; Accepted: 10 June 2019; Published: 19 June 2019

Abstract: Small intestinal Paneth cells, enteric glial cells (EGC), and goblet cells maintain gut mucosal integrity, homeostasis, and influence host physiology locally and through the gut-brain axis. Little is known about their roles during pregnancy, or how maternal malnutrition impacts these cells and their development. Pregnant mice were fed a control diet (CON), undernourished by 30% vs. control (UN), or fed a high fat diet (HF). At day 18.5 (term = 19), gut integrity and function were assessed by immunohistochemistry and qPCR. UN mothers displayed reduced mRNA expression of Paneth cell antimicrobial peptides (AMP; *Lyz2*, *Reg3g*) and an accumulation of villi goblet cells, while HF had reduced *Reg3g* and mucin (*Muc2*) mRNA and increased lysozyme protein. UN fetuses had increased mRNA expression of gut transcription factor *Sox9*, associated with reduced expression of maturation markers (*Cdx2*, *Muc2*), and increased expression of tight junctions (TJ; *Cldn-7*). HF fetuses had increased mRNA expression of EGC markers (*S100b*, *Bfabp*, *Plp1*), AMP (*Lyz1*, *Defa1*, *Reg3g*), and TJ (*Cldn-3*, *Cldn-7*), and reduced expression of an AMP-activator (*Tlr4*). Maternal malnutrition altered expression of genes that maintain maternal gut homeostasis, and altered fetal gut permeability, function, and development. This may have long-term implications for host-microbe interactions, immunity, and offspring gut-brain axis function.

Keywords: malnutrition; gut barrier; development; pregnancy

---

1. Introduction

The gut is critical to host health and disease development through interactions with microbes that colonize the gut and establishment of the gut epithelial barrier [1,2]. Major gut functions include absorption of water and nutrients; secretion of digestive enzymes, bile, and mucus; and motility of luminal contents down the gastrointestinal tract and in a mixing motion to ensure adequate contact with the epithelium for absorption [3]. In health, the gut is involved in vitamin synthesis [4], nutrient metabolism [5], and protection against pathogens [1,6]. It is also important for brain function: a direct and bidirectional channel of communication exists between the gut and the brain, dubbed the gut-brain axis, which includes a connection between the enteric and central nervous systems (CNS) [7]. Microbes that reside in the gut, their metabolites [8], hormones [9], and immune factors [10], have been shown to affect gut and brain function through this axis in humans and animals [7]. Since microbes and the metabolites they produce can activate host immune, metabolic, and stress pathways once they leave the gut environment [11], an important part of the gut-microbe-host relationship is the separation between the gut tissue, the gut ecosystem, and the rest of the host.

This separation is mediated by the intestinal epithelial barrier, which lies at the interface between exogenous host factors and the internal gut microenvironment and helps to regulate microbe-host interactions. During periods of optimal nutrition and host health, two key cell types are involved in supporting and maintaining this gut barrier: Paneth cells and enteric glial cells (EGCs). Paneth cells reside in the epithelium of the small intestine (SI), maintain gut integrity and prevent microbial translocation through production of tight junction (TJ) proteins [12], and produce antimicrobial peptides (AMPs) which regulate the host-microbe relationship [13]. As well, due to their proximity to crypt stem cells [14,15], Paneth cells can affect gut epithelial cell differentiation and gut maturation [14,15]. Toll-like receptors (TLRs), which activate the innate immune response when bacterial components are recognized in the gut environment, are purported to play a key role in inducing Paneth AMP production [16]. In addition, EGCs are part of the enteric nervous system (ENS) and gut-brain axis [17], and respond to and control gut inflammation [18,19]. EGCs also influence gut integrity and permeability [20] through their long cytoplasmic processes which make direct contact with the barrier [21]. Other enteric cells, such as goblet cells, maintain gut barrier integrity through their production of mucus [12]. Importantly, Paneth cells and EGCs are established in early development (Figure S1) [14,22], suggesting they are key for early and lifelong gut and brain health.

The integrity of the gut barrier and the composition and function of the gut microbiome can be greatly affected by the diet of the host [23–25], since the indigestible polysaccharides in the host diet become gut bacterial substrates and nutrients [26]. Indeed, adaptations to gut bacterial metabolism and transcription occur within days of dietary changes in humans [24]. In response to both over- [27] and undernutrition [28], collectively known as malnutrition, gut barrier function and integrity can become dysregulated, leading to gut microbial dysbiosis, altered gut function, and a leaky gut barrier. Yet, few studies have elucidated the effects of malnutrition on Paneth cell [29–33] and EGC [34,35] development and function (indeed, none have examined EGCs during undernutrition to our knowledge), or how either of these cells are affected by and/or perpetuate gut dysfunction. Since the rates of both undernutrition and underweight [36], as well as over-nutrition and obesity [36], are increasing worldwide, and a compromised gut barrier is both caused by, and leads to, a variety of immune-related, and chronic diseases, the effects of malnutrition on the gut-host relationship are important to understand.

During pregnancy, the intestinal epithelial barrier and the microbes contained within the gut are doubly important, as they protect both the mother and, by association, the fetus from harmful bacteria and xenobiotics [1,37], produce nutrients required for pregnancy health [38], and absorb nutrients into the blood stream that are vital to fetal development [39,40]. We and others have shown that maternal malnutrition impacts the maternal gut microbiome [41–44] and is associated with increased levels inflammation in the maternal gut and peripheral circulation [41,44]. In offspring, a mature intestinal epithelial barrier ensures a healthy and homeostatic gut environment [45], which allows the offspring to appropriately respond to infections [45], absorb and produce nutrients [16], and likely establish optimal communication with the brain and other organs [7]. Still, little is known about how malnutrition impacts the maternal epithelial barrier during pregnancy, a 'stress-test' in itself, or whether maternal malnutrition adversely programmes fetal gut development and function. Additionally, although we know that Paneth cells [14] and EGCs [46,47] are laid down and functional in early life, we know less about how early life adversity, including poor nutrition, or gut microbes, may influence their development and function.

We therefore sought to answer two questions: how does malnutrition during pregnancy affect maternal gut barrier function, and does maternal malnutrition impact fetal gut integrity, function, and development? We focused on Paneth cells and EGCs to answer these questions as these cells may be critical for long-term gut function and communication with the brain. We hypothesized that malnutrition would lead to an adverse maternal gut environment, and that mothers fed a high fat (HF) diet and their offspring would display the most affected gut function, since our previous work has demonstrated that the maternal gut is especially affected by a HF diet [41]. We also hypothesized that, due to the changes in maternal diets and gut environments, the fetal gut would display aberrant gut integrity and function and EGC development, with different outcomes in fetuses from undernourished (UN) and HF mothers.

2. Materials and Methods

2.1. Animal Model

All housing and breeding procedures were approved by the Animal Care Committee at Mount Sinai Hospital (Toronto, ON, Canada; AUP 16-0091H). Male and female C57BL/6J mice (Jackson Laboratories, Bar Harbor, ME, USA) were housed using a 12:12 light:dark cycle at 25 °C, with free access to food and water. Females were randomized into three diet groups: mice fed control diet ad libitum before and throughout pregnancy (CON, $n = 7$); mice fed control diet ad libitum before mating and undernourished by 30% from gestational day (d) 5.5 to 17.5 (UN, $n = 7$); and mice fed high fat diet (60% of calories as fat) ad libitum from 8 weeks prior to mating and through pregnancy (HF, $n = 8$). Males were fed control diet ad libitum and mated with females at ~10 weeks of age. Females were housed individually following confirmation of pregnancy status (presence of vaginal sperm plug). Dams were weighed weekly before and daily during pregnancy.

2.2. Maternal SI and Fetal Gut Collection

Dams were sacrificed by cervical dislocation at d18.5 (term = d19). Fetuses were collected, weighed, and at random, one male and female fetus from each litter was used for fetal biospecimen collections. Maternal and fetal gastrointestinal (GI) tracts were dissected as detailed previously [41]. A 2–5 mm piece of maternal SI from the mid portion of the SI (representing the jejunum) and the entire fetal gut were flash frozen in liquid nitrogen then stored at −80 °C for later molecular analyses. Another 2–5 mm of maternal SI from the mid portion was flushed with buffered 4% paraformaldehyde (PFA), cut longitudinally, and cut into two pieces for fixing in 4% PFA at 4 °C overnight. Fixed SI were washed thrice with 1× PBS and stored in 70% ethanol until paraffin embedded for later immunohistochemical analyses.

2.3. RNA Extraction and mRNA Expression Analysis

Total RNA was extracted from maternal SI and fetal guts using the Tissue Lyser II (Qiagen, Hilden, NRW, Germany) and RNA extraction kits following manufacturer's instructions (QIAGEN RNeasy Plus Mini Kit, Toronto, ON, Canada). Eluted RNA quality and quantity were assessed by spectrophotometry (DeNovix, Wilmington, DE, USA), and 1 µg of RNA was reverse transcribed using 5× iScript Reverse Transcription Supermix (Bio-Rad, Mississauga, ON, Canada).

We focused on genes involved in gut barrier function, integrity, and development to establish how maternal malnutrition may impact the maternal gut and, by consequence, fetal gut and ENS development (Table 1). mRNA expression data were normalized to the geometric mean of the three stably-expressed reference genes: TATA-Box Binding Protein (*Tbp*), Tyrosine 3-Monooxygenase/Tryptophan 5-Monooxygenase Activation Protein Zeta (*Ywhaz*), and Beta-actin (*Actb*). Primers were designed from gene sequences found in the NCBI Nucleotide Database or taken from the literature and analyzed using NCBI Primer-BLAST and Oligo Calc (Northwestern University, Evanston, IL, USA) for appropriate gene targeting and properties. Amplification and detection of mRNA expression was measured using CFX384 Touch Real-Time PCR Detection System (Bio-Rad). Samples, standards, and controls were pipetted in triplicate. Inter-run calibrators and non-template controls were run alongside each gene to normalize between plates and to assess contamination, respectively. The PCR cycling conditions were: 30 s at 95 °C, 40 × 5 s at 95 °C, 20 s at 60 °C. Data were analyzed applying the Pfaffl method [48].

Table 1. Primer sequences for quantitative PCR.

Gene	Sequence 5′–3′	Source
\multicolumn{3}{c}{Gut barrier function markers}		
Lyz1	F-GGGAACCTGTGACCTGTCTT R-GCCTCATGACACTGGGAACA	Accession: NM_013590.4
Lyz2	F-TCTACTGCAGCTCATTCGGT R-CTTAGAGGGGAAATCGAGGGAA	Accession: NM_017372.3
Pla2g2	F-AGGATTCCCCCAAGGATGCCAC R-CAGCCGTTTCTGACAGGAGTTCTGG	PMID: 19855381
Defa1	F-TCAAGAGGCTGCAAAGGAAGAGAAC R-TGGTCTCCATGTTCAGCGACAGC	PMID: 19855381
Defa5	F-CTTGTCCTGCTGGCCTTCC R-TAGACACAGCCTGGTCCTCT	Accession: NM_007851.2
Reg3g	F-CCATCTTCACGTAGCAGC R-CAAGATGTCCTGAGGGC	PMID: 22723890
Tlr4	Bio-Rad (Mississauga, ON, Canada)	Assay ID: qMmuCID0023548
	Gut barrier integrity markers	
Muc2	F-ACCTGGAAGGCCCAATCAAG R-CAGCGTAGTTGGCACTCTCA	Accession: NM_023566.3
Cldn-3	F-GCACCCACCAAGATCCTCTA R-AGGCTGTCTGTCCTCTTCCA	PMID: 17383680
Cldn-7	F-CATTGTGGCAGGTCTTGCTG R-CATGGGCGTCAAGGGGTTAT	Accession: NM_016887.6
	EGC maturation and function markers	
Bfabp	F-GGTTCGGTTGGATGGAGACA R-AGTCACGACCATCTTGCCAT	Accession: NM_021272.3
S100b	F-TGGCTGCGGAAGTTGAGATT R-ATGGCTCCCAGCAGCTAAAG	Accession: NM_009115.3
Gdnf	F-ACCAGTGACTCCAATATGCCTG R-CTGCCGCTTGTTTATCTGGTG	Accession: NM_001301357.1
Sox10	Bio-Rad (Mississauga, ON, Canada)	Assay ID: qMmuCID0007045
Plp1	Bio-Rad (Mississauga, ON, Canada)	Assay ID: qMmuCED0061105
	Fetal gut development markers	
Cdx2	F-AGCCAAGTGAAAACCAGGAC R-AGTGAAACTCCTTCTCCAGCTC	Accession: NM_007673.3
Sox9	F-GCCACGGAACAGACTCACAT R-AGATTGCCCAGAGTGCTCG	Accession: NM_011448.4
	Reference genes	
Ywhaz	F-GCAACGATGTACTGTCTCTTTTGG R-GTCCACAATTCCTTTCTTGTCATC	Accession: NM_011740.3
Tbp	F-CGGACAACTGCGTTGATTTTC R-AGCCCAACTTCTGCACAACTC	Accession: NM_013684.3
Actb	F-TCGTGCGTGACATCAAAGAGA R-GAACCGCTCGTTGCCAATA	Accession: NM_007393.5

Lyz, Lysozyme; *Pla2g2*, Phospholipase A2 Group II; *Defa*, Alpha Defensin; *Reg3g*, Regenerating Family Member 3 Gamma; *Tlr4*, Toll-Like Receptor 4; *Muc2*, Mucin 2; *Cldn*, Claudin; *Bfabp*, Brain-Type Fatty Acid-Binding Protein; *S100b*, S100 Calcium Binding Protein B; *Gdnf*, Glial-Derived Neurotrophic Factor; *Sox10*, SRY-Box 10; *Plp1*, Proteolipid Protein 1; *Cdx2*, Caudal Type Homeobox 2; *Sox9*, SRY-Box 9; *Ywhaz*, Tyrosine 3-Monooxygenase/Tryptophan 5-Monooxygenase Activation Protein Zeta; *Tbp*, TATA-Box Binding Protein; *Actb*, Beta-actin; F, Forward primer; R, Reverse primer; PMID, PubMed Identification.

2.4. Expression and Localization of Lysozyme Protein and Quantification of Goblet Cells in Maternal SI

Immunohistochemistry (IHC) was used to localize and semi-quantify immunoreactive (ir) staining of lysozyme (Lyz) and to quantify goblet cell number. Five millimeter sections of maternal SI were cut from paraffin-embedded blocks and mounted onto glass slides. For lysozyme staining, sections were deparaffinized and rehydrated in descending alcohol series and quenched with 3% hydrogen peroxide in 90% ethanol for 20 min at room temperature. Sections underwent antigen retrieval by sodium citrate/citric acid solution and microwaved for 10 min, then blocked in serum-free protein blocking solution (Agilent Dako, Santa Clara, CA, USA) for 1 h at room temperature, and incubated overnight at 4 °C with rabbit anti-lysozyme antibody (1:200 dilution; product #PA1-29680, Thermo Fisher Scientific, Waltham, MA, USA). Negative control sections were probed with normal rabbit IgG (0.4 µg/µL, 1:200; #sc-2027, Santa Cruz Biotechnology, Dallas, TX, USA). Sections were then incubated with goat biotinylated anti-rabbit secondary antibody (1:200; #BA-1000, Vector Labs, Burlingame, CA, USA) for 1 h at room temperature, followed by 1 h incubation with streptavidin-horseradish peroxidase (1:2000 in 1× PBS; Invitrogen, Carlsbad, CA, USA). Antibody-antigen interactions were visualized using DAB for 40 s (Vectastain DAB ABC kit, Vector Labs). Sections were counterstained with Gills I haematoxylin to provide greater contrast and to stain nuclei. Semi-quantitative analyses of ir-lysozyme staining intensity was performed using computerized image analysis (Image-Pro Plus 4.5, Media Cybernetics, Rockville, MD, USA; and Olympus BX61 microscope, Shinjuku, Tokyo, Japan). A superimage was composited from individual images (at 20× magnification) captured along the entire SI section from each dam. To semi-quantitatively measure staining intensity in each image (where higher levels of staining intensity may represent higher Lyz protein expression), an algorithm was developed (using Visiopharm NewCAST Analysis software, Hørsholm, Denmark) to detect intensity of DAB staining for ir-Lyz in four levels of intensity: low, moderate, high, and strong. Staining intensity for each of the four levels was summed across all images within a superimage for each animal. CON mothers had a mean ± SEM (range) of 10.3 ± 0.7 (7–13) images per superimage, UN had 10 ± 0.8 (7–13) images, and HF had 9 ± 0.6 (7–12) images. There was no statistical difference between diet groups in the number of images per superimage ($p = 0.37$).

For goblet cell quantification, sections were stained with alcian blue, which stains acetic mucins and acid mucosubstances. Sections were deparaffinized and rehydrated in descending alcohol series then incubated in 3% glacial acetic acid, followed by alcian blue (Sigma-Aldrich, Oakville, ON, Canada) and 0.1% nuclear fast red counterstain (Electron Microscopy Sciences, Hatfield, PA, USA). Eight images were randomly captured at 20× for each section (Leica DMIL LED inverted microscope, Wetzlar, Germany; and QCapture Pro software, Surrey, BC, Canada). We first counted the number of alcian blue positive cells (goblet cells) in one villus and one crypt nearest to the villus in each of the eight images and took the average across the images to determine the mean number of goblet cells in a villus and crypt for each animal. We also counted the number of alcian blue positive cells in two villi and two crypts nearest to these villi in each of four images (randomly selected from the eight images). The total number of epithelial cells in the same villi and crypts were also counted. The number of alcian blue positive cells and total epithelial cells across the two villi and two crypts in each image were summed. Then, the number of alcian blue positive cells and total epithelial cells were averaged across the four images (separately for villi and crypts) to obtain the average number of goblet cells and epithelial cells and the percentage of goblet cells per two villi or crypts for each animal. A researcher blinded to the experimental groups performed the counting.

2.5. Data Analysis

Data were checked for normality using the Shapiro–Wilk test and equal variance using Levene's test. Outliers were excluded from analyses. Non-parametric data were transformed by applying either logarithmic, square root, or cube root transformations. Differences between dietary groups for outcome measures were analyzed using: (1) ANOVA with Tukey's post hoc, (2) Kruskal–Wallis/Wilcoxon test with Steel–Dwass post hoc for non-parametric data, or (3) Welch's test with Games–Howell post hoc for normal data with unequal variance ($p < 0.05$) using JMP 13 software (SAS Institute, Cary, NC, USA). For qPCR and IHC data, biological replicates were CON $n = 6$–7 ($n = 10$–14 fetuses); UN $n = 6$–7 ($n = 7$–9 fetuses); and HF $n = 7$–8 ($n = 13$–15 fetuses). Data are shown as quantile box plots with 95% confidence diamonds; $p < 0.05$.

3. Results

3.1. Malnutrition Was Associated with Reduced Gut Barrier Function and Integrity

In UN mothers, SI mRNA expression levels of AMP genes *Lyz2* ($p = 0.02$, Figure 1B) and *Reg3g* ($p = 0.003$, Figure 1C) were decreased compared to CON. HF mothers had reduced mRNA expression levels of *Reg3g* ($p = 0.003$, Figure 1C) and *Muc2* ($p = 0.001$, Figure 1G) in SI compared to CON. There was no effect of maternal diet on the expression of AMP genes *Lyz1*, *Defa1*, *Defa5*, and *Pla2g2* (Figure 1A,D–F). The average number of goblet cells, sites of mucus secretion, was higher in SI villi, but not crypts, of UN mothers compared CON (Figure 2). Proportion of goblet cells relative to total number of epithelial cells in both villi and crypts were not different between dietary groups (Figure 2).

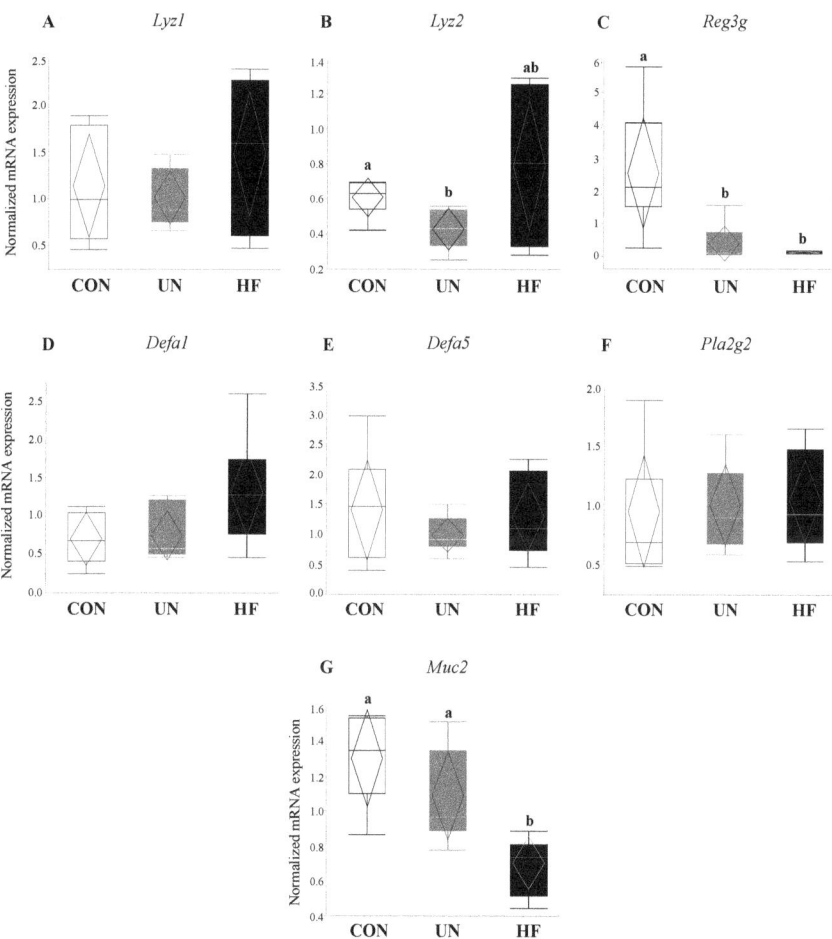

Figure 1. Maternal malnutrition was associated with altered gene expression of antimicrobial peptides and mucin. Maternal UN was associated with decreased mRNA expression of antimicrobial peptide genes *Lyz2* ($p = 0.02$) and *Reg3g* ($p = 0.003$) vs. CON, while HF diet was associated with decreased *Reg3g* ($p = 0.003$) and mucin (*Muc2*; $p = 0.001$) mRNA expression vs. CON ($n = 6$–8/group). Groups with different letters are significantly different ($p < 0.05$). UN, undernourished; HF, high fat; CON, control.

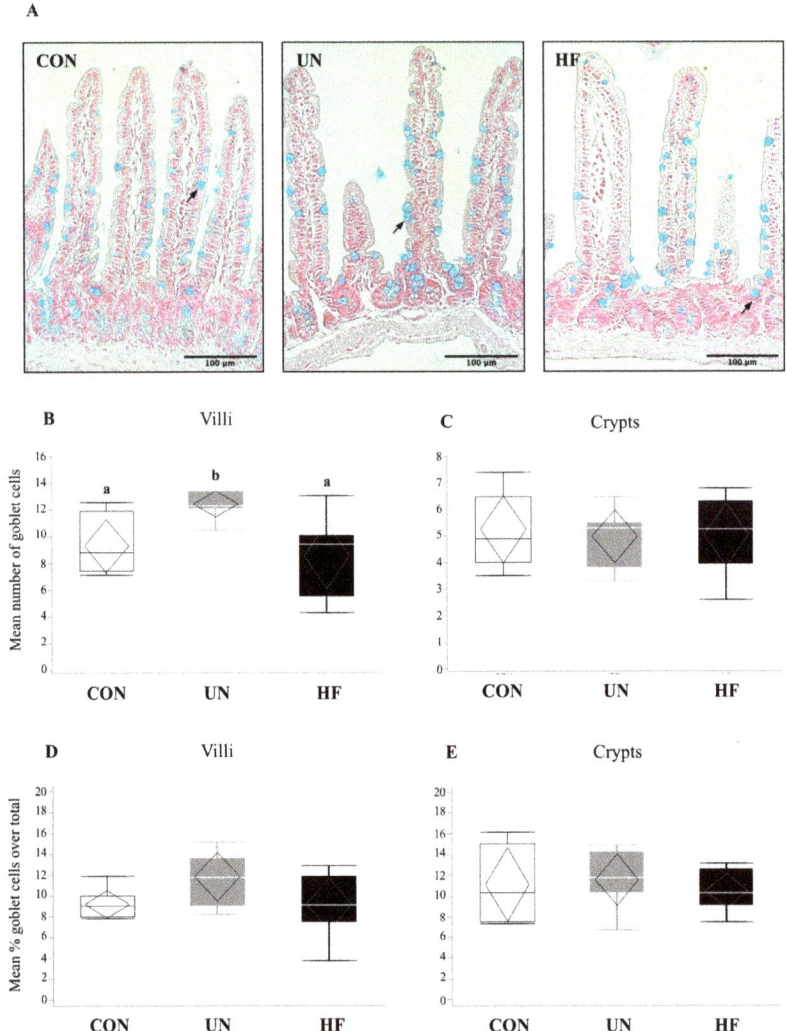

Figure 2. Maternal malnutrition may influence small intestinal goblet cell number. (**A**) Staining of goblet cells by alcian blue in small intestine (20× magnification). Arrows indicate goblet cells. Mean number of goblet cells across 8 villi (**B**) or crypts (**C**). Mean percentage of goblet cells (proportion of total number of epithelial cells) in villi (**D**) or crypts (**E**). There were a greater number of goblet cells in UN villi vs. CON ($p = 0.008$; $n = 6$–8/group). Groups with different letters are significantly different ($p < 0.05$). UN, undernourished; HF, high fat; CON, control.

3.2. Maternal HF Diet May Be Associated with Increased Lyz Production

Lyz protein was localized to Paneth cells in the crypts of the maternal SI (Figure 3A). In the SI of HF dams, semi-quantitative analyses revealed a significant reduction in the levels of low-intensity Lyz staining ($p = 0.03$, Figure 3B) compared to CON, which may suggest an overall greater production of this AMP in HF mothers. Further, an overall difference in moderate-intensity staining ($p = 0.04$; Figure 3C) was detected, but there was no difference between groups with post hoc testing, and there were no differences between diet groups in high- or strong-intensity staining levels (Figure 3D,E).

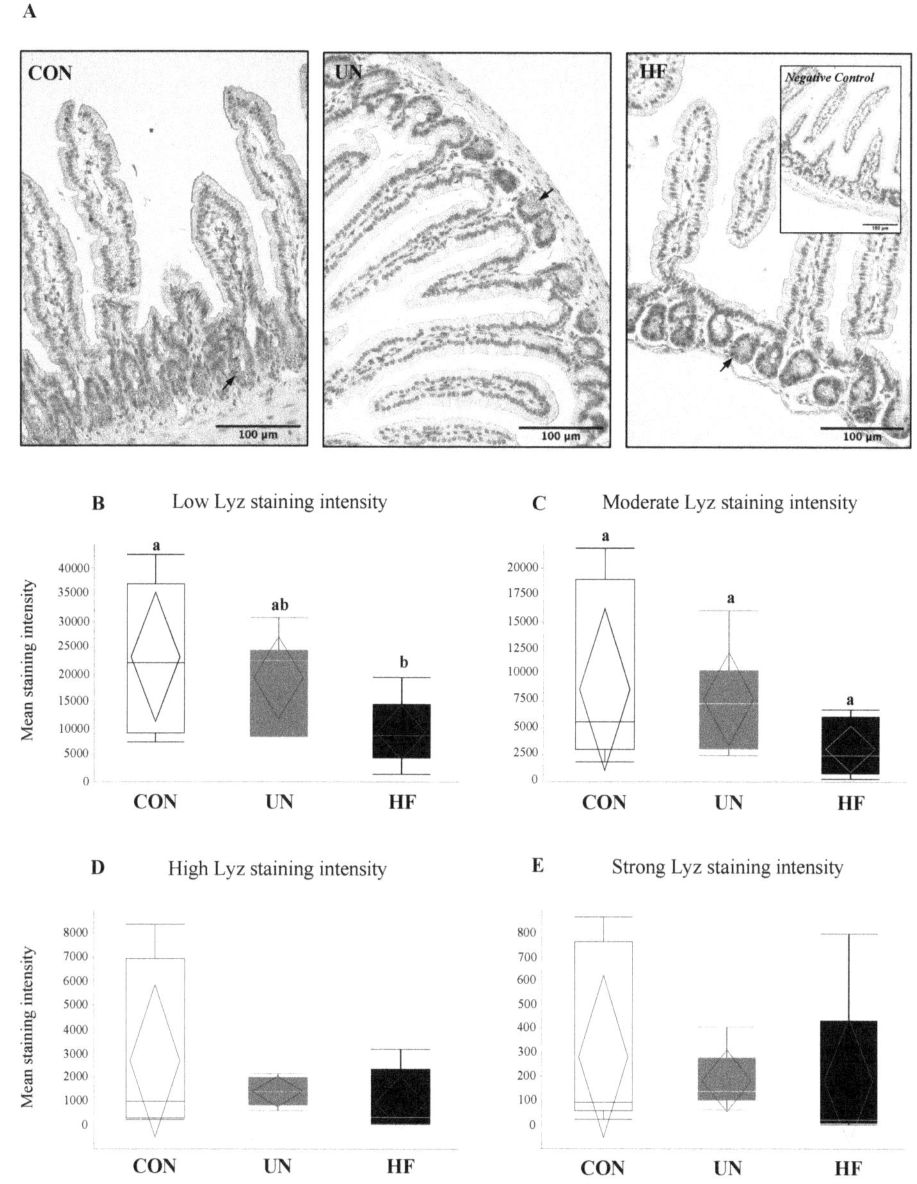

Figure 3. Maternal HF diet was associated with less low-intensity lysozyme staining in intestinal crypts. (**A**) Representative images of lysozyme protein immunoreactivity (ir) staining show localization to the crypts of the maternal small intestine (SI) at d18.5, with negative control inset (40× magnification). Arrows indicate lysozyme proteins within Paneth cells. (**B–E**) Lyz staining was quantified into low, moderate, high, and strong intensities, representing increasing levels of protein expression ($n = 6–8$/group). Semi-quantitative analysis revealed less low-intensity Lyz staining in SI from HF mothers ($p = 0.03$) vs. CON, and an overall difference in moderate-intensity Lyz staining ($p = 0.04$), but no difference between groups with post hoc testing. Groups with different letters are significantly different ($p < 0.05$). UN, undernourished; HF, high fat; CON, control.

3.3. Maternal UN Was Associated with Delayed Fetal Gut Development and Reduced Mucus Production

UN, but not HF, fetuses showed increased mRNA expression of the gut transcription factor *Sox9* ($p = 0.02$, Figure 4A) compared to CON, and an associated decrease in the mRNA expression levels of *Muc2* ($p = 0.002$, Figure 4B) and *Cdx2* ($p = 0.003$, Figure 4C) compared to CON.

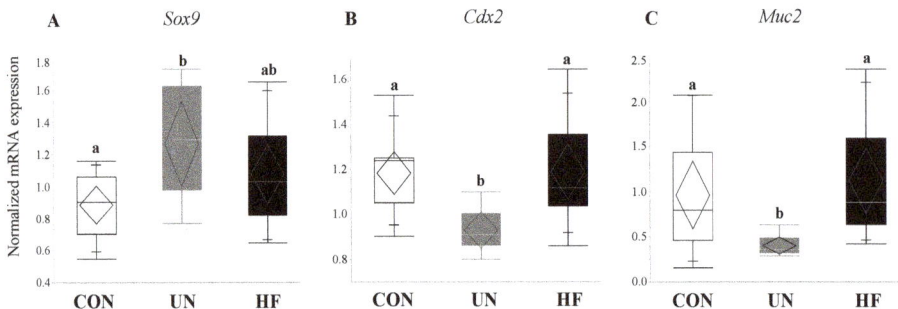

Figure 4. UN fetuses displayed activation of a gut transcription factor that represses gut barrier development and mucus production. Maternal UN was associated with increased fetal gut mRNA expression of *Sox9* ($p = 0.02$) vs. CON, and decreased *Muc2* ($p = 0.002$) and *Cdx2* ($p = 0.003$) vs. CON ($n = 8$–15/group). Groups with different letters are significantly different ($p < 0.05$). UN, undernourished; HF, high fat; CON, control.

3.4. Maternal HF Diet Was Associated with Increased Fetal EGC Development

Fetuses from HF-fed mothers had increased mRNA expression levels of EGC markers *Bfabp* (0.003, Figure 5A) and *S100b* ($p < 0.001$, Figure 5B) compared to CON, and increased *Plp1* ($p = 0.04$, Figure 5C) compared to UN. There were no differences in gut mRNA expression levels of the EGC marker *Sox10* and EGC neurotrophic factor *Gdnf* in fetuses exposed to different maternal diets (Figure 5D,E).

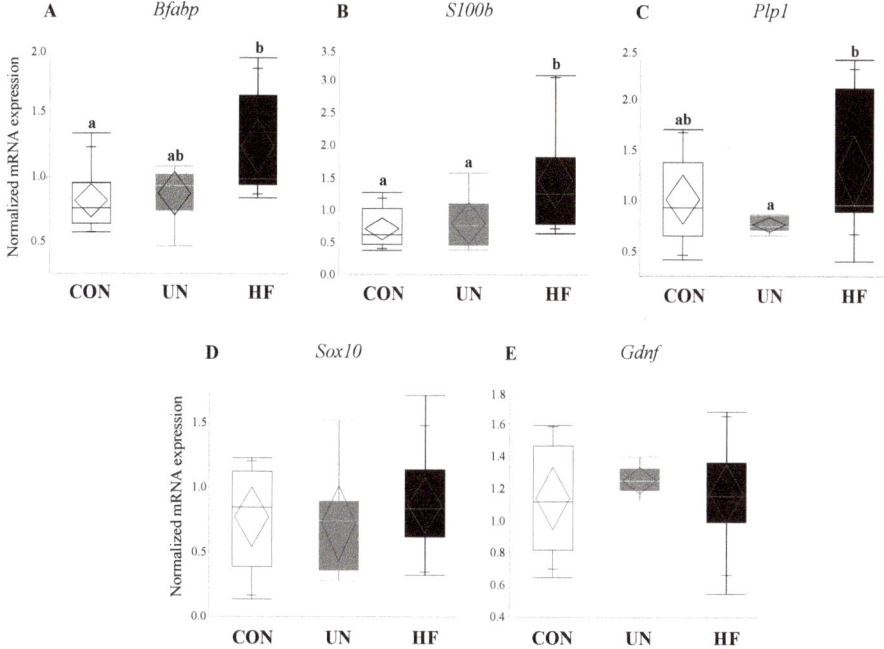

Figure 5. HF fetuses showed activation of enteric glial cell markers in the gut. Maternal HF diet was associated with increased fetal gut mRNA expression of enteric glial cells (EGC) markers *S100b* ($p < 0.001$) and *Bfabp* ($p = 0.003$) vs. CON, and *Plp1* ($p = 0.04$) vs. UN ($n = 7–15$/group). Groups with different letters are significantly different ($p < 0.05$). UN, undernourished; HF, high fat; CON, control.

3.5. Maternal HF Diet Was Associated with Increased Fetal Gut Barrier Function and Integrity

HF, but not UN, fetuses had increased mRNA expression levels of Paneth AMPs *Lyz1* ($p = 0.007$, Figure 6A), *Reg3g* ($p = 0.01$, Figure 6C), and *Defa1* ($p = 0.001$, Figure 6D), and lower gut mRNA expression levels of *Tlr4* ($p = 0.02$, Figure 6E) compared to CON. No between-group differences were detected in *Lyz2* fetal gut mRNA expression levels (Figure 6B). Maternal HF diet was also associated with an increase in fetal gut mRNA expression of TJ genes *Cldn-3* ($p = 0.008$, Figure 7A) and *Cldn-7* ($p < 0.001$, Figure 7B) compared to CON, while UN fetuses only displayed increased mRNA expression of *Cldn-7* compared to CON ($p < 0.001$, Figure 7B).

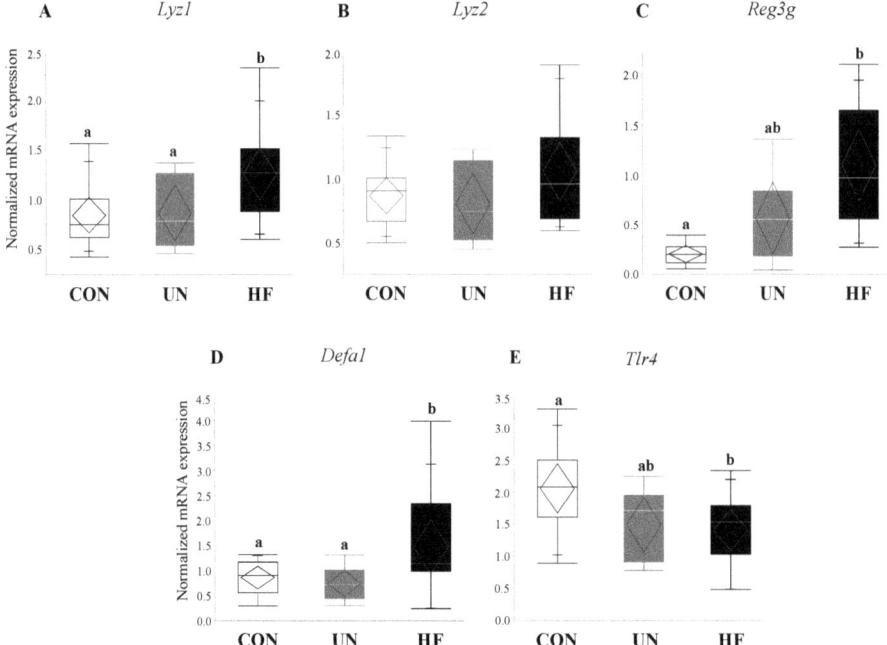

Figure 6. HF fetuses showed an upregulation of gut barrier function genes and downregulation of a microbe-sensing receptor. Maternal HF diet was associated with increased mRNA expression of antimicrobial peptide (AMP) genes *Lyz1* ($p = 0.007$), *Reg3g* ($p = 0.01$), and *Defa1* ($p = 0.001$) in the fetal gut vs. CON, though mRNA expression of the purported AMP-activating receptor *Tlr4* was decreased in these fetuses ($p = 0.02$) vs. CON ($n = 9$–15/group). Groups with different letters are significantly different ($p < 0.05$). UN, undernourished; HF, high fat; CON, control.

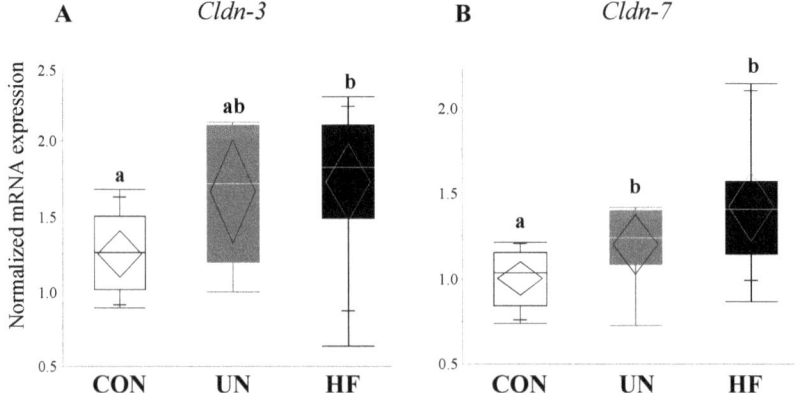

Figure 7. Maternal malnutrition altered fetal gut tight junction gene expression. Maternal UN was associated with increased mRNA expression of *Cldn-7* ($p < 0.001$) in fetal gut vs. CON, while fetuses from HF mothers increased *Cldn-3* ($p = 0.008$) and *Cldn-7* ($p < 0.001$) mRNA expression vs. CON ($n = 9$–15/group). Groups with different letters are significantly different ($p < 0.05$). UN, undernourished; HF, high fat; CON, control.

3.6. Maternal Malnutrition Altered Fetal Gut mRNA Expression in Males More than in Females

Data on fetal outcomes were also stratified by sex (Figures 8–11). We found that maternal HF diet was associated with increased mRNA expression of genes involved in EGC maturation (*Bfabp*, *Plp1*, and *S100b*), gut barrier function (*Lyz1* and *Lyz2*), and gut barrier integrity (*Cldn-3* and *Cldn-7*) in the male fetal gut ($p < 0.05$, Figures 9–11) compared to CON. Maternal HF diet was also associated with reduced male fetal gut mRNA expression of the microbe-sensing toll-like receptor *Tlr4* ($p = 0.03$, Figure 10E) compared to CON. Male UN fetuses showed a reduction in fetal gut mRNA expression of gut differentiation and maturation markers *Cdx2* ($p = 0.02$, Figure 8B) and *Muc2* ($p < 0.001$, Figure 8C), and an upregulation of EGC marker *Bfabp* ($p < 0.001$, Figure 9A) and TJ gene *Cldn-7* ($p = 0.004$, Figure 11B).

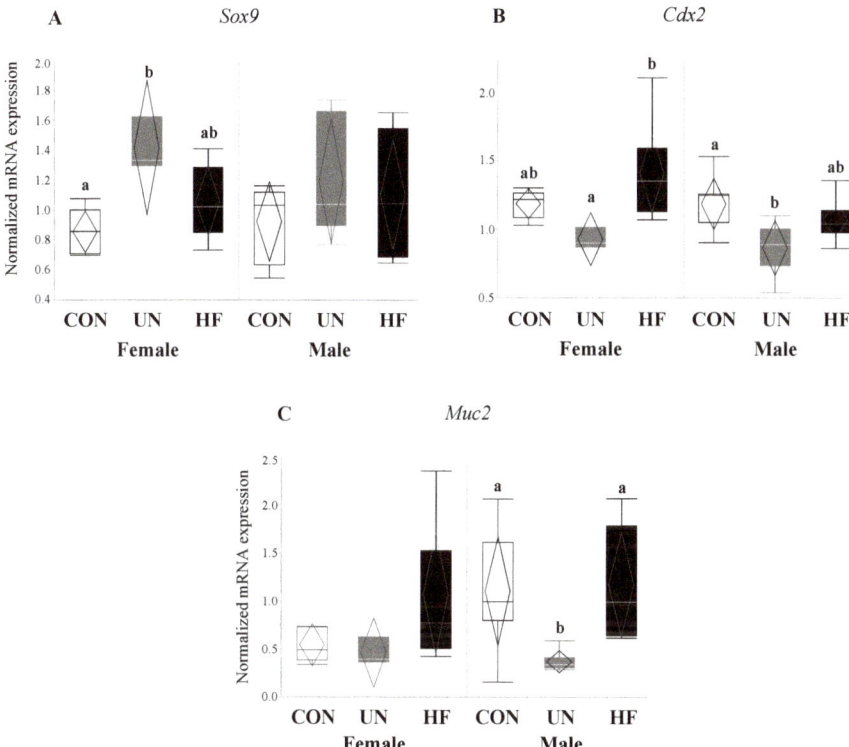

Figure 8. Maternal UN affected fetal gut barrier maturity in both sexes and mucus layer maturity in male fetuses only. Maternal UN was associated with increased gut transcription factor *Sox9* ($p = 0.004$) vs. CON and decreased *Cdx2* ($p = 0.02$) vs. HF in female fetal guts, and reduced *Cdx2* ($p = 0.02$) and *Muc2* ($p < 0.001$) vs. CON in male fetal guts ($n = 3$–8/group). Groups with different letters are significantly different ($p < 0.05$). UN, undernourished; HF, high fat; CON, control.

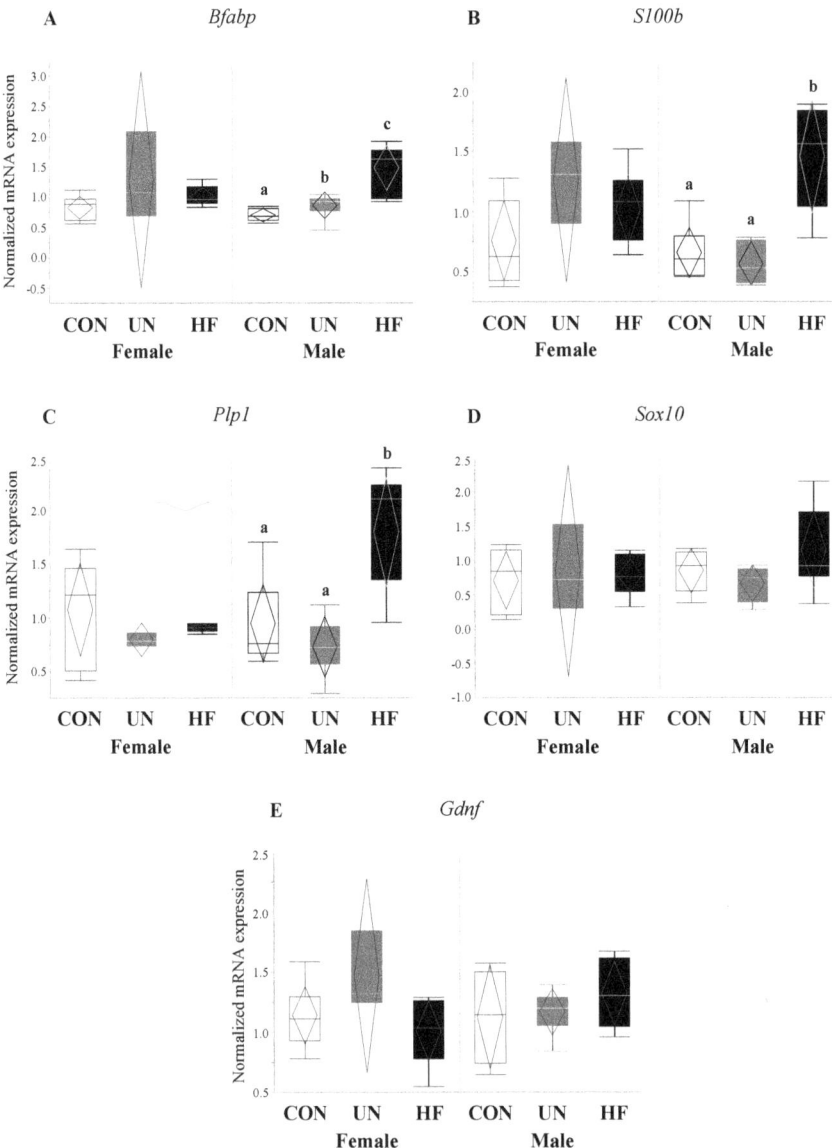

Figure 9. Maternal HF diet was associated with increased mRNA expression of enteric glial cell markers in male fetuses. In male fetal guts, maternal HF diet was associated with increased *Bfabp* ($p < 0.001$), *S100b* ($p < 0.001$), and *Plp1* ($p = 0.001$), while maternal UN was associated with increased *Bfabp* ($p < 0.001$). Maternal malnutrition (UN and HF) did not affect enteric glial cell development in female fetal guts (n = 3–8/group). Groups with different letters are significantly different ($p < 0.05$). UN, undernourished; HF, high fat; CON, control.

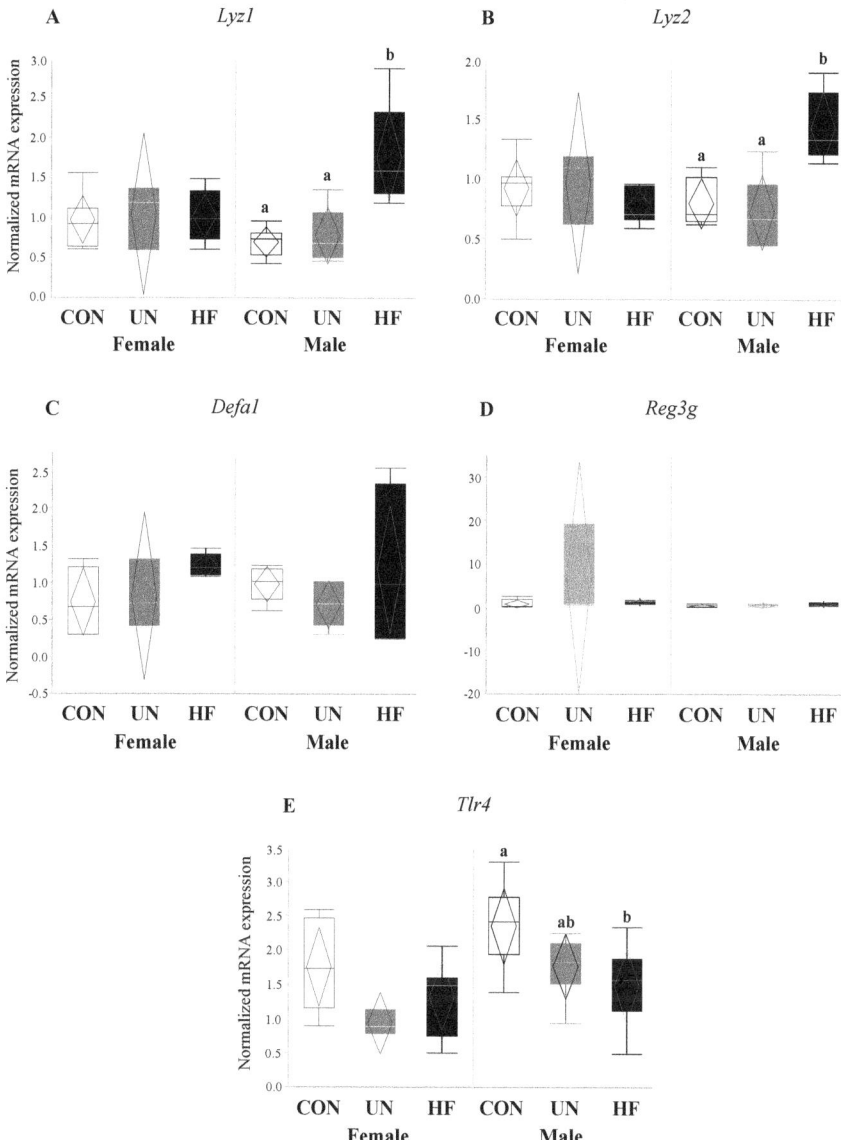

Figure 10. Maternal HF diet was associated with increased expression of antimicrobial peptides and decreased expression of microbe-sensing receptor in male fetuses. In male fetal guts, maternal HF diet was associated with increased *Lyz1* ($p < 0.001$) and *Lyz2* ($p < 0.001$), and decreased *Tlr4* ($p = 0.03$) mRNA expression levels. Maternal malnutrition (UN and HF) did not affect *Defa1* or *Reg3g* expression levels in male fetuses or antimicrobial peptide levels in female fetal guts (n = 3–8/group). Groups with different letters are significantly different ($p < 0.05$). UN, undernourished; HF, high fat; CON, control.

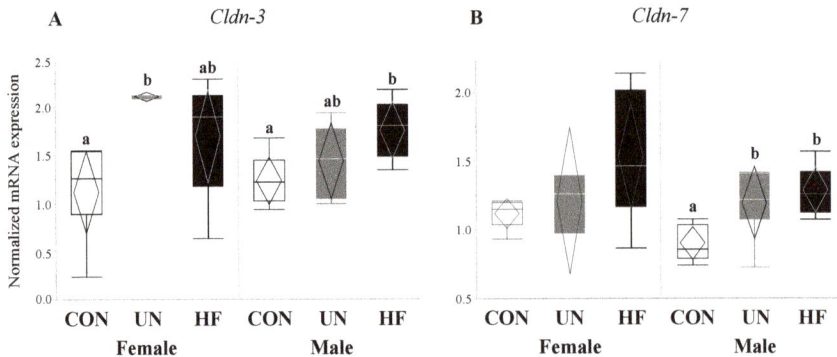

Figure 11. Maternal malnutrition was associated with an increase in gut tight junction gene expression in both fetal sexes. In male fetal guts, maternal HF diet was associated with increased *Cldn-3* ($p = 0.02$) and *Cldn-7* ($p = 0.004$), while maternal UN was associated with increased *Cldn-7* ($p = 0.004$); in female fetal guts, maternal UN was associated with increased *Cldn-3* ($p = 0.03$) mRNA expression ($n = 3$–8/group). Groups with different letters are significantly different ($p < 0.05$). UN, undernourished; HF, high fat; CON, control.

In female fetal guts, maternal UN was associated with increased mRNA expression of gut differentiation transcription factor *Sox9* ($p = 0.004$, Figure 8A) compared to CON, and decreased gut maturation marker *Cdx2* ($p = 0.02$, Figure 8B), though only compared to HF. Maternal UN was also associated with increased TJ gene *Cldn-3* ($p = 0.03$, Figure 11A) in female fetuses compared to CON.

4. Discussion

Despite the growing body of evidence linking the gut and its resident microbes to health [2,7], few studies have investigated these relationships in the mother during pregnancy or the developing fetus. Additionally, although we know that key enteric cells that support the gut barrier and establish communication with the brain are laid down and functional in early life [49–51], we know less about how early life nutritional adversity or altered gut microbes may influence their development and function [52,53]. Since malnutrition is an important insult to pregnancies and their outcomes [54], our study is the first to determine the effect of maternal malnutrition on maternal and fetal intestinal barrier integrity and function, and its implications for fetal development.

Extensive research has demonstrated that high fat diets markedly alter the diversity and composition of gut microbes [32,55–59], resulting in long-term aberrations in gut barrier integrity [32,55–57,60] and function [32,55,61] and chronic inflammation [32,56,57]. As a result, we hypothesized that mothers fed a HF diet would show pronounced changes in gut function. We found that maternal HF diet was associated with reduced mRNA expression of Paneth cell AMP *Reg3g*, which targets Gram-positive bacteria [62], no change in mRNA expression of AMPs *Lyz1* and *Lyz2*, and less low-intensity Lyz staining, which may suggest an overall greater production of Lyz protein, which can target both Gram-positive and Gram-negative bacteria [63]. These results are analogous to those where HF feeding for 8 weeks in non-pregnant mice was associated with a reduction in *Reg3g* SI mRNA expression and no expression changes in *Lyz1* [61]. Maternal HF diet was also associated with lower mRNA expression levels of the goblet cell-produced *Muc2*, which when translated, becomes the main component of the mucus layer: the first line of defense against gut infections and inflammation [64]. Although this was not associated with an increase in the number of goblet cells in HF SI, our mRNA results align with a study where male and female mice fed a HF diet from weeks 6–22 of life had a reduction in mRNA expression levels of *Muc2* and cryptidins (Paneth AMPs) [65]. Since Reg3g is critical to Muc2 distribution and spatial segregation of the gut epithelium and bacteria [62,66,67], concurrent downregulation of these genes due to a HF diet may further alter the mucus layer and reduce gut barrier defenses against microbes, resulting in bacterial contact with the epithelium, increased gut inflammation, gut tissue damage, and bacterial translocation [62].

Similar to high fat diets, undernutrition is known to be a significant insult to gut barrier integrity and function, increasing gut permeability [68–70], gut and peripheral inflammation [68,69], and bacterial translocation [71], and altering gut barrier structure [72] and enteric cell function [73,74]. Despite this, and to the best of our knowledge, our study is the first to investigate the effects of undernutrition on the maternal gut barrier during pregnancy. We hypothesized that undernutrition would lead to adverse changes in the maternal gut environment. We found that moderate maternal UN was associated with lower mRNA expression levels of Paneth cell AMPs (*Lyz2* and *Reg3g*), suggesting reduced gut barrier function and ability to maintain gut-microbe homeostasis. This is consistent with findings from a study where 48 h of starvation in non-pregnant mice led to a reduction in SI mRNA and protein expression of Reg3g and Lyz, and a hyper-permeable barrier [33]. Since AMPs limit bacterial contact with the gut barrier [75], a reduction in their expression has been associated with increased bacterial adherence to the barrier [76] and bacterial translocation [1,33,77]. Moreover, AMPs have been shown to block the release of IL-1β from activated immune cells [78], a pro-inflammatory cytokine which exacerbates gut barrier permeability by creating gaps between TJ proteins [78–80]. These data suggest that undernutrition may compromise the host's ability to mount an appropriate immune response through AMP pathways, resulting in increased susceptibility to infections and a leaky gut. In fact, malnutrition is known to impair gut immune functioning, causing increased susceptibility to environmental insults due to altered cytokine production [81]. This may be of particular concern in pregnancies where undernutrition/underweight and infection often coexist, such as in populations with low socioeconomic status [23,70]. Yet, even prior to bacteria reaching the gut barrier, a mucosal layer produced by goblet cells provides protection. We found that UN mothers had an increased number of villus-residing goblet cells, which may suggest an attempt to strengthen the gut barrier to offset the consequences of reduced AMP expression, such as the heightened propensity for leaky gut.

Though early life development is key to setting healthy trajectories throughout life [82], little is known about how perinatal events shape fetal gut development. Therefore, we were interested in the effect of maternal malnutrition on fetal gut and ENS development and function. We focused on the consequence of this nutritional adversity on genes involved in gut maturation and differentiation, EGC development, Paneth AMP production, and TJ formation. Consistent with our hypothesis, fetuses from HF mothers seemed to be most affected by an adverse nutritional exposure, as evidenced by significantly altered expression levels of nine of the 15 genes tested. HF diet was associated with changes in fetal EGC development, gut AMP production, and TJ expression. In fetuses from HF-fed dams, we found increased mRNA expression of specific and widely-expressed [17,46,83] markers of EGCs *S100b*, *Bfabp*, and *Plp1*, the latter of which to our knowledge has not been previously examined in the mouse fetal gut. This increase in EGC markers may be in response to the higher inflammatory environment in HF pregnancies that we have previously described [41], since EGCs are important for regulating inflammatory pathways [84] and gut barrier integrity [20,84]. During intestinal inflammation in adult animals, EGCs respond to pro-inflammatory signals and reverse inflammation-induced ENS damage [85] by driving enteric neuron death through nitric oxide production [41,43,44] and triggering ENS neurogenesis [86]. Previously, EGC proliferation was shown to occur in infant rats whose mothers were fed HF diets perinatally, though mRNA and protein expression of local pro-inflammatory cytokines were unchanged at all postnatal timepoints examined (2, 4, 6, and 12 weeks) [35]. Together with our results, this may suggest that maternal HF diet exposure in utero may reprogram mechanisms in the fetus that establish gut-brain connections and communication and increase EGC development to reduce gut inflammation by the time of birth, thereby negating some of the fetal ENS damage incurred during development. Future experiments should examine whether these changes in the fetal gut are associated with altered brain development and function.

Additionally, fetuses from HF mothers showed an increase in gut mRNA expression levels of AMP genes *Lyz1* (but not *Lyz2*), *Defa1*, and *Reg3g*. This is in contrast to studies in adult mice [27,32,55,65] that have demonstrated that HF diets decrease gut levels of AMPs, though one study [29] found differing results between the mRNA and protein levels. Nevertheless, our study is the first to our knowledge to uncover how maternal malnutrition alters expression levels of AMPs in near-term fetuses and is the first to assess *Reg3g* mRNA expression in fetal tissues. Due to the lack of other developed immune mechanisms, fetuses may be increasing the expression of AMP genes to regulate and protect themselves from the pro-inflammatory fetal environment observed in HF pregnancies [87–90]. In parallel to the upregulation of AMPs, we observed reduced expression of *Tlr4* mRNA in HF fetal gut, despite that, at least in adult models, *Tlr4* is known to initiate AMP production [91,92]. Nonetheless, depending on tissue type, the same TLR can downregulate, upregulate, or not affect AMP production [93], adding to the complexity of the TLR-AMP relationship which is further compounded by the lack of work on TLR function in fetal gut tissue.

In mammals, 24 members of the claudin family exist, each with unique charge-selectivity preferences that dictate barrier permeability and tight junction structure and function [94]. Of the claudin family, *Cldn-3* and *Cldn-7* are the amongst the most highly expressed in all sections of the mouse GI tract [95]. *Cldn-3* has previously been shown to be affected by HF diets [60] and commensal bacterial colonization [96], while *Cldn-7* has been found to have functions outside of barrier integrity, such as maintenance of intestinal homeostasis [97], and both its mRNA and protein are highly expressed in the small and large intestine [97]. Still, expression levels of neither *Cldn-3* nor *Cldn-7* have been recorded in mouse fetal guts. In our study, fetuses from HF mothers showed an increase in mRNA expression of both TJs *Cldn-3* and *Cldn-7*, however these data run counter to the effects of HF diets on TJ expression in adult mice and rats [56,57,60]. Research on the ontogeny and development of tight junction proteins in the fetal mouse gut is scarce, though in human fetuses, tight junction proteins appear at 8–10 weeks of gestation and begin to assemble junctional complexes (continuous, belt-like structures around cells) at 10–12 weeks [98]. Nevertheless, most of the development of the tight junction proteins and gut integrity occurs in postnatal life [94], indicating that the prenatal fetal gut may be particularly sensitive and vulnerable to the effects of adverse in utero exposures due to a permeable gut barrier. As HF fetuses showed increased expression of TJ proteins, which is consistent with their increased expression of EGC markers, it may be that HF fetuses are compensating for a highly inflammatory maternal environment by tightening the gut barrier.

Similar to HF fetuses, we hypothesized that fetal growth restriction would lead to aberrant gut barrier development in fetuses from UN mothers. Accordingly, we found that UN fetuses had increased expression of the gut transcription factor *Sox9*, which is known to repress the expression of *Muc2* and *Cdx2* through activation of the Wnt-β-Catenin-TCF4 pathway [99], and reduced mRNA expression of mucus (*Muc2*) and gut differentiation (*Cdx2*) genes. These data are indicative of immature gut barrier development, and consistent with the reduced weight of these fetuses compared to CON and HF fetuses. Importantly, an immature gut may be associated with increased susceptibility to inflammatory and infectious insults, due to reduced gut barrier integrity and function, gut dysbiosis [100–102], and irreversible [103] aberrant nutrient absorption [103–106]. This has long-term implications for growth-restricted offspring, including those born too soon, such as preterm infants, who are at greater risk for necrotizing enterocolitis [107–109] and nutrient malabsorption [103,105,106] due to poor gut development and function. Despite the changes in gut maturation and mucus production, fetuses from UN mothers showed an increase in gut *Cldn-7*, but not *Cldn-3*, mRNA expression, which might suggest an attempt to increase barrier integrity to compensate for the gut barrier immaturity. Lastly, our sex-stratified data suggest that the majority of observed group differences were driven by male fetuses, which is consistent with numerous reports which demonstrate that male fetuses are more susceptible to perinatal insults, especially due to maternal malnutrition [110–115].

One limitation of our study is our focus on gene expression, especially in fetal samples where sample volume is extremely limited. Findings from qPCR data can direct future experiments that examine changes at the protein level. Another limitation is the low n-number in the female fetal UN group; thus, sex-stratified changes in mRNA expression should be interpreted with caution. Lastly, although our study is cross-sectional, we focused on an important developmental time (d18.5) that can serve as an indicator of embryonic/fetal experiences and provide information that could explain neonatal development and adaptations. Future studies should examine when changes in Paneth cell and EGC development and function are initiated in the pregnant mother and developing offspring, and how long they persist, which could point to critical windows for intervention to correct adverse health and developmental trajectories set by malnutrition.

5. Conclusions

Our study is the first to examine the effect of both over- and undernutrition in parallel, during gestation in mice, and reveal the impact of maternal malnutrition on fetal gut Paneth cell function and EGC development—cells vital for gut barrier function and gut-brain axis connection. Our results indicate that malnutrition before and during gestation has adverse consequences for fetal gut development, maternal and fetal gut function, and potentially long-term programming of gut and brain function and gut immunity. If our findings are applicable to humans, this work may help inform research on dietary interventions that aim to prevent or mitigate the effects of being exposed to suboptimal nutrition in pregnancy (for mothers) and in early life (for offspring). Future work should place special focus on vulnerable populations wherein malnutrition and infection are more likely to coexist during pregnancy, exacerbating the negative repercussions of both noxious states on maternal and offspring wellbeing.

Supplementary Materials: The following are available online at http://www.mdpi.com/2072-6643/11/6/1375/s1, Figure S1: Earliest recorded expression of gut integrity, function, and development genes in the mouse fetal gut.

Author Contributions: Conceptualization, methodology, K.L.C.; data curation, formal analysis, K.L.C. and S.A.S.; investigation, K.L.C., S.A.S., T.T.-T.N.N., E.B.; writing—original draft preparation, S.A.S.; writing—review and editing, S.A.S. and K.L.C.; supervision, project administration, resources, K.L.C.

Funding: This research was funded by the Faculty of Science, Carleton University, and the Canadian Institutes of Health Research (animal model MOP-81238).

Acknowledgments: We deeply thank Stephen Lye for resource contributions towards the animal model; Tina Tu-Thu Ngoc Nguyen for the Lyz image capture and data collation; Ryszard Bielecki for the Lyz algorithm set up; and Enrrico Bloise for the goblet cell quantification.

Conflicts of Interest: The authors declare no conflict of interest.

References

1. Vaishnava, S.; Behrendt, C.L.; Ismail, A.S.; Eckmann, L.; Hooper, L.V. Paneth cells directly sense gut commensals and maintain homeostasis at the intestinal host-microbial interface. *Proc. Natl. Acad. Sci. USA* **2008**, *105*, 20858–20863. [CrossRef] [PubMed]
2. Kinross, J.M.; Darzi, A.W.; Nicholson, J.K. Gut microbiome-host interactions in health and disease. *Genome Med.* **2011**, *3*, 14. [CrossRef] [PubMed]
3. Greenwood-Van Meerveld, B.; Johnson, A.C.; Grundy, D. Gastrointestinal physiology and function. In *Handbook of Experimental Pharmacology*; Springer: Cham, Switzerland, 2017; Volume 239, pp. 1–16.
4. LeBlanc, J.G.; Milani, C.; de Giori, G.S.; Sesma, F.; van Sinderen, D.; Ventura, M. Bacteria as vitamin suppliers to their host: A gut microbiota perspective. *Curr. Opin. Biotechnol.* **2013**, *24*, 160–168. [CrossRef] [PubMed]
5. Nicholson, J.K.; Holmes, E.; Kinross, J.; Burcelin, R.; Gibson, G.; Jia, W.; Pettersson, S. Host-gut microbiota metabolic interactions. *Science* **2012**, *336*, 1262–1267. [CrossRef] [PubMed]
6. Stecher, B.; Hardt, W.-D. Mechanisms controlling pathogen colonization of the gut. *Curr. Opin. Microbiol.* **2011**, *14*, 82–91. [CrossRef]
7. Carabotti, M.; Scirocco, A.; Maselli, M.A.; Severi, C. The gut-brain axis: Interactions between enteric microbiota, central and enteric nervous systems. *Ann. Gastroenterol.* **2015**, *28*, 203–209. [PubMed]

8. De Vadder, F.; Kovatcheva-Datchary, P.; Goncalves, D.; Vinera, J.; Zitoun, C.; Duchampt, A.; Bäckhed, F.; Mithieux, G. Microbiota-generated metabolites promote metabolic benefits via gut-brain neural circuits. *Cell* **2014**, *156*, 84–96. [CrossRef]
9. Field, B.C.T.; Chaudhri, O.B.; Bloom, S.R. Bowels control brain: Gut hormones and obesity. *Nat. Rev. Endocrinol.* **2010**, *6*, 444–453. [CrossRef]
10. Tracey, K.J. The inflammatory reflex. *Nature* **2002**, *420*, 853–859. [CrossRef]
11. Kane, A.V.; Dinh, D.M.; Ward, H.D. Childhood malnutrition and the intestinal microbiome. *Pediatr. Res.* **2015**, *77*, 256–262. [CrossRef]
12. Pearce, S.C.; Al-Jawadi, A.; Kishida, K.; Yu, S.; Hu, M.; Fritzky, L.F.; Edelblum, K.L.; Gao, N.; Ferraris, R.P. Marked differences in tight junction composition and macromolecular permeability among different intestinal cell types. *BMC Biol.* **2018**, *16*, 19. [CrossRef] [PubMed]
13. Ayabe, T.; Satchell, D.P.; Wilson, C.L.; Parks, W.C.; Selsted, M.E.; Ouellette, A.J. Secretion of microbicidal α-defensins by intestinal Paneth cells in response to bacteria. *Nat. Immunol.* **2000**, *1*, 113–118. [CrossRef] [PubMed]
14. Bry, L.; Falk, P.; Huttner, K.; Ouellette, A.; Midtvedt, T.; Gordon, J.I. Paneth cell differentiation in the developing intestine of normal and transgenic mice. *Proc. Natl. Acad. Sci. USA* **1994**, *91*, 10335–10339. [CrossRef] [PubMed]
15. Sato, T.; van Es, J.H.; Snippert, H.J.; Stange, D.E.; Vries, R.G.; van den Born, M.; Barker, N.; Shroyer, N.F.; van de Wetering, M.; Clevers, H. Paneth cells constitute the niche for Lgr5 stem cells in intestinal crypts. *Nature* **2011**, *469*, 415–418. [CrossRef] [PubMed]
16. Said, H.M. Intestinal absorption of water-soluble vitamins in health and disease. *Biochem. J.* **2011**, *437*, 357–372. [CrossRef] [PubMed]
17. Rühl, A. Glial cells in the gut. *Neurogastroenterol. Motil.* **2005**, *17*, 777–790. [CrossRef] [PubMed]
18. Rühl, A.; Franzke, S.; Stremmel, W. IL-1beta and IL-10 have dual effects on enteric glial cell proliferation. *Neurogastroenterol. Motil.* **2001**, *13*, 89–94. [CrossRef]
19. Murakami, M.; Ohta, T.; Ito, S. Lipopolysaccharides enhance the action of bradykinin in enteric neurons via secretion of interleukin-1beta from enteric glial cells. *J. Neurosci. Res.* **2009**, *87*, 2095–2104. [CrossRef]
20. Savidge, T.C.; Newman, P.; Pothoulakis, C.; Ruhl, A.; Neunlist, M.; Bourreille, A.; Hurst, R.; Sofroniew, M.V. Enteric glia regulate intestinal barrier function and inflammation via release of S-nitrosoglutathione. *Gastroenterology* **2007**, *132*, 1344–1358. [CrossRef]
21. Bernstein, C.N.; Vidrich, A. Isolation, identification, and culture of normal mouse colonic glia. *Glia* **1994**, *12*, 108–116. [CrossRef]
22. Rothman, T.P.; Tennyson, V.M.; Gershon, M.D. Colonization of the bowel by the precursors of enteric glia: Studies of normal and congenitally aganglionic mutant mice. *J. Comp. Neurol.* **1986**, *252*, 493–506. [CrossRef] [PubMed]
23. Subramanian, S.; Huq, S.; Yatsunenko, T.; Haque, R.; Mahfuz, M.; Alam, M.A.; Benezra, A.; DeStefano, J.; Meier, M.F.; Muegge, B.D.; et al. Persistent gut microbiota immaturity in malnourished Bangladeshi children. *Nature* **2014**, *510*, 417–421. [CrossRef] [PubMed]
24. David, L.A.; Maurice, C.F.; Carmody, R.N.; Gootenberg, D.B.; Button, J.E.; Wolfe, B.E.; Ling, A.V.; Devlin, A.S.; Varma, Y.; Fischbach, M.A.; et al. Diet rapidly and reproducibly alters the human gut microbiome. *Nature* **2014**, *505*, 559–563. [CrossRef] [PubMed]
25. Turnbaugh, P.J.; Ley, R.E.; Mahowald, M.A.; Magrini, V.; Mardis, E.R.; Gordon, J.I. An obesity-associated gut microbiome with increased capacity for energy harvest. *Nature* **2006**, *444*, 1027–1031. [CrossRef] [PubMed]
26. Tremaroli, V.; Bäckhed, F. Functional interactions between the gut microbiota and host metabolism. *Nature* **2012**, *489*, 242–249. [CrossRef]
27. Araújo, J.R.; Tomas, J.; Brenner, C.; Sansonetti, P.J. Impact of high-fat diet on the intestinal microbiota and small intestinal physiology before and after the onset of obesity. *Biochimie* **2017**, *141*, 97–106. [CrossRef] [PubMed]
28. Genton, L.; Cani, P.D.; Schrenzel, J. Alterations of gut barrier and gut microbiota in food restriction, food deprivation and protein-energy wasting. *Clin. Nutr.* **2015**, *34*, 341–349. [CrossRef]
29. Hodin, C.M.; Verdam, F.J.; Grootjans, J.; Rensen, S.S.; Verheyen, F.K.; Dejong, C.H.C.; Buurman, W.A.; Greve, J.W.; Lenaerts, K. Reduced Paneth cell antimicrobial protein levels correlate with activation of the unfolded protein response in the gut of obese individuals. *J. Pathol.* **2011**, *225*, 276–284. [CrossRef]

30. Lee, J.-C.; Lee, H.-Y.; Kim, T.K.; Kim, M.-S.; Park, Y.M.; Kim, J.; Park, K.; Kweon, M.-N.; Kim, S.-H.; Bae, J.-W.; et al. Obesogenic diet-induced gut barrier dysfunction and pathobiont expansion aggravate experimental colitis. *PLoS ONE* **2017**, *12*, e0187515. [CrossRef]
31. Guo, X.; Tang, R.; Yang, S.; Lu, Y.; Luo, J.; Liu, Z. Rutin and its combination with inulin attenuate gut dysbiosis, the inflammatory status and endoplasmic reticulum stress in Paneth cells of obese mice induced by high-fat diet. *Front. Microbiol.* **2018**, *9*, 2651. [CrossRef]
32. Guo, X.; Li, J.; Tang, R.; Zhang, G.; Zeng, H.; Wood, R.J.; Liu, Z. High fat diet alters gut microbiota and the expression of Paneth cell-antimicrobial peptides preceding changes of circulating inflammatory cytokines. *Mediat. Inflamm.* **2017**, *2017*, 9474896. [CrossRef] [PubMed]
33. Hodin, C.M.; Lenaerts, K.; Grootjans, J.; de Haan, J.J.; Hadfoune, M.; Verheyen, F.K.; Kiyama, H.; Heineman, E.; Buurman, W.A. Starvation compromises Paneth cells. *Am. J. Pathol.* **2011**, *179*, 2885–2893. [CrossRef] [PubMed]
34. Baudry, C.; Reichardt, F.; Marchix, J.; Bado, A.; Schemann, M.; des Varannes, S.B.; Neunlist, M.; Moriez, R. Diet-induced obesity has neuroprotective effects in murine gastric enteric nervous system: Involvement of leptin and glial cell line-derived neurotrophic factor. *J. Physiol.* **2012**, *590*, 533–544. [CrossRef] [PubMed]
35. McMenamin, C.A.; Clyburn, C.; Browning, K.N. High-fat diet during the perinatal period induces loss of myenteric nitrergic neurons and increases enteric glial density, prior to the development of obesity. *Neuroscience* **2018**, *393*, 369–380. [CrossRef] [PubMed]
36. NCD Risk Factor Collaboration (NCD-RisC). Worldwide trends in body-mass index, underweight, overweight, and obesity from 1975 to 2016: A pooled analysis of 2416 population-based measurement studies in 128·9 million children, adolescents, and adults. *Lancet (Lond. Engl.)* **2017**, *390*, 2627–2642. [CrossRef]
37. Ruiter-Ligeti, J.; Vincent, S.; Czuzoj-Shulman, N.; Abenhaim, H.A. Risk factors, incidence, and morbidity associated with obstetric clostridium difficile infection. *Obstet. Gynecol.* **2018**, *131*, 387–391. [CrossRef] [PubMed]
38. Pompei, A.; Cordisco, L.; Amaretti, A.; Zanoni, S.; Matteuzzi, D.; Rossi, M. Folate production by bifidobacteria as a potential probiotic property. *Appl. Environ. Microbiol.* **2007**, *73*, 179–185. [CrossRef]
39. Wikoff, W.R.; Anfora, A.T.; Liu, J.; Schultz, P.G.; Lesley, S.A.; Peters, E.C.; Siuzdak, G. Metabolomics analysis reveals large effects of gut microflora on mammalian blood metabolites. *Proc. Natl. Acad. Sci. USA* **2009**, *106*, 3698–3703. [CrossRef]
40. Reid, J.N.S.; Bisanz, J.E.; Monachese, M.; Burton, J.P.; Reid, G. The rationale for probiotics improving reproductive health and pregnancy outcome. *Am. J. Reprod. Immunol.* **2013**, *69*, 558–566. [CrossRef]
41. Connor, K.L.; Chehoud, C.; Altrichter, A.; Chan, L.; DeSantis, T.Z.; Lye, S.J. Maternal metabolic, immune, and microbial systems in late pregnancy vary with malnutrition in mice. *Biol. Reprod.* **2018**, *98*, 579–592. [CrossRef]
42. Gohir, W.; Whelan, F.J.; Surette, M.G.; Moore, C.; Schertzer, J.D.; Sloboda, D.M. Pregnancy-related changes in the maternal gut microbiota are dependent upon the mother's periconceptional diet. *Gut Microbes* **2015**, *6*, 310–320. [CrossRef] [PubMed]
43. Million, M.; Diallo, A.; Raoult, D. Gut microbiota and malnutrition. *Microb. Pathog.* **2017**, *106*, 127–138. [CrossRef] [PubMed]
44. Wankhade, U.D.; Zhong, Y.; Kang, P.; Alfaro, M.; Chintapalli, S.V.; Thakali, K.M.; Shankar, K. Enhanced offspring predisposition to steatohepatitis with maternal high-fat diet is associated with epigenetic and microbiome alterations. *PLoS ONE* **2017**, *12*, e0175675. [CrossRef]
45. Peterson, L.W.; Artis, D. Intestinal epithelial cells: Regulators of barrier function and immune homeostasis. *Nat. Rev. Immunol.* **2014**, *14*, 141–153. [CrossRef]
46. Young, H.M.; Bergner, A.J.; Müller, T. Acquisition of neuronal and glial markers by neural crest-derived cells in the mouse intestine. *J. Comp. Neurol.* **2003**, *456*, 1–11. [CrossRef]
47. Burns, A.J.; Thapar, N. Advances in ontogeny of the enteric nervous system. *Neurogastroenterol. Motil.* **2006**, *18*, 876–887. [CrossRef]
48. Pfaffl, M.W. A new mathematical model for relative quantification in real-time RT-PCR. *Nucleic Acids Res.* **2001**, *29*, e45. [CrossRef] [PubMed]
49. Grubišić, V.; Gulbransen, B.D. Enteric glia: The most alimentary of all glia. *J. Physiol.* **2017**, *595*, 557–570. [CrossRef]

50. Kabouridis, P.S.; Lasrado, R.; McCallum, S.; Chng, S.H.; Snippert, H.J.; Clevers, H.; Pettersson, S.; Pachnis, V. Microbiota controls the homeostasis of glial cells in the gut lamina propria. *Neuron* **2015**, *85*, 289–295. [CrossRef]

51. Stockinger, S.; Hornef, M.W.; Chassin, C. Establishment of intestinal homeostasis during the neonatal period. *Cell. Mol. Life Sci.* **2011**, *68*, 3699–3712. [CrossRef]

52. Santer, R.M.; Conboy, V.B. Prenatal undernutrition permanently decreases enteric neuron number and sympathetic innervation of Auerbach's plexus in the rat. *J. Anat.* **1990**, *168*, 57–62. [PubMed]

53. Stenkamp-Strahm, C.; Patterson, S.; Boren, J.; Gericke, M.; Balemba, O. High-fat diet and age-dependent effects on enteric glial cell populations of mouse small intestine. *Auton. Neurosci. Basic Clin.* **2013**, *177*, 199–210. [CrossRef] [PubMed]

54. Coad, J.; Al-Rasasi, B.; Morgan, J. Nutrient insult in early pregnancy. *Proc. Nutr. Soc.* **2002**, *61*, 51–59. [CrossRef] [PubMed]

55. Tomas, J.; Mulet, C.; Saffarian, A.; Cavin, J.-B.; Ducroc, R.; Regnault, B.; Tan, C.K.; Duszka, K.; Burcelin, R.; Wahli, W.; et al. High-fat diet modifies the PPAR-γ pathway leading to disruption of microbial and physiological ecosystem in murine small intestine. *Proc. Natl. Acad. Sci. USA* **2016**, *113*, E5934–E5943. [CrossRef] [PubMed]

56. Kim, K.-A.; Gu, W.; Lee, I.-A.; Joh, E.-H.; Kim, D.-H. High fat diet-induced gut microbiota exacerbates inflammation and obesity in mice via the TLR4 signaling pathway. *PLoS ONE* **2012**, *7*, e47713. [CrossRef] [PubMed]

57. de La Serre, C.B.; Ellis, C.L.; Lee, J.; Hartman, A.L.; Rutledge, J.C.; Raybould, H.E. Propensity to high-fat diet-induced obesity in rats is associated with changes in the gut microbiota and gut inflammation. *Am. J. Physiol. Gastrointest. Liver Physiol.* **2010**, *299*, G440–G448. [CrossRef] [PubMed]

58. Ley, R.E.; Bäckhed, F.; Turnbaugh, P.; Lozupone, C.A.; Knight, R.D.; Gordon, J.I. Obesity alters gut microbial ecology. *Proc. Natl. Acad. Sci. USA* **2005**, *102*, 11070–11075. [CrossRef]

59. Turnbaugh, P.J.; Backhed, F.; Fulton, L.; Gordon, J.I. Marked alterations in the distal gut microbiome linked to diet-induced obesity. *Cell Host Microbe* **2008**, *3*, 213–223. [CrossRef]

60. Suzuki, T.; Hara, H. Dietary fat and bile juice, but not obesity, are responsible for the increase in small intestinal permeability induced through the suppression of tight junction protein expression in LETO and OLETF rats. *Nutr. Metab.* **2010**, *7*, 19. [CrossRef]

61. Everard, A.; Lazarevic, V.; Gaïa, N.; Johansson, M.; Ståhlman, M.; Backhed, F.; Delzenne, N.M.; Schrenzel, J.; François, P.; Cani, P.D. Microbiome of prebiotic-treated mice reveals novel targets involved in host response during obesity. *ISME J.* **2014**, *8*, 2116–2130. [CrossRef]

62. Loonen, L.M.P.; Stolte, E.H.; Jaklofsky, M.T.J.; Meijerink, M.; Dekker, J.; van Baarlen, P.; Wells, J.M. REG3γ-deficient mice have altered mucus distribution and increased mucosal inflammatory responses to the microbiota and enteric pathogens in the ileum. *Mucosal Immunol.* **2014**, *7*, 939–947. [CrossRef] [PubMed]

63. Markart, P.; Faust, N.; Graf, T.; Na, C.-L.; Weaver, T.E.; Akinbi, H.T. Comparison of the microbicidal and muramidase activities of mouse lysozyme M and P. *Biochem. J.* **2004**, *380*, 385–392. [CrossRef] [PubMed]

64. Hansson, G.C. Role of mucus layers in gut infection and inflammation. *Curr. Opin. Microbiol.* **2012**, *15*, 57–62. [CrossRef] [PubMed]

65. Schulz, M.D.; Atay, C.; Heringer, J.; Romrig, F.K.; Schwitalla, S.; Aydin, B.; Ziegler, P.K.; Varga, J.; Reindl, W.; Pommerenke, C.; et al. High-fat-diet-mediated dysbiosis promotes intestinal carcinogenesis independently of obesity. *Nature* **2014**, *514*, 508–512. [CrossRef] [PubMed]

66. Wang, L.; Fouts, D.E.; Stärkel, P.; Hartmann, P.; Chen, P.; Llorente, C.; DePew, J.; Moncera, K.; Ho, S.B.; Brenner, D.A.; et al. Intestinal REG3 lectins protect against alcoholic steatohepatitis by reducing mucosa-associated microbiota and preventing bacterial translocation. *Cell Host Microbe* **2016**, *19*, 227–239. [CrossRef] [PubMed]

67. Vaishnava, S.; Yamamoto, M.; Severson, K.M.; Ruhn, K.A.; Yu, X.; Koren, O.; Ley, R.; Wakeland, E.K.; Hooper, L.V. The antibacterial lectin RegIIIgamma promotes the spatial segregation of microbiota and host in the intestine. *Science* **2011**, *334*, 255–258. [CrossRef] [PubMed]

68. Reynolds, J.; O'Farrelly, C.; Feighery, C.; Murchan, P.; Leonard, N.; Fulton, G.; O'Morain, C.; Keane, F.; Tanner, W. Impaired gut barrier function in malnourished patients. *Br. J. Surg.* **2005**, *83*, 1288–1291. [CrossRef]

69. Welsh, F.K.S.; Farmery, S.M.; MacLennan, K.; Sheridan, M.B.; Barclay, G.R.; Guillou, P.J.; Reynolds, J.V. Gut barrier function in malnourished patients. *Gut* **1998**, *42*, 396–401. [CrossRef]

70. Mondal, D.; Minak, J.; Alam, M.; Liu, Y.; Dai, J.; Korpe, P.; Liu, L.; Haque, R.; Petri, W.A. Contribution of enteric infection, altered intestinal barrier function, and maternal malnutrition to infant malnutrition in Bangladesh. *Clin. Infect. Dis.* **2012**, *54*, 185–192. [CrossRef]
71. Deitch, E.A.; Winterton, J.; Li, M.; Berg, R. The gut as a portal of entry for bacteremia. Role of protein malnutrition. *Ann. Surg.* **1987**, *205*, 681–692. [CrossRef]
72. Oriá, R.B.; Vieira, C.M.G.; Pinkerton, R.C.; de Castro Costa, C.M.; Lopes, M.B.; Hussaini, I.; Shi, W.; Brito, G.A.C.; Lima, A.A.M.; Guerrant, R.L. Apolipoprotein E knockout mice have accentuated malnutrition with mucosal disruption and blunted insulin-like growth factor I responses to refeeding. *Nutr. Res.* **2006**, *26*, 427–435. [CrossRef] [PubMed]
73. Guerrant, R.L.; Oriá, R.B.; Moore, S.R.; Oriá, M.O.; Lima, A.A. Malnutrition as an enteric infectious disease with long-term effects on child development. *Nutr. Rev.* **2008**, *66*, 487–505. [CrossRef] [PubMed]
74. Dock-Nascimento, D.B.; Junqueira, K.; de Aguilar-Nascimento, J.E. Rapid restoration of colonic goblet cells induced by a hydrolyzed diet containing probiotics in experimental malnutrition. *Acta Cir. Bras.* **2007**, *22*, 72–76. [CrossRef] [PubMed]
75. Wehkamp, J.; Schauber, J.; Stange, E.F. Defensins and cathelicidins in gastrointestinal infections. *Curr. Opin. Gastroenterol.* **2007**, *23*, 32–38. [CrossRef] [PubMed]
76. Nieuwenhuis, E.E.S.; Matsumoto, T.; Lindenbergh, D.; Willemsen, R.; Kaser, A.; Simons-Oosterhuis, Y.; Brugman, S.; Yamaguchi, K.; Ishikawa, H.; Aiba, Y.; et al. Cd1d-dependent regulation of bacterial colonization in the intestine of mice. *J. Clin. Investig.* **2009**, *119*, 1241–1250. [CrossRef] [PubMed]
77. Grootjans, J.; Hodin, C.M.; de Haan, J.-J.; Derikx, J.P.M.; Rouschop, K.M.A.; Verheyen, F.K.; van Dam, R.M.; Dejong, C.H.C.; Buurman, W.A.; Lenaerts, K. Level of activation of the unfolded protein response correlates with Paneth cell apoptosis in human small intestine exposed to ischemia/reperfusion. *Gastroenterology* **2011**, *140*, 529–539. [CrossRef]
78. Shi, J.; Aono, S.; Lu, W.; Ouellette, A.J.; Hu, X.; Ji, Y.; Wang, L.; Lenz, S.; van Ginkel, F.W.; Liles, M.; et al. A novel role for defensins in intestinal homeostasis: Regulation of IL-1beta secretion. *J. Immunol.* **2007**, *179*, 1245–1253. [CrossRef]
79. Al-Sadi, R.; Ye, D.; Dokladny, K.; Ma, T.Y. Mechanism of IL-1beta-induced increase in intestinal epithelial tight junction permeability. *J. Immunol.* **2008**, *180*, 5653–5661. [CrossRef]
80. Al-Sadi, R.M.; Ma, T.Y. IL-1beta causes an increase in intestinal epithelial tight junction permeability. *J. Immunol.* **2007**, *178*, 4641–4649. [CrossRef]
81. Cunningham-Rundles, S. Malnutrition and gut immune function. *Curr. Opin. Gastroenterol.* **1994**, *10*, 664–670. [CrossRef]
82. Barker, D.J.P.; Osmond, C.; Kajantie, E.; Eriksson, J.G. Growth and chronic disease: Findings in the Helsinki birth cohort. *Ann. Hum. Biol.* **2009**, *36*, 445–458. [CrossRef] [PubMed]
83. Rao, M.; Nelms, B.D.; Dong, L.; Salinas-Rios, V.; Rutlin, M.; Gershon, M.D.; Corfas, G. Enteric glia express proteolipid protein 1 and are a transcriptionally unique population of glia in the mammalian nervous system. *GLIA* **2015**, *63*, 2040–2057. [CrossRef] [PubMed]
84. Cabarrocas, J.; Savidge, T.C.; Liblau, R.S. Role of enteric glial cells in inflammatory bowel disease. *GLIA* **2003**, *41*, 81–93. [CrossRef] [PubMed]
85. Margolis, K.G.; Gershon, M.D. Enteric neuronal regulation of intestinal inflammation. *Trends Neurosci.* **2016**, *39*, 614–624. [CrossRef] [PubMed]
86. Laranjeira, C.; Sandgren, K.; Kessaris, N.; Richardson, W.; Potocnik, A.; Van den Berghe, P.; Pachnis, V. Glial cells in the mouse enteric nervous system can undergo neurogenesis in response to injury. *J. Clin. Investig.* **2011**, *121*, 3412–3424. [CrossRef] [PubMed]
87. Challier, J.C.; Basu, S.; Bintein, T.; Minium, J.; Hotmire, K.; Catalano, P.M.; Hauguel-de Mouzon, S. Obesity in pregnancy stimulates macrophage accumulation and inflammation in the placenta. *Placenta* **2008**, *29*, 274–281. [CrossRef] [PubMed]
88. Bilbo, S.D.; Tsang, V. Enduring consequences of maternal obesity for brain inflammation and behavior of offspring. *FASEB J.* **2010**, *24*, 2104–2115. [CrossRef]
89. Roberts, K.A.; Riley, S.C.; Reynolds, R.M.; Barr, S.; Evans, M.; Statham, A.; Hor, K.; Jabbour, H.N.; Norman, J.E.; Denison, F.C. Placental structure and inflammation in pregnancies associated with obesity. *Placenta* **2011**, *32*, 247–254. [CrossRef]

90. Freeman, L.R.; Small, B.J.; Bickford, P.C.; Umphlet, C.; Granholm, A.-C. A high-fat/high-cholesterol diet inhibits growth of fetal hippocampal transplants via increased inflammation. *Cell Transpl.* **2011**, *20*, 1499–1514. [CrossRef]
91. Wu, Y.-Y.; Hsu, C.-M.; Chen, P.-H.; Fung, C.-P.; Chen, L.-W. Toll-like receptor stimulation induces nondefensin protein expression and reverses antibiotic-induced gut defense impairment. *Infect. Immun.* **2014**, *82*, 1994–2005. [CrossRef]
92. Hu, G.; Gong, A.-Y.; Roth, A.L.; Huang, B.Q.; Ward, H.D.; Zhu, G.; Larusso, N.F.; Hanson, N.D.; Chen, X.-M. Release of luminal exosomes contributes to TLR4-mediated epithelial antimicrobial defense. *PLoS Pathog.* **2013**, *9*, e1003261. [CrossRef] [PubMed]
93. De Oca, E.P.M. Antimicrobial peptide elicitors: New hope for the post-antibiotic era. *Innate Immun.* **2013**, *19*, 227–241. [CrossRef] [PubMed]
94. Lu, Z.; Ding, L.; Lu, Q.; Chen, Y.-H. Claudins in intestines. *Tissue Barriers* **2013**, *1*, e24978. [CrossRef] [PubMed]
95. Holmes, J.L.; Van Itallie, C.M.; Rasmussen, J.E.; Anderson, J.M. Claudin profiling in the mouse during postnatal intestinal development and along the gastrointestinal tract reveals complex expression patterns. *Gene Expr. Patterns* **2006**, *6*, 581–588. [CrossRef] [PubMed]
96. Patel, R.M.; Myers, L.S.; Kurundkar, A.R.; Maheshwari, A.; Nusrat, A.; Lin, P.W. Probiotic bacteria induce maturation of intestinal Claudin 3 expression and barrier function. *Am. J. Pathol.* **2012**, *180*, 626–635. [CrossRef]
97. Ding, L.; Lu, Z.; Foreman, O.; Tatum, R.; Lu, Q.; Renegar, R.; Cao, J.; Chen, Y.-H. Inflammation and disruption of the mucosal architecture in claudin-7-deficient mice. *Gastroenterology* **2012**, *142*, 305–315. [CrossRef] [PubMed]
98. Polak-Charcon, S.; Shoham, J.; Ben-Shaul, Y. Tight junctions in epithelial cells of human fetal hindgut, normal colon, and colon adenocarcinoma. *J. Natl. Cancer Inst.* **1980**, *65*, 53–62.
99. Blache, P.; van de Wetering, M.; Duluc, I.; Domon, C.; Berta, P.; Freund, J.-N.; Clevers, H.; Jay, P. SOX9 is an intestine crypt transcription factor, is regulated by the Wnt pathway, and represses the CDX2 and MUC2 genes. *J. Cell Biol.* **2004**, *166*, 37–47. [CrossRef]
100. Sherman, M.P. New concepts of microbial translocation in the neonatal intestine: Mechanisms and prevention. *Clin. Perinatol.* **2010**, *37*, 565–579. [CrossRef]
101. Mai, V.; Torrazza, R.M.; Ukhanova, M.; Wang, X.; Sun, Y.; Li, N.; Shuster, J.; Sharma, R.; Hudak, M.L.; Neu, J. Distortions in development of intestinal microbiota associated with late onset sepsis in preterm infants. *PLoS ONE* **2013**, *8*, e52876. [CrossRef]
102. Deitch, E.A. Gut-origin sepsis: Evolution of a concept. *Surgeon* **2012**, *10*, 350–356. [CrossRef] [PubMed]
103. Buddington, R.K.; Diamond, J.M. Ontogenetic development of intestinal nutrient transporters. *Annu. Rev. Physiol.* **1989**, *51*, 601–619. [CrossRef] [PubMed]
104. Rouwet, E.V.; Heineman, E.; Buurman, W.A.; ter Riet, G.; Ramsay, G.; Blanco, C.E. Intestinal permeability and carrier-mediated monosaccharide absorption in preterm neonates during the early postnatal period. *Pediatr. Res.* **2002**, *51*, 64–70. [CrossRef] [PubMed]
105. Clifford, S.H.; Weller, K.F. The absorption of vitamin A in prematurely born infants: With experience in the use of absorbable vitamin A in the prophylaxis of retrolental fibroplasia. *Pediatrics* **1948**, *1*, 505–511. [PubMed]
106. Siggers, R.H.; Siggers, J.; Thymann, T.; Boye, M.; Sangild, P.T. Nutritional modulation of the gut microbiota and immune system in preterm neonates susceptible to necrotizing enterocolitis. *J. Nutr. Biochem.* **2011**, *22*, 511–521. [CrossRef] [PubMed]
107. Gephart, S.M.; McGrath, J.M.; Effken, J.A.; Halpern, M.D. Necrotizing enterocolitis risk: State of the science. *Adv. Neonatal Care* **2012**, *12*, 77–87. [CrossRef]
108. Gregory, K.E.; Deforge, C.E.; Natale, K.M.; Phillips, M.; Van Marter, L.J. Necrotizing enterocolitis in the premature infant: Neonatal nursing assessment, disease pathogenesis, and clinical presentation. *Adv. Neonatal Care* **2011**, *11*, 155–164. [CrossRef] [PubMed]
109. Luig, M.; Lui, K.; NSW & ACT NICUS Group. Epidemiology of necrotizing enterocolitis—Part II: Risks and susceptibility of premature infants during the surfactant era: A regional study. *J. Paediatr. Child Health* **2005**, *41*, 174–179. [PubMed]
110. Di Renzo, G.C.; Rosati, A.; Sarti, R.D.; Cruciani, L.; Cutuli, A.M. Does fetal sex affect pregnancy outcome? *Gend. Med.* **2007**, *4*, 19–30. [CrossRef]

111. Zeitlin, J.; Saurel-Cubizolles, M.-J.; De Mouzon, J.; Rivera, L.; Ancel, P.-Y.; Blondel, B.; Kaminski, M. Fetal sex and preterm birth: Are males at greater risk? *Hum. Reprod.* **2002**, *17*, 2762–2768. [CrossRef] [PubMed]
112. Mayoral, S.R.; Omar, G.; Penn, A.A. Sex differences in a hypoxia model of preterm brain damage. *Pediatr. Res.* **2009**, *66*, 248–253. [CrossRef] [PubMed]
113. Eriksson, J.G.; Kajantie, E.; Osmond, C.; Thornburg, K.; Barker, D.J.P. Boys live dangerously in the womb. *Am. J. Hum. Biol.* **2010**, *22*, 330–335. [CrossRef] [PubMed]
114. Van Abeelen, A.F.M.; de Rooij, S.R.; Osmond, C.; Painter, R.C.; Veenendaal, M.V.E.; Bossuyt, P.M.M.; Elias, S.G.; Grobbee, D.E.; van der Schouw, Y.T.; Barker, D.J.P.; et al. The sex-specific effects of famine on the association between placental size and later hypertension. *Placenta* **2011**, *32*, 694–698. [CrossRef] [PubMed]
115. Aiken, C.E.; Ozanne, S.E. Sex differences in developmental programming models. *Reproduction* **2013**, *145*, R1–R13. [CrossRef] [PubMed]

© 2019 by the authors. Licensee MDPI, Basel, Switzerland. This article is an open access article distributed under the terms and conditions of the Creative Commons Attribution (CC BY) license (http://creativecommons.org/licenses/by/4.0/).

Article

Resistant Starch Is Actively Fermented by Infant Faecal Microbiota and Increases Microbial Diversity

Geetha Gopalsamy [1,2], Elissa Mortimer [2], Paul Greenfield [3], Anthony R. Bird [4], Graeme P. Young [2] and Claus T. Christophersen [5,6,*]

1. Eastern Health Clinical School, Monash University, Box Hill, VIC 3128, Australia; gopa0006@gmail.com
2. Flinders Centre for Innovation in Cancer, College of Medicine and Public Health, Flinders University, Bedford Park, SA 5042, Australia; elissa.mortimer@flinders.edu.au (E.M.); graeme.young@flinders.edu.au (G.P.Y.)
3. CSIRO Environomics Future Science Platform, North Ryde, NSW 2113, Australia; Paul.Greenfield@csiro.au
4. CSIRO Health and Biosecurity, Adelaide, SA 5000, Australia; Tony.Bird@csiro.au
5. School of Medical & Health Sciences, Edith Cowan University, Joondalup, WA 6027, Australia
6. School of Molecular & Life Sciences, Curtin University, Bentley, WA 6102, Australia
* Correspondence: c.christophersen@ecu.edu.au; Tel.: +61-8-6304-5278

Received: 13 May 2019; Accepted: 12 June 2019; Published: 14 June 2019

Abstract: In adults, fermentation of high amylose maize starch (HAMS), a resistant starch (RS), has a prebiotic effect. Were such a capacity to exist in infants, intake of RS might programme the gut microbiota during a critical developmental period. This study aimed to determine if infant faecal inocula possess the capacity to ferment HAMS or acetylated-HAMS (HAMSA) and characterise associated changes to microbial composition. Faecal samples were collected from 17 healthy infants at two timepoints: Preweaning and within 10 weeks of first solids. Fermentation was assessed using in vitro batch fermentation. Following 24 h incubation, pH, short-chain fatty acid (SCFA) production and microbial composition were compared to parallel control incubations. In preweaning infants, there was a significant decrease at 24 h in pH between control and HAMS incubations and a significant increase in the production of total SCFAs, indicating fermentation. Fermentation of HAMS increased further following commencement of solids. Fermentation of RS with weaning faecal inocula increased Shannon's diversity index (H) and was associated with increased abundance of *Bifidobacterium* and *Bacteroides*. In conclusion, the faecal inocula from infants is capable of RS fermentation, independent of stage of weaning, but introduction of solids increases this fermentation capacity. RS may thus function as a novel infant prebiotic.

Keywords: short-chain fatty acid (SCFA); pH; dietary fibre; gut health; prebiotic

1. Introduction

The initiation of solid foods in early infant feeding represents a dynamic period of change in the composition of the gut microbiota [1]. This period is also critical to child development with implications for general nutrition and immune development, amongst other benefits [1]. The administration of prebiotics during this period could have profound health consequences [2] as they are non-digestible, generally safe and inexpensive food ingredients that selectively stimulate the growth and/or activity of one or a limited number of bacterial species that already reside in the colon [3].

In adults, high amylose maize starch (HAMS), a cultivar obtained through selective breeding, may function as a prebiotic [4]. It is a form of resistant starch (RS), which is defined as the sum of starch and products of starch degradation which have not been absorbed in the small intestine of healthy individuals and which become available for microbial fermentation in the colon [5]. RS is classified into five types and several of these starches have been shown to alter gut fermentation and change gut microbial composition [5]. HAMS is an example of a type 2 RS. Starches may also be chemically modified (type 4 RS) to reduce their digestibility and obtain favourable physicochemical properties. Acylated type 4 RSs may also increase the delivery of short-chain fatty acids (SCFAs) directly to the colon. One example of this is acetylated-HAMS (HAMSA), which is esterified with acetyl groups.

Due to significant differences in the composition and function of infant and adult gut microbiota, the potential of RS to function as a prebiotic in infants cannot be extrapolated from adult studies. It is possible that the relatively immature gut microbial ecosystem of the infant may not have acquired the necessary diversity of bacteria to ferment a complex carbohydrate such as RS [6,7].

Due to the relative inaccessibility of the proximal colon and portal vein, in vivo measurement of substrate digestion and fermentation requires highly invasive procedures and feeding studies to preweaning infants that are not practical. In vitro static batch fermentation is a rapid, inexpensive method to initially assess substrate fermentation and has been widely used to evaluate the fermentation capacity of both adult and infant faecal inocula [8,9]. Substrate fermentation during in vitro batch fermentation studies is evidenced by substrate disappearance, increased production of SCFAs, a decrease in pH, gas generation, or differences in microbial composition in those ferments containing the substrate when compared to a control. The control incubation does not have an added substrate.

In the present study, we used an in vitro batch fermentation system that simulates human colonic fermentation to determine: (1) The capacity of infant faecal inocula, pre-weaning and weaned, to ferment pre-digested HAMS and HAMSA by monitoring changes in the activity (fermentation) and community structure of the faecal microbiota of preweaning and early weaning infants; and (2) whether the introduction of solids into the diet (weaning) influences these variables.

2. Materials and Methods

2.1. Study Design

We conducted an observational study in which faecal samples were collected from infants, prior to and following weaning (commencement of first solid foods). The first faecal sample was collected from 8 weeks of age until weaning, while the second sample was collected within 10 weeks of weaning commencing. Collected samples were incubated in vitro with fermentable substrates to address the aims. Each infant was expected to provide two samples, one for each time point. This study complied with National Health and Medical Research Council (Australia) guidelines relating to ethical conduct in human research and was reviewed and approved by the Southern Adelaide Clinical Human Research Ethics Committee (September 2013, 339.13).

2.2. Participants and Intervention

Caregivers of preweaning infants were recruited either by direct approach following childbirth and prior to hospital discharge or through advertisements in a local parenting magazine. Inclusion criteria included that the infant was full term at birth with a gestational age of more than 38 weeks; at or above the 10th percentile for weight at birth and with no known cardiac, respiratory or gastrointestinal disease. Infants were excluded if they or their mothers received probiotic supplementation or antibiotics post-delivery and prior to stool collection. The mode of infant feeding, breastfed or bottle-fed or mixed (not exclusively breast fed), was recorded. There were 17 preweaning infants and 16 of these infants went on to provide weaning samples.

2.3. Faecal Sample Collection and Processing

Within 15 min of an infant passing a motion, stool was collected from a disposable nappy using a sterile container, placed into an airtight bag and put into a portable freezer set at −20 °C. Within two hours, the sample was processed in the laboratory. Working within an anaerobic cabinet containing 5% H_2, 5% CO_2, and 90% N_2, the faecal samples were homogenised in 50% glycerol (1:1 dilution) and stored at −80 °C until further analysis. The use of frozen samples compared to fresh samples in fermentation experiments has been previously validated [10].

2.4. Carbohydrates and Chemicals

The two RSs, HAMS (Hylon VII) and modified HAMS (HAMSA-Crispfilm) were obtained from Ingredion, USA. Hylon VII is composed of 70% amylose and 30% amylopectin. It is estimated that Hylon VII contains 50 g of RS per 100 g [11]. Crispfilm (CF) has a Hylon VII backbone and has undergone a further esterification process to form a starch acetate. The degree of acetylation of CF is less than 2.5%, the limit imposed by U.S. Food and Drug Administration (FDA) food regulation and CF has Generally Recognized as Safe (GRAS) status. Lactulose was obtained from Sigma-Aldrich, Australia.

2.5. In Vitro Fermentation

An in vitro pre-digestion step was performed to simulate the digestive action of the infant small intestine prior to the fermentation experiments [12]. See Appendix A.1, for further detail of the pre-digestion method. The in vitro fermentation method was based on the technique described by Edwards et al. [13] and Goni et al. [14]. For each preweaning infant sample there were three groups of incubations: HAMS, lactulose and the control. The control incubation had no additional substrate and lactulose was added to confirm that viable bacteria were present in the faecal inocula of each subject. For weaning infant samples there were four incubations, with HAMSA as the additional group, along with HAMS, lactulose and the control. HAMSA was not tested in preweaning infants due to small sample volumes. All fermentations were performed in triplicate for each donor/substrate at 0 and 24 h time points. Briefly, 100 mg of pre-digested starch residue was weighed into triplicate 15 mL sterile culture tubes together with 8–10 sterile 2.5 mm glass beads and 9 mL of autoclaved fermentation media. Frozen homogenised faecal material from participants was thawed and a 10% w/v faecal slurry was prepared by homogenisation and dilution in pre-reduced phosphate-buffered saline (PBS) (0.1 M, pH 7.2). Working within an anaerobic chamber, 1 mL of faecal slurry was added to each fermentation tube (1% w/v). Controls were incubated in parallel with incubations containing HAMS, lactulose or HAMSA. Tubes were incubated under anaerobic conditions with gentle agitation for 24 h. Fermentation was terminated at 0 h (Blanks) and at 24 h samples by centrifugation at 13,000× g, 4 °C for 10 min and the supernatant was stored at −80 °C for further measurements.

2.6. SCFA and pH

SCFA concentration was determined using capillary column gas chromatography according to McOrist et al. [15]. A digital pH meter was used to measure pH. Fermentation of the substrate was deemed to have occurred if there was a statistically significant decrease in pH and an increase in production of SCFA at 24 h in ferments containing added RS when compared to the respective control.

2.7. Molecular and Sequence Analysis

2.7.1. DNA Extraction

DNA was extracted from ferments using bead beating followed by the MoBio PowerMag Microbiome RNA/DNA isolation kit (Qiagen, Hilden, Germany) optimised for epMotion (Eppendorf, Hamburg, Germany) platforms (see Appendix A.2). Due to the recognised variation in fermentation between replicates during in vitro batch fermentation [16], DNA was extracted from a pooled sample containing 1 mL from each ferment replicate.

2.7.2. Real-time Quantitative Polymerase Chain Reaction (qPCR)

Selected bacteria including *bifidobacterium*, *lactobacillus* and total bacteria, were quantified by specific primers targeting the 16S rRNA gene using qPCR. See Appendix A, Table A1, for primer sequences and the optimised qPCR conditions. The ability of a substrate to selectively stimulate the growth of a given bacterial taxon was determined by comparing incubations with either HAMS or HAMSA to the 24 h control. All qPCR analysis was performed on the CFX 384TM real-time PCR detection system (Bio-Rad, Hercules, CA, USA) (See Appendix A.3). Absolute abundance was estimated according to Christophersen et al. [17].

2.7.3. Sequencing of 16S Ribosomal RNA Encoding Gene Amplicons

16S ribosomal DNA gene sequencing was performed on DNA extracted from each participant's 24 h fermentation samples (preweaning control, preweaning HAMS, weaning HAMS, weaning HAMSA, weaning control). The 24 h samples were chosen as the fermentation is not only affected by the added substrate but also by the remaining substrates in the faecal slurry. We therefore believe the true control for each subject and substrate is a control fermentation with no added substrate to take into account the available substrate in the faecal slurry. The methods outlined in Illumina's 16S Metagenomic Sequencing Library Preparation protocol (Illumina, San Diego, CA, USA) were followed with minor adjustments made to PCR thermal cycle conditions, as described in Appendix A.4.

2.7.4. Taxonomic Assignments to 16S Reads

An in-house (CSIRO) amplicon clustering and classification pipeline (GHAP) based on tools from Usearch [18] and a Ribosomal Database Project (RDP) classifier [19] combined with locally written tools for demultiplexing and generating Operational Taxonomic Unit (OTU) tables were used to process the amplicon sequence data. Following the merging of paired reads, dereplication, clustering at 97% and chimera checking were also performed using the pipeline. Classification of the reads was then performed by using the RDP to assign taxonomy and by finding the closest match to the OTU from a set of reference 16S sequences [19]. OTUs were defined at a 97% sequence similarity level and classified to genus level. Sequences which were not classified using the pipeline were manually blasted against the NCBI database.

2.8. Statistical Analysis

For SCFA and pH results, data normality was assessed using the Shapiro–Wilk test using SPSS Version 22.0. A boxplot of the dataset was used to identify outliers within preweaning and weaning groups, Univariate ANOVA with Bonferroni correction was used to analyse for differences in starting pH and total SCFA of the different groups within the weaning and preweaning infants. Due to differences in the number of formula and breastfed infants, a general linear mixed model was used to determine if within the preweaning group, the method of feeding influenced the effect of incubation with HAMS on both change in pH and total SCFA production. A repeated measures two-factor ANOVA was used to determine if there was an effect of weaning on parameters of HAMS fermentation (pH and total SCFA) when compared to controls. Values are presented as means ± their standard errors. Statistical significance was accepted as $p < 0.05$.

For analysis of the molecular results, the qPCR values were log10 transformed and the means were compared using Student's *t*-test. Microbial abundance at 24 h was compared with those at 0 h for each substrate. Multivariate analysis of the sequencing data was performed using PRIMER 7 with PERMANOVA (PRIMER-e, Auckland, New Zealand). Statistical analysis of Bray-Curtis dissimilarities were calculated using relative abundance of bacterial genera at the family level following 24 h of fermentation. Alpha diversity index was also calculated at the family level. Principal coordinate analysis (PCOA) was used to visualise the dissimilarity data.

3. Results

The seventeen participants provided a preweaning sample, with all but one of these infants also providing a second (weaning) sample (see Table 1).

Table 1. Description of infants used as faecal donors for in vitro fermentation experiments.

Participants	Preweaning (*n*)	Weaning (*n*)	Age Preweaning Sample (Months) Mean ± SE	Age Weaning Sample (Months) Mean ± SE
Exclusively Breast-Fed	10	10	3.32 ± 0.37 [a]	7.03 ± 0.27 [b]
Mixed	7	6	3.29 ± 0.22 [a]	7.39 ± 0.21 [b]
Total	17	16	3.31 ± 0.23	7.16 ± 0.18

SE: Standard error. Means in a row without a common superscript letter differ significantly ($p < 0.05$).

3.1. SCFA and pH Levels

The Shapiro–Wilk test confirmed a normal distribution of faecal pH and SCFA data in both preweaning and weaning incubation samples. One participant, in the preweaning exclusively breast-fed group, was noted to have an outlier for total SCFA for each substrate. Parallel series of calculations performed with and without the inclusion of this participant's data did not alter the final statistical conclusions. Within the preweaning group, linear mixed model analysis revealed no effect of mode of feeding (exclusively breast fed or mixed) on SCFA ($p = 0.754$) or pH ($p = 0.809$), following incubation with either substrate (HAMS or lactulose).

Results for initial, final and change in pH at 24 h following incubation are presented in Table 2. All incubations, including the lactulose, resulted in a significant decrease in pH at 24 h. In both weaning and preweaning groups the decrease in pH was significantly greater following incubation with HAMS than in the controls, consistent with active fermentation of HAMS. It was expected to observe a decrease in pH for control samples due to residual substrate in the faecal inoculate. In the weaning group, incubation with HAMSA also led to a decrease in pH when compared to the control.

Table 2. Initial and final change in pH following incubation of infant faecal microbiota with high amylose maize starch (HAMS) and other substrates. Mean ± SE.

	Preweaning			Weaning			
	Control	HAMS	Lactulose	Control	HAMS	HAMSA	Lactulose
Initial pH (0 h)	7.64 ± 0.03 [a]	7.63 ± 0.03 [a]	7.69 ± 0.03 [a]	7.66 ± 0.03 [a]	7.65 ± 0.02 [a]	7.65 ± 0.03 [a]	7.72 ± 0.04 [a]
Final pH (24 h)	6.72 ± 0.08 [a]	6.40 ± 0.09 [b,*]	4.73 ± 0.12 [c,*]	6.77 ± 0.05 [a]	6.43 ± 0.05 [b,*]	6.37 ± 0.04 [b,*]	4.65 ± 0.08 [c,*]
Δ pH	−0.93 ± 0.07 [a]	−1.21 ± 0.06 [b,*]	−3.03 ± 0.06 [c,*]	−0.82 ± 0.07 [a]	−1.14 ± 0.08 [b,*]	−1.09 ± 0.10 [b,*]	−2.89 ± 0.20 [c,*]

Within the preweaning and weaning groups a one-way ANOVA was used. Unlike superscript letters within each row are significantly different (Bonferroni adjusted $p < 0.05$). * defined as different from control ($p < 0.05$). HAMSA: High Amylose Maize Starch Acetylated.

Analysis of variance showed no effect of the substrates on initial pH, $p = 0.36$. A repeated measures two-factor ANOVA revealed an effect of HAMS $p < 0.001$, but not stage of weaning on change in pH, $p = 0.34$. There was no interaction between the substrate and stage of weaning $p = 0.54$, indicating that the introduction of solids did not influence the effect of incubation with HAMS on change in pH when compared to controls. The means and standard error thereof for total and major individual production of SCFAs at 24 h are shown in Table 3. In the faecal inocula of the preweaning group there was a significant increase in the concentration of SCFAs, acetate and butyrate when HAMS was added compared to the controls, indicating that the preweaning faecal inocula had some capacity to utilise HAMS as a substrate.

In the faecal inocula from weaned infants, there was a significant increase in the production of total SCFAs, acetate and propionate for both HAMSA and HAMS. A significant increase in butyrate was seen only with HAMS and not HAMSA. These changes confirmed fermentation of the RSs by the infant faecal inocula. The molar ratios for all the substrates confirmed previous findings that, in young infants, acetate is by far the dominant SCFA during fermentation.

In relation to total SCFA concentration, a two-way repeated measures ANOVA demonstrated a significant interaction between incubation with HAMS and stage of weaning, $p = 0.015$, suggesting that the fermentation of HAMS is enhanced post the introduction of solids. A pair-wise comparison with Bonferroni correction revealed no statistically significant effect of weaning on total SCFA concentration, $p = 0.06$. However, there was a statistically significant effect for HAMS $p < 0.001$.

Table 3. Short-chain fatty acid (SCFA) concentrations (mmol/L) following incubation with test substrates using pre- and weaning infant faecal inoculum. Mean ± SE.

	Preweaning				Weaning			
	Control	HAMS	Lactulose	Control	HAMS	HAMSA	Lactulose
Total SCFA [#]	16.68 ± 1.7 [a]	23.70 ± 1.7 [b,*]	61.84 ± 5.4 [c,*]	20.11 ± 2.3 [a]	34.85 ± 4.0 [b,*]	37.81 ± 4.2 [b,*]	78.27 ± 4.5 [c,*]
Acetate	14.70 ± 1.5 [a]	20.93 ± 1.5 [b,*]	56.11 ± 6.5 [c,*]	16.98 ± 2.1 [a]	27.73 ± 3.3 [b,*]	30.27 ± 3.1 [b,*]	73.42 ± 5.0 [c,*]
Propionate	0.83 ± 0.17 [a]	1.33 ± 0.33 [a]	1.13 ± 0.80 [a]	1.60 ± 0.32 [a]	4.14 ± 0.91 [b*]	5.56 ± 1.2 [c*]	2.89 ± 0.74 [a,b,c,*]
Butyrate	0.36 ± 0.13 [a]	0.74 ± 0.26 [b]	0.27 ± 0.12 [a]	1.09 ± 0.26 [a]	2.20 ± 0.59 [b*]	1.23 ± 0.33 [a]	1.81 ± 1.1 [a]

[#] Total SCFA (mmol/L) = sum of acetate, propionate, butyrate and minor SCFAs (valeric, caproic, isobutyric, isovaleric). Within the preweaning and weaning groups a one-way ANOVA was used. Unlike superscript letters within each row are significantly different (Bonferroni adjusted $p < 0.05$). * defined as different from control ($p < 0.05$).

3.2. Microbial Community Analysis

Across all samples collected at 24 h there were 321 operational taxonomic units identified, with an average of 94 OTUs per sample (59–213). For preweaning faecal fermentation samples, a total of 632,032 usable reads were obtained and for weaning, 873,706 reads were obtained for downstream analysis. DNA from two samples from the preweaning group failed to amplify during library preparation for unknown reasons and therefore these participants were omitted from sequence analysis. Sequences were classified at each phylogenetic level from phylum to genus.

3.2.1. Alpha Diversity

Alpha diversity was calculated in this study, although the study uses an in vitro model (closed system), because changes in diversity can still be observed as the Shannon index combines species richness and their relative abundances. In the preweaning group at 24 h of incubation, there was no significant difference in the Shannon Index between bacterial communities following incubation with HAMS or the controls. For example, at the family level, a one-way ANOVA determined that, in the preweaning group, the overall mean of the log(e) of the Shannon Index did not differ between the control and the HAMS incubations ($p = 0.10$). The Shannon Diversity Index boxplot based on OTU abundance at the family level is presented in Figure 1A. In the weaning group, at the family level, there was a significant increase in the Shannon index in the RS groups compared to the controls ($p = 0.05$). However, the type of RS (HAMSA or HAMS) did not have an effect on this measure of diversity. The Shannon Diversity Index boxplot based on OTU abundance at the family level for the weaning infants is presented in Figure 1B.

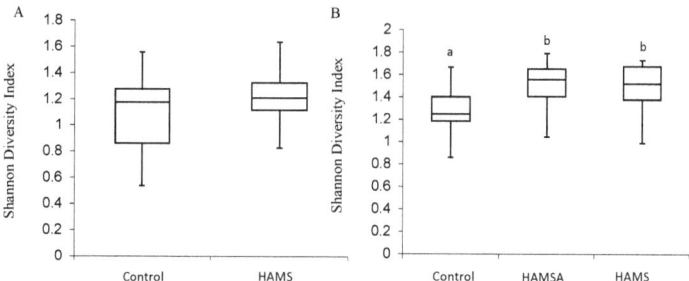

Figure 1. Boxplot of diversity at family level in (**A**) preweaning and (**B**) weaning infant faecal inocula following 24 h in vitro fermentation. Boxes indicate 25th to 75th percentiles, with mean values marked as a line and whiskers indicating minimum and maximum values. Different letters mean significantly different from each other ($p < 0.05$).

3.2.2. Beta Diversity

Among the preweaning samples, multivariate analysis did not reveal an effect of HAMS on microbial community structure ($p > 0.05$). However, in the weaning samples, the effect of both HAMS and HAMSA were significant ($p < 0.05$), compared to the controls. Pairwise comparison was performed to investigate the differences between the two test groups. This demonstrated that both RSs had a similar effect on community structure ($p = 0.97$) (see Figure 2). It is apparent that samples from the same individual clustered closely.

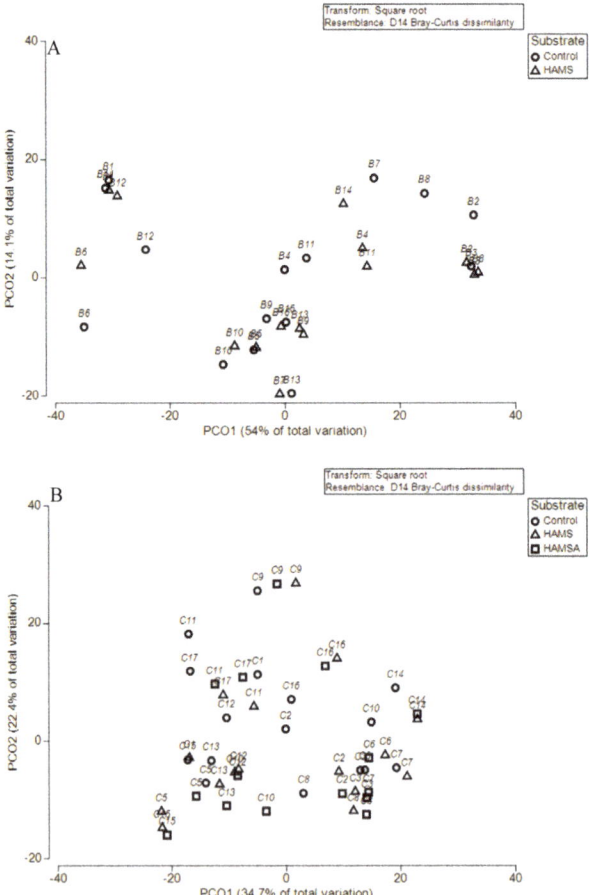

Figure 2. Principal coordinate analysis of the Bray-Curtis dissimilarity matrix for (**A**) preweaning and (**B**) weaning samples, calculated at the family level. PCO: Principal Coordinate Analysis.

3.2.3. Relative Abundance

In order to identify the changes in the composition of bacterial communities that might utilise the RSs, the relative abundance of bacterial groups in fermentation fluid at 24 h of incubation was assessed. The relative abundance was calculated and is presented in Appendix A, Table A2. Across all levels of classification, there were a number of statistically significant differences in the relative abundance of bacteria between HAMS and controls in the weaning group when compared to the preweaning group. At the phyla level in weaning infants, there was a significant increase in the proportion of

Actinobacteria and Bacteroidetes at 24 h of fermentation for both RSs in comparison to the controls ($p < 0.05$). There was also a significant reduction in Proteobacteria following incubation with both HAMS and HAMSA ($p < 0.05$). At a genus level in weaning infants, the abundances of *Bacteroides* and *Bifidobacterium* were significantly increased following incubation with the RSs when compared to controls ($p < 0.05$).

Incubation with both HAMS and HAMSA led to a concomitant reduction in the relative abundance of *Enterobacter*. Compared to the control, the relative abundance of *Ruminococcus* was significantly increased following incubation with HAMSA ($p = 0.01$), but not HAMS ($p = 0.58$). Figure 3 illustrates microbial composition after 24 h of fermentation at a genus level for each substrate, with a relative abundance threshold of 0.1%.

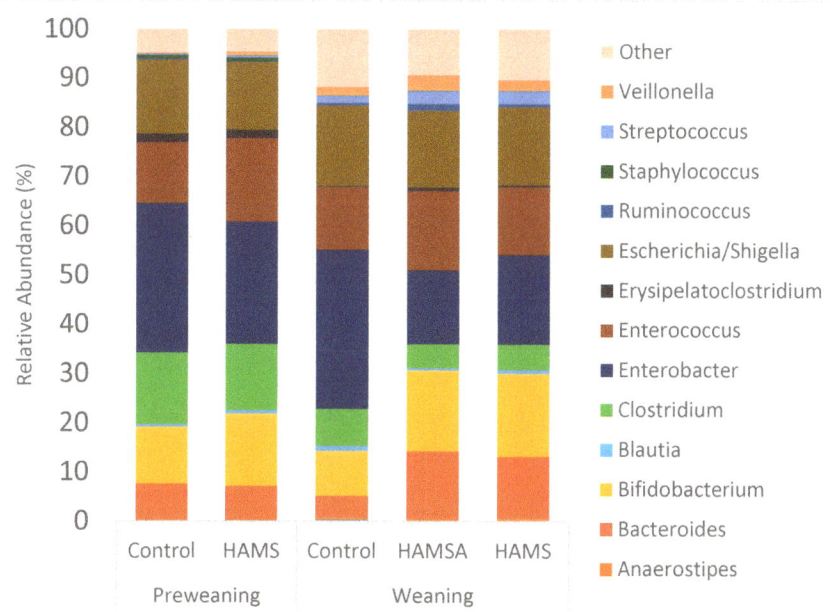

Figure 3. Genus-level composition of the microbial community after sequencing. DNA from 24 h in vitro fermentations of infant faecal inocula. Bacterial genus with a relative abundance of less than 1% are grouped as "other".

3.2.4. Quantitative PCR

All qPCR assays had previously been verified using single cell colony sequencing and they were found to be 100% specific to the bacterial group assigned. In preweaning infants, the absolute abundance for total bacteria increased after 24 h of fermentation in HAMS and controls (see Table 4) compared to time 0 h in the controls (representative of the baseline). However, at 24 h the absolute abundance of total bacteria, bifidobacteria and *Lactobacillus* did not differ between the control and the HAMS incubations. In the weaning infants, following 24 h of in vitro fermentation, the absolute abundance of total bacteria, bifidobacteria and *Lactobacillus* increased in all three groups (control, HAMS and HAMSA) compared to time 0 h controls. While there was no difference in the absolute abundance of total bacteria or *Lactobacillus* at 24 h compared to the control, the absolute abundance of *Bifidobacterium* was significantly greater in HAMS and HAMSA compared to controls. This indicates that both RS's can stimulate the growth of *Bifidobacterium* in the faecal microbiota of weaning infants.

Table 4. 16S rDNA copy numbers (log 10 copy numbers mL^{-1} fermentation effluent) of specific bacterial groups before (0 h) and after 24 h of in vitro fermentation with infant faecal inocula and test substrates, as determined by qPCR. Mean ± SE.

		Total Bacteria		Lactobacillus		Bifidobacterium	
	Substrate	0 h	24 h	0 h	24 h	0 h	24 h
Preweaning	control (n = 17)	6.6 ± 0.15	7.18 ± 0.24 [a,*]	4.26 ± 0.4	4.19 ± 0.23 [a]	6.06 ± 0.09	6.08 ± 0.32 [a]
	HAMS (n = 17)		7.21 ± 0.1 [a,*]		4.02 ± 0.28 [a]		6.39 ± 0.25 [a]
Weaning	Control (n = 16)	6.65 ± 0.07	7.32 ± 0.06 [a,*]	2.9 ± 0.27	3.34 ± 0.86 [a]	5.09 ± 0.31	5.56 ± 0.21 [a]
	HAMSA (n = 16)		7.42 ± 0.06 [a,*]		3.52 ± 0.91 [a]		6.21 ± 0.10 [b,*]
	HAMS (n = 15)		7.47 ± 0.05 [a,*]		3.49 ± 0.89 [a]		6.04 ± 0.19 [b,*]

In the preweaning and weaning groups, within column values which do not share a superscript letter are significantly different ($p < 0.05$). * Significant difference between 0 and 24 h values for each substrate, $p < 0.05$.

4. Discussion

A divergence in the intake of fibre between those living in Western countries and those consuming a predominantly agrarian diet, as occurs in many rural parts of the world, emerges as soon as solids are introduced into the infant diet [1]. Thus, targeted manipulation of fibre content in the early diet may affect the emerging gut microbiota during a critical period and lead to functional and compositional changes which could benefit the host's developing immune system [20,21]. It is within such a context that we were interested in whether forms of RS, which function as a prebiotic in adults, could have a role in infancy.

In both the weaning and preweaning groups, the production of SCFAs by the faecal inocula from these infants was significantly greater in the presence of HAMS when compared to controls. While fermentation might be expected in faeces from the weaning cohort, it is surprising that the microbiota of preweaning infants already possess capacity to ferment starch. However, this capacity did increase significantly post commencement of complementary feeds.

Previous results regarding the potential of young infant faecal bacteria to ferment complex carbohydrates, are conflicting. An early study by Parrett et al. suggested that the capacity of human faecal microbiota to ferment complex carbohydrates does not emerge for several months after solids are commenced [6]. In contrast, Christian et al. found that the faecal microbiota of early weaning infants is highly efficient at fermenting the digestible waxy maize starch [8]. Heterogeneity in methodologies across in vitro fermentation studies, including the inclusion or exclusion of substrate pre-digestion, the age of study participants and the preparation of faecal inocula, limits comparison of results across studies. In this study, we used a substrate pre-digestion which reflects the low level of alpha-amylase in the infant gut.

There was no significant difference in the production of total SCFAs, especially acetate, between HAMS and HAMSA (which is acetylated HAMS). It might be that the degree of substitution was insufficient for cleavage of the esterified acetate to achieve a statistically significant increase in the overall pool of acetate. It could also be that infant microbiota might not have yet acquired the esterase capacity needed to cleave the additional acetate.

Surprisingly, and despite the fermentation findings, our molecular findings showed little difference in bacterial composition following incubation of HAMS with the preweaning infant inocula in comparison to the controls. There were no selective differences in the abundance of groups of bacteria following incubation of HAMS with the preweaning infant inocula, in comparison to the controls. However, within ten weeks of commencing solids, the molecular changes within the weaning infant faecal incubations suggest emergence of such a capacity. If selective stimulation of *Bifidobacterium* and an increase in the production of SCFAs are components of the prebiotic definition [2] then, based on these findings, HAMS and HAMSA may well function as a novel prebiotics during infancy.

A significant number of *Bifidobacterium* spp. possess genes belonging to the GH13 family of glycosyl hydrolases (GH) [22,23]. Enzymes belonging to this family are heavily involved in the degradation of starch and starch related-substrates [24]. Despite the relatively high abundance of *Bifidobacterium* in preweaning infant inocula, several possibilities exist for why there were no significant differences in microbial profile between controls and substrate incubations. It could be that the significant inter-individual variation in the gut microbial composition of preweaning infants might have masked the capacity to establish substrate level differences in the microbial profile [25]. Another explanation rests upon differences in metabolic function amongst various members of the *Bifidobacterium* genus. Depending on the particular ecological niche, given the myriad of metabolic pathways available to *Bifidobacterium* spp., selective pressure will support the growth of those *Bifidobacterium* spp. that are able to utilise the substrates that are most available [22]. Metagenomic studies have demonstrated differences in which bifidobacterial GH genes are transcribed between infants and adults [26]. For example, in breast feeding infants, there is greater transcription of bifidobacterial GH genes that are involved in the degradation of Human Milk Oligosaccharides (HMO) and mucin, compared to adults which reveals a greater predominance of bifidobacterial GH encoding genes involved in the breakdown of complex plant-derived carbohydrates [26]. Following the introduction of solids, when more complex carbohydrates escape the digestive enzymes of the host and are encountered by the infant gut microbiota, those microbial genes that are involved in the utilisation of these novel nutrients will be switched on [27]. Given the immaturity of pancreatic function in the young infant, following the initiation of solids, significant amounts of dietary starch will enter the infant's colon. This will favour a selective increase in the expression of genes associated with starch utilisation and may account for the findings in the weaning faecal inocula.

Despite the recognised relevance of *Bifidobacterium* to starch utilisation, it had been suggested that *Ruminococcus bromii*, a member of the Firmicutes, is a keystone species in the degradation of HAMS [28]. Our results cannot confirm this assertion. While there was a selective increase in the relative abundance of *Ruminococcus* spp. following incubation with HAMSA, there was no such increase following incubation of HAMS with the weaning infant inocula. This was despite evidence of starch utilization, as seen in the increase in the relative abundance of other groups of bacteria in comparison to the controls. This suggests that many strains of bacteria are able to utilise HAMS, which aligns with studies conducted in pigs [29,30]. Members of Bacteroidetes, a dominant bacterial phylum in the mammalian gut, also encode numerous discrete polysaccharide utilisation loci (PULs) that may facilitate starch utilisation. Incubation of weaning infant faecal inocula with both RSs led to an increase in the relative abundance of Bacteroidetes and an increase in the ratio of Bacteroidetes:Firmicutes. An increase in the ratio of Bacteroidetes:Firmicutes may be associated with several positive health outcomes, including reduced risks of obesity and intestinal inflammation [31].

The concordance between HAMS and HAMSA in relation to the effects on microbial composition and diversity observed in this study could be due to the minimal nature of the chemical modification. The HAMSA starch used in this study is approved for consumption in infant foods. For starch acetate, the FDA and the WHO have stipulated an acetyl group's percentage below 2.5 g/100 g (corresponding to a maximum Degree of Substitution (DS) of 0.1) for a food application. Although HAMS with higher degrees of acetylation have been shown to facilitate increased delivery of SCFA into the human colon, their comparable effects on the gut microbiota have not yet been examined [32].

In a study feeding modified and unmodified RS to adult participants, different effects on faecal microbiota composition between the two starches were found [33]. The authors found the intake of the modified, but not the unmodified starch, over a three-week period led to a significant increase in the abundance of Actinobacteria and *Bacteroides*. Using our in vitro model, we demonstrated a similar taxonomic effect following incubation of the weaning infant feces with the modified RS. However, in contrast to Martinez et al. (2010) we found incubation of the unmodified RS led to similar taxonomic changes at the phylum level. From a translational viewpoint, if HAMS is eventually used in infant nutrition, the unmodified form may have greater appeal to caregivers due to its lack of chemical modification.

Being an in vitro design, our study has inherent limitations. The high relative abundance of Proteobacteria at 24 h of incubation, particularly in the controls, reflects a difference between in vitro and in vivo conditions and suggests that in vitro conditions might have favoured the growth of Proteobacteria. This also reflects that not all bacteria grow in vitro and faeces contain a large proportion of dead bacteria which may not be representative of the living mucosal-adherent and/or luminal bacteria in the gut. The limited availability of infant faecal inocula also precluded comparative evaluation of the fermentation of the chosen RS to other substrates which may already be recognised as prebiotics in infants, e.g., fructo-oligosaccharides (FOS). To circumvent this, it might have been possible to pool the faecal samples from different donors. However, individual variation in the microbiota might be greater than the effects of treatment.

A further limitation was that the actual degradation of the substrate in each ferment was not measured. It is not uncommon for in vitro fermentation studies to omit this measurement, instead relying, as we have done, on the products of fermentation and changes in microbial composition to provide an estimate of fermentation. It should also be noted that, irrespective of the benefits to gut health, RS may impact the nutritional status of the infant diet. Therefore, further studies are required to determine a safe dose of RS for highly vulnerable populations, such as growing babies and children.

Ultimately, any claim to prebiotic status for HAMS during the infant period must be supported by well-designed human studies. Inulin, fructo-oligosaccharides and galacto-oligosaccharides, the subjects of several randomised controlled trials, are generally held to be the main substrates to have prebiotic potential during infancy. However, these agents are costly to produce and, particularly in the case of FOS and galacto-oligosaccharides (GOS) due to their small size, confer a luminal osmotic effect and are rapidly fermented. These features carry the risk of precipitating undesirable clinical effects, such as diarrhoea and abdominal discomfort. HAMS has several advantages. It can be readily cultivated, is slowly fermented and does not produce an osmotic effect in the large intestine. It can also be readily incorporated into foods without altering the processing or organoleptic properties [34].

5. Conclusions

This study confirms that faecal inocula, whether preweaning or at weaning infants has the capacity to utilise HAMS and HAMSA as a potential substrate. Incubation with both starches selectively stimulated *Bifidobacterium* copy numbers and increased the Bacteroidetes:Firmicutes ratio, outcomes that, if replicated in direct feeding studies, may be associated with beneficial health outcomes. These findings justify further in vivo infant studies to examine the short- and long-term effects of different doses of HAMS during weaning on the composition and function of the emerging gut microbiota and clinical outcomes.

Author Contributions: Conceptualization, G.G., C.T.C. and G.P.Y.; methodology, G.G., A.R.B. and C.T.C.; formal analysis, G.G., C.T.C. and P.G.; writing—original draft preparation, G.G. and C.T.C.; writing—review and editing, all authors; supervision, C.T.C., G.P.Y. and A.R.B.; project administration, G.G. and E.M.; funding acquisition, G.G.

Funding: This report is based on research funded by an Australian NHMRC grant (GNT1074403) to Dr Geetha Gopalsamy. The authors also thank the Department of Health, WA, for funding through the Telethon—Perth Children's Hospital Research Fund 2015, Stream 2 to support the time of CTC.

Conflicts of Interest: The authors declare no conflict of interest

Appendix A

Appendix A.1. Pre-Digestion Method

5 g of test starch was accurately weighted into tubes (8) and incubated with 25 mL of pepsin (1 g/mL, made up in 0.02 mol HCl/L, pH 2, for 30 min at 37 °C. The next phase of digestion was replicated by neutralisation with sodium hydroxide (0.02 M) and incubation with 25 mL of amyloglucosidase (200 U/mL) and alpha-amylase (1.5 U/mL) solution for 5 h in a shaking water bath at 37 °C. Following the incubation, contents were pooled and transferred into a beaker containing 80% ethanol and left

overnight. The supernatant was then removed and the residue was washed with 80% ethanol. This was then centrifuged (2000× g for 10 min), followed by the removal of the supernatant. Washing with alcohol, centrifuging and removal of the supernatant was repeated four times. Following air-drying in the fume hood for another 24 h, the residue was collected for use in the small-scale batch in vitro fermentation method. Only one batch of pre-digestion was performed for each substrate and this provided sufficient amounts for all the required incubations.

Appendix A.2. DNA Extraction Method

DNA was extracted using the PowerMag Microbiome DNA Isolation kit (Qiagen, Hilden, Germany) from a pooled sample containing 1 mL from each ferment replicate. No DNA was extracted from ferments incubated with lactulose as human faecal microbiota of all age groups are recognised to readily ferment lactulose. The purpose of incubation with the lactulose was only to confirm the viability of the faecal microbiota from the particular participant. A sample of 1 mL from each of triplicate ferments was transferred into three separate 2 mL Eppendorf tubes and centrifuged at 13,000× g at 4 °C for 5 min using a bench centrifuge (Eppendorf, Hamburg, Germany). Each resulting cell pellet was retained and the supernatant discarded. Heated PowerMag Microbiome lysis solution (650 µL) was added to the first of the three tubes (each containing the cell pellet) and mixed thoroughly. All the solution was then transferred into the second tube, mixed, and the contents transferred into the third tube, which was also subjected to vigorous mixing. Sterile 0.4 g PowerMag glass beads were added to the final tube and the sample was homogenised for 3 min at maximum speed on a Mini-Beadbeater (BioSpec Products, Bartlesville, OK, USA). The tube was then centrifuged for 5 min at 13,000× g at room temperature. The supernatant was transferred into fresh PCR-grade 1.75 mL tubes. 30 µL of Proteinase K was added to the supernatant, mixed and kept for 10 min at 70 °C. 30 µL of PowerMag Inhibitor Removal Solution was then added to each sample and mixed well. The samples were incubated at room temperature for 5 min and then were again centrifuged at 13,000× g and the supernatants transferred to a fresh 2 mL Deep Well Plate, ensuring no transfer of any residual pellet. 5 µL of RNAse was added to each well of this Deep Well Plate.

The extraction method was completed on the Eppendorf epMotion 5075TMX platform following the manufactures instructions using the PowerMag Microbiome DNA Isolation kit. Reagents used in this programme included Clear Mag Wash Solution, the Clear Mag Binding Solution and Clear Mag Beads. Briefly, using the automated wash and elution protocol, beads and a binding solution were added to each well and mixed together. In this process, the DNA would adhere to the beads after which the plate was transported to a magnet and the waste removed. This process was repeated 3 times with the lysate and binding solution followed by an ethanol wash. Low heat was applied to remove traces of ethanol and an elution buffer was added. This released the nucleic acids from the beads and the elute was separated from the beads and stored at −80 °C.

Table A1. PCR primers and their amplification conditions.

Target	Primers	Sequence (5'–3')	Conc (nM)	Annealing Temp (°C)	Annealing Time (s)	References
Total Bacteria	UnivF UnivR	TCCTACGGGAGGCAGCAGT GGACTACCAGGGTATCTATCCTGTT	500	60	45	[35]
Lactobacillus Spp	Lacto-F Lacto-R	AGCAGTAGGGAATCTTCCA CACCGCTACACATGGAG	500	56	20	[36]
Bifidobacterium Spp	Bifi-F Bifi-R	TCGCGTCCGGTGTGAAAG CCACATCCAGCGTCCAC	500	56	20	[37]

Appendix A.3. qPCR

All qPCR analysis was performed on the CFX 384TM real-time PCR detection system (Bio-Rad, Hercules). Reactions were performed in triplicate, with a total reaction volume of 10 µL. Each reaction consisted of 3 µL (2.8 ng/µL) of DNA template and 7 µL PCR mixture containing 5µL of SYBR mix, bovine serum albumin (0.2 µL), forward (0.1 µL) and reverse primers (0.1 µL) (2.5 ng/µL) and PCR-grade water (1.6 µL). The qPCR cycling conditions had an initial hot start at 98 °C for 3 min, followed by 35 cycles of two step qPCR with denaturing at 98 °C for 15 s using the annealing/elongation temperatures as in Table 1. Fluorescence intensities were detected during the last step of each cycle. qPCR melting curves were obtained after amplification by continuously collecting fluorescence intensity measurements as the reactions were slowly heated from 55 to 95 °C in increments of 0.50 °C/s.

Appendix A.4. Next Generation Sequencing

The methods outlined in Illumina's "16S Metagenomic Sequencing Library Preparation" protocol were followed [38] with adjustments made to PCR thermal cycle conditions, as detailed below. The hypervariable region V4 of the 16SrRNA gene was amplified from the extracted DNA using modified primer pairs with Illumina adapter overhang sequences. The full-length primer sequences using standard IUPAC nucleotide nomenclature were:

16S Amplicon PCR Forward Primer =
5'-TCGTCGGCAGCGTCAGATGTGTATAAGAGACAGGTGCCAGCMGCCGCGGTAA-3'
Amplicon PCR Reverse Primer =
5'-TCTCGTGGGCTCGGAGATGTGTATAAGAGACAGGGACTACHVGGGTWTCTAAT-3'

A two-step PCR process was required. The PCR reaction contained 5 µL of forward primer (1 nM), 5 µL of reverse primer (1 nM) and 12.5 µL of 2 × KAPA H-iFi Hotstart ReadyMix (KAPA Biosystems, Wilmington, MA, USA) in a total volume of 25 µL. The PCR reaction was performed on a Veriti Thermal Cycler (Thermo Fisher Scientific, Waltham, MA, USA) using the following programme: 25 cycles of 95 °C for 30 s, 55 °C for 30 s, 72 °C for 30 s followed by holding at 72 °C for a further 5 min. Following a clean-up of the PCR product, indexing PCR was performed. This step used the Nextera XT Index Kit to attach the dual indices and the Illumina sequencing adapters. The manufacturer's instructions were followed as described in the Illumina library preparation protocol, as mentioned above. The final library was pair-end sequenced using a MiSeq Reagent Kit v3 on the Illumina MiSeq platform. Library preparation and sequencing was performed at Flinders University, South Australia. A standard *t*-test was used to compare the relative abundances of control and test groups at the different taxonomic levels (see Table A2).

Table A2. Percent abundance of the dominant major bacterial taxa at each taxonomic level.

Taxonomy	Preweaning (n = 15)			Weaning (n = 14)				
	Control	HAMS	p Value	Control	HAMS	p Value	mHAMS	p Value
Phylum								
Actinobacteria	12.50	15.48	0.35	10.16	18.55	0.004	17.88	0.004
Bacteroidetes	7.87	7.29	0.86	4.73	13.04	0.001	15.56	0.03
Firmicutes	33.33	37.74	0.09	33.5	32.51	0.79	34.37	0.11
Proteobacteria	46.23	39.39	0.17	51.27	35.76	0.001	32.06	0.004
Class/subclass								
Actinobacteria	12.51	15.48	0.35	10.17	18.56	0.004	17.88	0.004
Bacilli	13.46	18.74	0.02	14.48	16.83	0.32	19.04	0.08
Bacteroidia	7.87	7.29	0.86	4.73	13.04	0.001	15.56	0.001
Clostridia	17.71	16.28	0.58	16.25	12.31	0.26	10.79	0.07
Gammaproteobacteria	46.10	39.13	0.17	50.69	35.19	<0.001	31.56	<0.001
Order								
Bacteroidales	7.87	7.29	0.86	4.73	13.04	0.001	15.56	0.001
Bifidobacteriales	11.64	14.73	0.32	9.12	16.83	0.01	16.33	0.01
Clostridiales	17.71	16.27	0.58	16.24	12.30	0.26	10.79	0.07
Enterobacteriales	45.98	39.04	0.17	50.61	35.16	<0.001	31.51	<0.001
Erysipelotrichales	1.77	1.73	0.93	0.25	0.54	0.07	0.77	0.09
Lactobacillales	12.81	17.67	0.03	14.44	16.58	0.36	19.03	0.08

Table A2. Cont.

Taxonomy	Preweaning (n = 15)			Weaning (n = 14)				
	Control	HAMS	p Value	Control	HAMS	p Value	mHAMS	p Value
Family								
Bifidobacteriaceae	11.67	14.82	0.32	9.31	17.54	0.007	16.86	0.004
Bacteroidaceae	7.49	7.06	0.89	4.60	12.91	<0.001	14.34	0.002
Enterococcaceae	12.42	17.04	0.03	12.79	13.96	0.61	16.32	0.16
Clostridiceae	14.58	13.44	0.67	7.71	5.22	0.27	4.98	0.13
Erysipelotrichaceae	1.77	1.73	0.94	0.26	0.55	0.07	0.78	0.08
Peptostreptococcaceae	1.99	1.57	0.45	3.14	0.94	0.04	0.54	0.02
Ruminococcaceae	0.42	0.34	0.34	1.59	1.46	0.62	2.02	0.28
Enterobacteriaceae	46.12	39.19	0.17	51.19	35.88	<0.001	32.20	<0.001
Genus								
Bacteroides	7.48	7.02	0.89	4.54	12.56	0.009	13.92	0.002
Bifidobacterium	11.64	14.73	0.32	9.12	16.83	0.006	16.33	0.006
Clostridium	14.53	13.38	0.67	7.44	5.09	0.27	4.78	0.11
Enterobacter	30.31	24.89	0.21	32.32	18.21	<0.001	15.02	<0.001
Enterococcus	12.40	16.99	0.03	12.72	13.75	0.65	16.12	0.18
Escherichia/Shigella	15.05	13.67	0.73	16.39	15.88	0.73	15.59	0.67
Ruminococcus	0.22	0.17	0.43	0.50	0.56	0.58	1.35	0.01

References

1. Albenberg, L.G.; Wu, G. Diet and the Intestinal microbiome: Associations, Functions, and Implications for Health and Disease. *Gastroenterology* **2014**, *146*, 1564–1572. [CrossRef] [PubMed]
2. Gibson, G.R.; Probert, H.M.; Loo, J.V.; Rastall, R.A.; Roberfroid, M.B. Dietary modulation of the human colonic microbiota: Updating the concept of prebiotics. *Nutr. Res. Rev.* **2004**, *17*, 259–275. [CrossRef] [PubMed]
3. Zaman, S.A.; Sarbini, S.R. The potential of resistant starch as a prebiotic. *Crit. Rev. Biotechnol.* **2015**, *36*, 578–584. [CrossRef] [PubMed]
4. Englyst, H.N.; Kingman, S.M.; Cummings, J.H. Classification and measurement of nutritionally important starch fractions. *Eur. J. Clin. Nutr.* **1992**, *46*, S33–S50. [PubMed]
5. Bird, A.R.; Conlon, M.A.; Christophersen, C.T.; Topping, D.L. Resistant starch, large bowel fermentation and a broader perspective of prebiotics and probiotics. *Benef. Microbes* **2010**, *1*, 423–431. [CrossRef]
6. Parrett, A.M.; Edwards, C.A.; Lokerse, E. Colonic fermentation capacity in vitro: Development during weaning in breast-fed infants is slower for complex carbohydrates than for sugars. *Am. J. Clin. Nutr.* **1997**, *65*, 927–933. [CrossRef]
7. Palmer, C.; Bik, E.M.; DiGiulio, D.B.; Relman, D.A.; Brown, P.O. Development of the human infant intestinal microbiota. *PLoS Biol.* **2007**, *5*, e177. [CrossRef]
8. Christian, M.T.; Edwards, C.A.; Preston, T.; Johnston, L.; Varley, R.; Weaver, L.T. Starch fermentation by faecal bacteria of infants, toddlers and adults: Importance for energy salvage. *Eur. J. Clin. Nutr.* **2003**, *57*, 1486–1491. [CrossRef]
9. Parrett, A.M.; Edwards, C.A. In vitro fermentation of carbohydrate by breast fed and formula fed infants. *Arch. Dis. Child.* **1997**, *76*, 249–253. [CrossRef]
10. Rose, D.J.; Venema, K.; Keshavarzian, A.; Hamaker, B.R. Starch-entrapped microspheres show a beneficial fermentation profile and decrease in potentially harmful bacteria during in vitro fermentation in faecal microbiota obtained from patients with inflammatory bowel disease. *Br. J. Nutr.* **2010**, *103*, 1514–1524. [CrossRef]
11. Vonk, R.J.; Hagedoorn, R.E.; De Graff, R.; Elzinga, H.; Tabak, S.; Yang, Y.X.; Stellaard, F. Digestion of so-called resistant starch sources in the human small intestine. *Am. J. Clin. Nutr.* **2000**, *72*, 432–438. [CrossRef] [PubMed]
12. Bird, A.R.; Usher, S.; May, B.; Topping, T.L.; Morrell, M.K. Resistant Starch: Measurements, Intakes, and Dietary Targets. In *Dietary Fiber and Health 2012*; CRC Press: Boca Raton, FL, USA, 2012; pp. 41–56.
13. Edwards, C.A.; Gibson, G.; Champ, M.; Jensen, B.B.; Mather, J.C.; Nagengast, F.; Rumney, C.; Quehl, A. In Vitro Method for Quantification of the Fermentation of Starch by Human Faecal Bacteria. *J. Sci. Food Agric.* **1996**, *71*, 209–217. [CrossRef]
14. Goñi, I.; Martin-Carrón, N. In vitro fermentation and hydration properties of commercial dietary fiber-rich supplements. *Nutr. Res.* **1998**, *18*, 1077–1089. [CrossRef]

15. McOrist, A.L.; Abell, G.C.; Cooke, C.; Nyland, K. Bacterial population dynamics and faecal short-chain fatty acid (SCFA) concentrations in healthy humans. *Br. J. Nutr.* **2008**, *100*, 138–146. [CrossRef] [PubMed]
16. Payne, A.N.; Zihler, A.; Chassard, C.; Lacroix, C. Advances and perspectives in in vitro human gut fermentation modeling. *Trends Biotechnol.* **2012**, *30*, 17–25. [CrossRef] [PubMed]
17. Christophersen, C.T.; Morrison, M.; Conlon, M.A. Overestimation of the Abundance of Sulfate-Reducing Bacteria in Human Feces by Quantitative PCR Targeting the Desulfovibrio 16S rRNA Gene. *Appl. Environ. Microbiol.* **2011**, *77*, 3544–3546. [CrossRef]
18. Edgar, R.C. Search and clustering orders of magnitude faster than BLAST. *Bioinformatics* **2010**, *26*, 2460–2461. [CrossRef]
19. Wang, Q.; Garrity, G.M.; Tiedje, J.M.; Cole, J.R. Naive Bayesian classifier for rapid assignment of rRNA sequences into the new bacterial taxonomy. *Appl. Environ. Microbiol.* **2007**, *73*, 5261–5267. [CrossRef]
20. Tang, M.K.; Lodge, C.J. Exmining the evidence for using synbiotics to treat or prevent atopic dermatitis. *JAMA Pediatr.* **2016**, *170*, 201–203. [CrossRef]
21. Boehm, G.; Moro, G. Structural and functional aspects of prebiotics used in infant nutrition. *J. Nutr.* **2008**, *138*, 1818S–1828S. [CrossRef]
22. Pokusaeva, K.; Fitzgerald, G.F.; Van Sinderen, D. Carbohydrate metabolism in Bifidobacteria. *Genes Nutr.* **2011**, *6*, 285–306. [CrossRef] [PubMed]
23. Odamaki, T.; Horiogome, A.; Sugahara, H.; Hashikura, N.; Minami, J.; Xiao, J.Z.; Abe, F. Comparative Genomics Revealed Genetic Diversity and Species/Strain-Level Differences in Carbohydrate Metabolism of Three Probiotic Bifidobacterial Species. *Int. J. Genom.* **2015**, *2015*, 12. [CrossRef] [PubMed]
24. Van der Maarel, M.J.; Leemhuis, H. Starch modification with microbial alpha-glucanotransferase enzymes. *Carbohydr. Polym.* **2013**, *93*, 116–121. [CrossRef] [PubMed]
25. Avershina, E.; Storro, O.; Oien, T.; Johnsen, R.; Pope, P.; Rudi, K. Major faecal microbiota shifts in composition and diversity with age in a geographically restricted cohort of mothers and their children. *FEMS Microbiol. Ecol.* **2014**, *87*, 280–290. [CrossRef] [PubMed]
26. Milani, C.; Lugi, G.A.; Duranti, S.; Turroni, F.; Mancabelli, L.; Ferraio, C.; Mangifesta, M.; Hevia, A.; Viappiani, A.; Scholz, M.; et al. Bifidobacteria exhibit social behavior through carbohydrate resource sharing in the gut. *Sci. Rep.* **2015**, *5*, 15782. [CrossRef] [PubMed]
27. Rakoff-Nahoum, S.; Kong, Y.; Kleinstein, S.H.; Subramanian, S.; Ahern, P.P.; Gordon, J.I.; Medzhitov, R. Analysis of gene–environment interactions in postnatal development of the mammalian intestine. *Proc. Natl. Acad. Sci. USA* **2015**, *112*, 1929–1936. [CrossRef] [PubMed]
28. Ze, X.; Duncan, S.H.; Louis, P.; Flint, J.H. Ruminococcus bromii is a keystone species for the degradation of resistant starch in the human colon. *ISME J.* **2012**, *6*, 1535–1543. [CrossRef] [PubMed]
29. Fouhse, J.M.; Ganzle, M.G.; Regmi, P.R.; Van Kempen, T.A.; Zijlstra, R.T. High Amylose Starch with Low In Vitro Digestibility Stimulates Hindgut Fermentation and Has a Bifidogenic Effect in Weaned Pigs. *J. Nutr.* **2015**, *145*, 2464–2470. [CrossRef]
30. Jiang, X.; Li, B.; Su, Y.; Weiyun, Z. Shifts in bacterial community compositions during in vitro fermentation of amylopectin and resistant starch by colonic inocula of pigs. *J. Food Nutr. Res.* **2013**, *1*, 156–163.
31. Turnbaugh, P.H.; Ley, R.E.; Mahowald, M.A.; Magrini, V.; Mardis, E.R.; Gordon, J.I. An obesity-associated gut microbiome with increased capacity for energy harvest. *Nature* **2006**, *444*, 1027–1031. [CrossRef]
32. Clarke, J.M.; Bird, A.R.; Topping, D.L.; Cobiac, L. Excretion of starch and esterified short-chain fatty acids by ileostomy subjects after the ingestion of acylated starches. *Am. J. Clin. Nutr.* **2007**, *86*, 1146–1151. [CrossRef] [PubMed]
33. Martinez, I.; Kim, J.; Duffy, P.R.; Schlegel, V.L.; Walter, J. Resistant starches types 2 and 4 have differential effects on the composition of the fecal microbiota in human subjects. *PLoS ONE* **2010**, *5*, e15046. [CrossRef] [PubMed]
34. Ashwar, B.A.; Gani, A.; Shah, A.; Wani, I.; Masoodi, F. Preparation, health benefits and applications of resistant starch—A review. *Starch Stärke* **2016**, *68*, 287–301. [CrossRef]
35. Nadkarni, M.A.; Martin, F.E. Determination of bacterial load by real time PCR using a broad range (universal) probe and primers set. *Microbiology* **2002**, *148*, 257–266. [CrossRef] [PubMed]
36. Walter, J.C.; Hertel, C.; Tannock, G.W.; Lis, C.M.; Munro, K.; Hammes, W.P. Detection of Lactobacillus, Pediococcus, Leuconostoc, and Weissella species in human feces by using group-specific PCR primers and denaturing gradient gel electrophoresis. *Appl. Environ. Microbiol.* **2001**, *67*, 2578–2585. [CrossRef] [PubMed]

37. Rinttila, T.; Kassinen, A.; Malinen, E.; Krogius, L.; Palva, A. Development of an extensive set of 16S rDNA targeted primers for quantification of pathogenic and indigenous bacteria in faecal samples by real time PCR. *J. Appl. Microbiol.* **2004**, *97*, 1166–1177. [CrossRef] [PubMed]
38. 16S Metagenomic Sequencing Library Preparation. Available online: http://www.illumina.com/content/dam/illumina-support/documents/documentation/chemistry_documentation/16s/16s-metagenomic-library-prep-guide-15044223-b.pdf (accessed on 1 May 2019).

© 2019 by the authors. Licensee MDPI, Basel, Switzerland. This article is an open access article distributed under the terms and conditions of the Creative Commons Attribution (CC BY) license (http://creativecommons.org/licenses/by/4.0/).

Article

Social Inequalities in Prenatal Folic Acid Supplementation: Results from the ELFE Cohort

Aurore Camier [1,2], Manik Kadawathagedara [1,2], Sandrine Lioret [1,2], Corinne Bois [3], Marie Cheminat [3], Marie-Noëlle Dufourg [3], Marie Aline Charles [1,2,3] and Blandine de Lauzon-Guillain [1,2,4,*]

1. INSERM, UMR1153 Center for Research in Epidemiology and StatisticS (CRESS), Research Team on Early Life Origins of Health (EAROH), 75004 Paris, France; aurore.camier@inserm.fr (A.C.); manik.kadawathagedara@inserm.fr (M.K.); sandrine.lioret@inserm.fr (S.L.); marie-aline.charles@inserm.fr (M.A.C.)
2. Université de Paris, UMR1153 Center for Research in Epidemiology and StatisticS (CRESS), Research Team on Early Life Origins of Health (EAROH), 75004 Paris, France
3. Unité Mixte Inserm-Ined-EFS Elfe, Ined, 75020 Paris, France; corinne.bois@ined.fr (C.B.); marie.cheminat@ined.fr (M.C.); marie-noelle.dufourg@ined.fr (M.-N.D.)
4. INRA, U1125 Center for Research in Epidemiology and StatisticS (CRESS), Research Team on Early Life Origins of Health (EAROH), 75004 Paris, France
* Correspondence: blandine.delauzon@inserm.fr; Tel.: +33-145-595-019; Fax: +33-147-269-454

Received: 23 April 2019; Accepted: 15 May 2019; Published: 18 May 2019

Abstract: Most professional and international organizations recommend folic acid supplementation for women planning pregnancy. Various studies have shown high levels of non-compliance with this recommendation. This study aimed to identify sociodemographic characteristics related to this compliance. The analyses were based on 16,809 women from the French nationwide ELFE cohort (Etude Longitudinale Française depuis l'Enfance). Folic acid supplementation was assessed at delivery, and sociodemographic characteristics were collected at two months postpartum. The association between sociodemographic characteristics and compliance with recommendations on folic acid supplementation (no supplementation, periconceptional supplementation, and supplementation only after the periconceptional period) was examined using multivariate multinomial logistic regression. Only 26% of French women received folic acid supplementation during the periconceptional period, 10% of women received supplementation after the periconceptional period, and 64% received no supplementation. Young maternal age, low education level, low family income, multiparity, single parenthood, maternal unemployment, maternal overweight, and smoking during pregnancy were related to lower likelihood of folic acid supplementation during the periconceptional period compared to no supplementation. These associations were not explained by unplanned pregnancy. Immigrant and underweight women were more likely to receive folic acid supplementation after the periconceptional period. Our study confirms great social disparities in France regarding the compliance with the recommendations on folic acid supplementation.

Keywords: folic acid supplementation; pregnancy; epidemiology; social inequalities

1. Introduction

Neural tube defects are one of the most common congenital diseases in Europe [1] and, in France, represent 1.3 cases (live births, fetal deaths, termination of pregnancy for fetal anomaly) per 1000 births [2]. Randomized trials have proved the efficacy of folic acid supplementation during the periconceptional period in neural tube defects prevention [3–5].

Most professional and international organizations, including the World Health Organization, recommend folic acid supplementation for women planning pregnancy [6,7]. However, the specifically targeted population groups (e.g., women of childbearing age, women planning pregnancy, women of childbearing age without safe contraception) and the timing or duration of the supplementation recommendations vary across countries. In North America, maternal folic acid intake is addressed by the folic acid fortification of food products (flours, cereals) [8], whereas, in Europe, the prevention policy is based on supplementation exclusively. In France, the latest guidelines from the Agency for Food, Environmental, and Occupational Health & Safety reinforced the need for folic acid supplementation for women within the periconceptional period (eight weeks before and eight weeks after conception) to achieve a daily intake of 400 µg dietary folate equivalents [9].

The neural tube defects prevention policy based on folic acid supplementation during the periconceptional period assumes that pregnancies are planned and that future mothers visit health professionals before pregnancy. However, in Europe, 45% of pregnancies were unintended in 2012 [10], and women from disadvantaged households tend to have lower access to health services, particularly for preventive care [11–13].

Within this context, the objective of the present study was to identify sociodemographic characteristics associated with compliance with recommendations on folic acid supplementation to prevent neural tube defects in France, with a special focus on the timing of this supplementation, on the basis of a national survey performed in 2011 on women giving birth.

2. Materials and Methods

2.1. Study Population

The present study was based on the ELFE study (Etude Longitudinale Française depuis l'Enfance), a multidisciplinary study comprising a nationally representative birth cohort, which included 18,258 children born in 349 randomly selected maternity units in France in 2011 [14]. Inclusion took place during 25 selected recruitment days over 4 waves encompassing 4 to 8 days each and all 4 seasons. Inclusion criteria were as follows: children born after 33 weeks of gestation, mothers aged 18 years or older and who were not planning to move outside of metropolitan France during the following three years. Foreign families could also participate in the study if mothers were able to read French, Arabic, Turkish, or English. Participating mothers signed informed consent for themselves and their child. A total of 51% of contacted parents agreed to participate. Data were collected through standardized interviews conducted by trained midwives and through self-completed questionnaires. The follow-up of this birth cohort is ongoing.

The ELFE study was given ethical approval by the Advisory Committee for the Treatment of Information on Health Research (Comité Consultatif sur le Traitement des Informations pour la Recherche en Santé), the National Agency Regulating Data Protection (Commission National Informatique et Libertés), and the National Statistics Council.

2.2. Maternal Characteristics

Mothers were first interviewed in the maternity ward after delivery, to collect information about their pregnancy, their newborn, and their general characteristics (employment status, education level, age). Two months post-partum, telephone interviews with mothers and fathers took place, which included additional questions on demographic and socioeconomic characteristics such as country of birth, educational level, employment, monthly income, and number of family members.

As family data were more comprehensively collected during the two-month-post-partum interview than during the maternity interview and because family sociodemographic characteristics only marginally evolved within two months, we prioritized data collected at two months in our analyses. The sociodemographic characteristics collected during the maternity stay were used only in the case of missing values in the two-months-post-partuminterview. Maternal characteristics included in the analyses were: migration status (native French, immigrant, descendant of immigrants), age at first delivery (18–25 years, 25–29 years, 30–34 years, ≥35 years), family type (traditional, stepfamily, one-parent family), educational level measured on the basis of the highest academic degree attained (<secondary school, secondary school, high school, two-year university degree, ≥three-year university degree), employment status (employed, housewife/parental leave, retired/disability/unemployed, student, other), monthly family income (≤€1500, €1501–2300, €2301–3000, €3001–4000, €4001–5000, >€5000), smoking status (never smoked, smoked only before pregnancy, smoked until early pregnancy, smoked during the whole pregnancy), planned pregnancy, and fertility treatment.

2.3. Folic Acid Supplementation

Information on maternal folic acid supplementation was collected retrospectively at the maternity unit during face-to-face interviews utilizing five questions: 'Have you taken folic acid (also called vitamin B9) before and/or during pregnancy (to prevent nervous system abnormalities)?' 'If yes, indicate the periods of time you took it: one to three months before pregnancy (yes/no), in the first two months of pregnancy (yes/no), between the second and sixth month of pregnancy (yes/no), beyond six months of pregnancy (yes/no)'. Women were divided into three groups according to their folic acid supplementation: supplementation during the periconceptional period (before pregnancy and/or first two months of pregnancy), late supplementation (only after the second month of pregnancy), and no supplementation. No information about family history of neural tube defects or folic acid supplements dosages was collected.

2.4. Sample Selection

Women who withdrew consent within the first year ($n = 128$) or for whom it was not possible to verify the eligibility criteria due to missing data ($n = 350$) were excluded from the study, resulting in 17,574 eligible mothers. We also excluded women with missing data ($n = 3418$), leaving a total of 14,156 women in the main analyses.

2.5. Statistical Analyses

Comparisons between excluded and included subjects were conducted with chi-square tests.

In order to provide representative descriptive statistics of births in 2011 in France, the descriptive data (rates) were weighted to consider the sampling design and biases related to non-consent. Weighting also included calibration on margins from the state register of statistical data and the 2010 French National Perinatal study [15] regarding the following variables: age, region, marital status, migration status, level of education, and primiparity.

Associations between sociodemographic variables and folic acid supplementation were tested by multivariable multinomial logistic regression, including maternal characteristics (maternal age at first delivery, parity, family composition, migration status, education level, employment status, family income, pre pregnancy body mass index (BMI), smoking status), and additionally adjusted for mothers' region of residence, size of maternity unit, and wave of recruitment.

Several sensitivity analyses were performed. First, analyses were performed only for women with planned pregnancy and without fertility treatment, in order to check that the associations between familial characteristics and compliance with folic acid supplementation guidelines were not driven by the specific cases of unplanned pregnancy or fertility treatment. Then, we calculated weighted multivariate models. Afterward, using multiple imputations to deal with missing data on sociodemographic variables, using the SAS software. We assumed that data were missing at random and generated five independent datasets using the fully conditional specification method (MI procedure, FCS statement, NIMPUTE option) and then calculated pooled effect estimates (MIANALYSE procedure). Imputation model variables included both the potentially predicting non-response and the outcomes. Categorical variables were imputed using a multinomial model, ordinal or binary variables using logistic regressions, and continuous variables using linear regressions. Further details are available in Supplementary Table S1.

3. Results

Women excluded from our analysis were younger, more likely to be single, born in a country other than France, had a lower education level, were less likely to be employed during pregnancy, and had quit smoking before pregnancy compared to women included in our analysis (Table 1).

Table 1. Study sample characteristics.

	Excluded Women	Included Women	p-Value
N	3418	14,156	
Age at delivery (years)	29.4 (5.7)	30.4 (4.9)	<0.0001
Birth order			<0.0001
First child	43.7% (1491)	45.0% (6365)	
Second child	33.5% (1142)	36.0% (5101)	
Third child	14.3% (488)	13.8% (1949)	
Fourth child or more	8.5% (291)	5.2% (741)	
Single parenthood			<0.0001
Yes	12.4% (414)	3.9% (548)	
No	87.6% (2928)	96.1% (13,549)	
Country of birth			<0.0001
France	78.5% (2599)	88.9% (12,516)	
Another country	21.5% (713)	11.1% (1568)	
Education level			<0.0001
Primary school	11.2% (376)	3.5% (491)	
Secondary school	20.0% (673)	12.6% (1783)	
General high school	11.2% (378)	7.2% (1023)	
Technical/professional high school	15.7% (530)	12.3% (1741)	
University	41.9% (1410)	64.4% (9116)	
Employment status			<0.0001
Employed	59.3% (1968)	72.7% (10,298)	
Housewife/parental leave	2.5% (83)	3.4% (485)	
Retired/disability/unemployed	11.9% (395)	12.0% (1702)	
Student	21.7% (721)	10.1% (1424)	
Other	4.6% (154)	1.7% (247)	
Pre-pregnancy body mass index (kg/m^2)	23.7 (5.1)	23.4 (4.7)	0.0138
Smoking status			<0.0001
Never smoked	59.0% (1902)	56.7% (8020)	
Smoked only before pregnancy	18.1% (585)	23.7% (3356)	
Smoked only during early pregnancy	4.0% (130)	3.9% (559)	
Smoker during the whole pregnancy	18.9% (608)	15.7% (2221)	

Values are % (n).

The weighted rate of folic acid supplementation was 26.0% during the periconceptional period and 9.9% for supplementation started only after the first two months of pregnancy. A total of 64.1% of women did not receive folic acid supplementation before or during pregnancy.

Bivariable associations between maternal characteristics and folic acid supplementation are shown in Table 2.

Table 2. Bivariate analyses between familial characteristics and folic acid supplementation.

	N	No Supplementation (N = 8328)	Periconceptional Supplementation (N = 4520)	Late Supplementation Only (N = 1308)	p-Value
Age at first delivery					<0.0001
<25 years	3611	74.9%	15.1%	9.6%	
25–29 years	6134	59.9%	31.3%	8.7%	
30–34 years	3412	50.7%	38.7%	9.8%	
≥35 years	999	49.9%	39.3%	9.6%	
Parity					<0.0001
First child	6365	55.2%	35.4%	9.0%	
Second child	5101	63.4%	26.9%	9.3%	
Third child	1949	67.9%	21.8%	9.7%	
Fourth child or more	741	78.4%	10.9%	10.1%	
Family composition					<0.0001
Traditional	12,499	59.7%	30.6%	9.2%	
Single parenthood	532	80.7%	11.1%	8.8%	
Stepfamily	1125	68.8%	20.9%	10.3%	
Migration					<0.0001
Native French	10,229	60.9%	30.1%	8.7%	
First-generation immigrant	1582	64.0%	23.5%	11.7%	
Second-generation immigrant	2345	61.7%	27.9%	9.7%	
Educational level					<0.0001
<Secondary school	999	76.9%	11.8%	10.1%	
Secondary school	2010	75.3%	16.3%	8.9%	
High school	2620	66.4%	23.9%	8.9%	
Two-year university degree	3199	58.6%	31.8%	9.5%	
Three-year university degree	2495	52.3%	37.8%	9.4%	
≥Five-year university degree	2833	46.5%	43.4%	9.0%	
Employment status					<0.0001
Employed	10,298	57.0%	33.7%	9.0%	
Student	485	61.8%	30.4%	8.7%	
Unemployed	1702	68.5%	20.6%	11.2%	
Housewife/parental leave	1424	74.6%	13.9%	9.6%	
Other	247	71.8%	21.9%	6.9%	
Family income					<0.0001
<€1500/month	1417	78.4%	12.3%	9.2%	
€1501–2300/month	2177	69.3%	19.5%	10.5%	
€2301–3000/month	3974	61.4%	29.3%	9.1%	
€3001–4000/month	3745	56.6%	33.8%	9.2%	
€4001–5000/month	1551	50.0%	40.1%	9.1%	
>€5000/month	1292	45.5%	45.9%	8.0%	
Smoking status					<0.0001
Never smoked	8020	59.7%	30.5%	9.3%	
Smoked only before pregnancy	3356	56.7%	33.7%	9.0%	
Smoked only during early pregnancy	559	68.8%	21.6%	10.7%	
Smoked during the whole pregnancy	2221	72.2%	18.3%	9.1%	
Planned pregnancy					<0.0001
No	1203	74.6%	15.3%	10.1%	
Yes	12,079	58.8%	31.6%	9.0%	
Previous treatment for infertility					<0.0001
No	12,915	63.0%	27.2%	9.4%	
Yes	1105	42.4%	49.6%	7.4%	

Values are n and weighted %. Chi-square tests on weighted data.

Women were less likely to take folic acid supplementation during the periconceptional period, compared to no pre-conceptional supplementation, when they were younger, multiparous, single, with low education level, low family income, unemployed during pregnancy, and smoked during pregnancy (Table 3). Compared to mothers who had never smoked, women who had quit smoking before pregnancy were more likely to take folic acid supplementation during the periconceptional period, whereas the relationship was inverse for those women who smoked during the whole pregnancy.

Table 3. Multivariate associations between sociodemographic characteristics and timing of folic acid supplementation (n = 14,156).

	Folic Acid Supplementation (Reference = No Supplementation)		
	Periconceptional Supplementation	Late Supplementation Only	p-Value
Age at first delivery			<0.0001
<25 years	0.72 [0.64–0.81]	1.04 [0.88–1.23]	
25–29 years	1 [Ref]	1 [Ref]	
30–34 years	1.23 [1.12–1.36]	1.32 [1.13–1.54]	
≥35 years	1.30 [1.12–1.51]	1.33 [1.04–1.71]	
Birth order			<0.0001
First child	1 [Ref]	1 [Ref]	
Second child	0.62 [0.57–0.68]	0.92 [0.80–1.05]	
Third child	0.58 [0.51–0.66]	0.96 [0.80–1.17]	
Fourth child or more	0.40 [0.32–0.51]	0.87 [0.66–1.16]	
Family composition			<0.0001
Traditional	1 [Ref]	1 [Ref]	
Single-parenthood	0.73 [0.56–0.96]	0.80 [0.57–1.12]	
Stepfamily	1.22 [1.04–1.43]	1.21 [0.97–1.51]	
Migration			<0.0001
Native French	1 [Ref]	1 [Ref]	
Immigrant	0.90 [0.79–1.04]	1.31 [1.08–1.59]	
Descendant of immigrant	0.92 [0.83–1.03]	1.07 [0.90–1.26]	
Education level			<0.0001
<Secondary school	0.47 [0.38–0.59]	0.73 [0.54–0.98]	
Secondary school	0.51 [0.43–0.61]	0.67 [0.52–0.87]	
High school	0.68 [0.60–0.79]	0.74 [0.59–0.93]	
Two-year university degree	0.76 [0.68–0.86]	0.89 [0.73–1.09]	
Three-year university degree	0.89 [0.79–1.00]	0.90 [0.74–1.11]	
≥Five-year university degree	1 [Ref]	1 [Ref]	
Employment status			<0.0001
Employed	1 [Ref]	1 [Ref]	
Retired/disability/unemployed	0.86 [0.75–0.98]	1.20 [1.00–1.45]	
Housewife/parental leave	0.82 [0.69–0.97]	0.93 [0.74–1.16]	
Other	0.84 [0.62–1.14]	0.64 [0.38–1.07]	
Student	1.11 [0.90–1.37]	1.00 [0.71–1.41]	
Monthly family income			<0.0001
<€1500	0.62 [0.51–0.75]	0.89 [0.69–1.14]	
€1501–2300	0.77 [0.68–0.89]	1.06 [0.87–1.28]	
€2301–3000	1 [Ref]	1 [Ref]	
€3001–4000	1.05 [0.94–1.16]	1.03 [0.87–1.22]	
€4001–5000	1.10 [0.95–1.26]	0.98 [0.77–1.23]	
>€5000	1.32 [1.13–1.54]	0.91 [0.69–1.19]	
Pre-pregnancy body mass index			<0.0001
<18.5 kg/m^2	1.06 [0.91–1.23]	1.31 [1.05–1.63]	
18.5–24.9 kg/m^2	1 [Ref]	1 [Ref]	
25.0–29.9 kg/m^2	0.79 [0.71–0.88]	0.93 [0.79–1.10]	
30 kg/m^2 or more	0.72 [0.62–0.84]	0.96 [0.78–1.18]	
Smoking status			<0.0001
Never smoked	1 [Ref]	1 [Ref]	
Only before pregnancy	1.10 [1.01–1.21]	1.04 [0.89–1.21]	
Only during early pregnancy	0.72 [0.58–0.88]	1.09 [0.82–1.47]	
During the whole pregnancy	0.68 [0.60–0.77]	0.89 [0.74–1.07]	

Values are adjusted OR [95% CI]. Multivariate multinomial logistic regression, also adjusted for maternal region of residence, size of maternity unit, and wave of recruitment.

Women were more likely to start folic acid supplementation after the periconceptional period when they were older, underweight before pregnancy, or immigrant.

The results were very similar when the specific weighting to deal with non-inclusion bias was used in sensitivity analyses (Supplementary Table S2). Moreover, the results were not modified by excluding women with an unplanned pregnancy and those with fertility treatment. In the analysis by multiple imputation with five independent datasets, the findings were very consistent with those described in our main analysis.

4. Discussion

The results from our study revealed that only 26% of French pregnant women received folic acid supplementation during the periconceptional period. Moreover, we found that all dimensions related to the socioeconomic level (young maternal age, low education level, employment status, income, and single parenthood) were independently related to a lower odds ratio of pre-conceptional folic acid supplementation. These associations remained after excluding women with an unplanned pregnancy or those who had received fertility treatment.

In most countries in Europe, the rates of periconceptional supplementation are still low, even though guidelines on this public health issue were introduced more than 15 years ago. Most countries report folic acid supplementation rates of 25–35% during the periconceptional period [16]. In the Dutch Generation R cohort, 37% of women received folic acid supplementation within the appropriate period [17]. The results were lower in Norway and Denmark, with 10% and 14% of women receiving folic acid supplementation during the periconceptional period, respectively [18,19]. In France, the results from the 2010 National Perinatal Survey indicated that 24% of women received folic acid supplementation in the periconceptional period [20], which is consistent with the rate highlighted in the present study. This low compliance with the folic acid supplementation guidelines could impair the effectiveness of the current neural tube defects prevention policy, and some studies have indicated that the implementation of the recommendations on folic acid supplementation is not clearly related to a decrease in the incidence of neural tube defects [21–23].

Folic acid supplementation rates vary according to country, but sociodemographic variables such as younger age of women and lower family income have consistently been associated with lower rates of folic acid supplementation [17,18,24–26]. European supplementation policies similar to those of France, without fortification policies, might, therefore, play a role in maintaining social inequalities [27,28]. In our study, social inequalities in periconceptional folic acid supplementation remained even when pregnancy planning was considered. This finding accentuates the poor preventive care during early pregnancy of the most socially disadvantaged women. The main reasons may be a decreased awareness in disadvantaged populations regarding this recommendation as well as their lower use of health care facilities [29,30]. To address this issue in the Netherlands, a mass media campaign was implemented in 1995 to increase the rate of folic acid supplementation [30].

Interestingly, in our study, women who stopped smoking before pregnancy were more likely to have folic acid supplementation during the periconceptional period, suggesting that they were probably more inclined to follow guidelines related to pregnancy and be involved in preventive care. This situation may give some leads for further folic acid supplementation strategies, such as considering the preventive prenatal care as a specific moment to address women's general health (nutrition, lifestyle, physical activity, smoking cessation).

Preventing neural tube defects by folic acid supplementation is a difficult goal to achieve, as changes in behavior are difficult to generate, and the human reproduction period can be long [31]. Indeed, the time between planning a pregnancy and becoming pregnant may take several months or even years, and women have first to achieve and maintain the objective of acid folic intake. To address this issue, more than 40 countries have decided to fortify foods with folic acid, and this strategy appears to be the most effective today [8,23,29,32]. This strategy could be particularly relevant for young and low-income women, because these groups of women are more likely to have unplanned pregnancies and less likely to receive or respond to health-promotion messages [16]. Studies in Canada suggest an improved efficacy of food fortification policies: while no effect on the incidence of neural tube defects was observed after recommendations of folic acid periconceptional supplementation between 1993 and 1997, a marked decrease in the incidence of neural tube defects was observed after food fortification implementation in 1997 [8,23]. The target group of this fortification policy are women of childbearing age, even though the general population has increased dietary folate intake [29]. This fortification strategy aims to supplement women continuously during their reproductive period, reaching women with unplanned pregnancies, and to limit social inequalities in health [8,33]. In 2017, a recommendation by the US Preventive Services Task Force (USPSTF) reinforced the importance of folic acid supplementation for all women of childbearing age (even in the absence of planned pregnancies) in addition to food fortification [34]. One of the disadvantages of this strategy concerns elderly people. Indeed, high folate intake may mask anemia resulting from vitamin B12 deficiency, which may cause neurological deterioration [35]. Moreover, there is a supposed relation between high intake of folic acid and cancer risk (especially colorectal cancer), even though human epidemiological data are inconclusive [36]. Even if voluntary fortification is very common in Europe (e.g., breakfast cereals, dairy products, fruit juices), mandatory food fortification is not practiced in any European countries mainly because of this cancer risk [36]. In France, systematic flour fortification with folic acid was considered. A pilot study to assess risks and benefits was proposed in early 2000s but ultimately was not launched. Our findings could contribute to this debate [37].

Some maternal characteristics, such as immigrant status and pre-pregnancy underweight status, were specifically related to folic acid supplementation only after the periconceptional period. This could be due to the fact that this late supplementation is not taken for the prevention of neural tube defects but is started following treatment for other pregnancy issues. The WHO recommends daily oral iron and folic acid supplementation among pregnant women to prevent maternal anaemia, puerperal sepsis, low birth weight, and preterm birth [38]. In the ELFE study, women who had started folic acid supplementation only after the periconceptional period were more likely to have haemoglobin levels below 11 g/100 mL during pregnancy (15% vs 20%, $p < 0.0001$), compared to women who received folic acid supplementation during the periconceptional period. Immigrant and underweight women are probably more at risk of maternal anaemia [39] and low birth weight.

The ELFE cohort is a representative study of births from the year 2011 in metropolitan France (excluding very premature babies), and descriptive statistics used weighted data to provide accurate prevalence of folic acid supplementation. In our multivariate analysis of the association between familial characteristics and folic acid supplementation, we excluded part of the sample because of missing data. The comparison of the characteristics of the women included versus those of the women excluded from the analysis showed that women with a higher educational level were overrepresented in the study, which may have implications for generalizing the findings. However, in sensitivity analyses based on multiple imputations of missing data, the results remained similar. Because women were recruited at delivery, it is necessary to acknowledge that miscarriages or medical abortions due to congenital malformations related to folic acid condition were not included in the present study. The main strengths of the ELFE study include the large sample size and the wide range of sociodemographic variables, which allowed us to demonstrate that various components of the social background were related to compliance with folic acid supplementation guidelines. Unfortunately, we were not able to distinguish between the timing patterns of folic acid supplementation related to neural tube defects prevention from supplementation for other purposes. Moreover, as we did not record the dose of folic acid consumed by women, we were not able to identify women with folic acid intake higher than the tolerable upper limit of 1000 µg/day [40]. Finally, the study design did not allow to assess the effect of such supplementation on birth outcomes as stillbirth, very premature birth, and termination of pregnancy.

5. Conclusions

Our study confirms low rates of folic acid supplementation during the periconceptional period as well as great social disparities concerning the use of maternal periconceptional folic acid supplementation. Therefore, it is important to find alternative methods, especially for vulnerable populations, to increase the effectiveness of prevention policies by considering media campaign.

Supplementary Materials: The following are available online at http://www.mdpi.com/2072-6643/11/5/1108/s1, Table S1: Type of variable, model used to predict missing data, and percentage of values missing for each variable included in the imputation model, Table S2: Sensitivity analyses on multivariate associations between familial characteristics and timing of folic acid supplementation in comparison to no supplementation

Author Contributions: The corresponding author affirms that all listed authors meet authorship criteria and that no others meeting these criteria have been omitted. A.C. conceptualized and designed the study, conducted the statistical analyses, interpreted the results, drafted the initial manuscript, and approved the final manuscript as submitted. M.K., S.L., C.B., M.C. and M.-N.D. critically reviewed the manuscript and approved the final version submitted. M.A.C. and B.d.L.-G. conceptualized and designed the study, contributed to the interpretation of the results, reviewed and revised the manuscript, and approved the final manuscript as submitted. A.C. and B.d.L.-G. are sponsors of the present work.

Funding: This analysis was funded by an ANR grant within the framework "Social determinants of health" (grant number: ANR-12-DSSA-0001). The Elfe survey is a joint project between the French Institute for Demographic Studies (INED) and the National Institute of Health and Medical Research (INSERM), in partnership with the French blood transfusion service (Etablissement français du sang, EFS), Santé publique France, the National Institute for Statistics and Economic Studies (INSEE), the Direction générale de la santé (DGS, part of the Ministry of Health and Social Affairs), the Direction générale de la prévention des risques (DGPR, Ministry for the Environment), the Direction de la recherche, des études, de l'évaluation et des statistiques (DREES, Ministry of Health and Social Affairs), the Département des études, de la prospective et des statistiques (DEPS, Ministry of Culture), and the Caisse nationale des allocations familiales (CNAF), with the support of the Ministry of Higher Education and Research and the Institut national de la jeunesse et de l'éducation populaire (INJEP). Via the RECONAI platform, it receives a government grant managed by the National Research Agency under the "Investissements d'avenir" programme (ANR-11-EQPX-0038).

Acknowledgments: We would like to thank the scientific coordinators (Marie Aline Charles, Bertrand Geay, Henri Léridon, Corinne Bois, Marie-Noëlle Dufourg, Jean-Louis Lanoé, Xavier Thierry, Cécile Zaros), IT and data managers, statisticians (A Rakotonirina, R Kugel, R Borges-Panhino, M Cheminat, H Juillard), administrative and family communication staff, and study technicians (C Guevel, M Zoubiri, L Gravier, I Milan, R Popa) of the ELFE coordination team as well as the families who gave their time for the study.

Conflicts of Interest: The authors have no conflicts of interest relevant to this study to disclose. The authors have no financial relationship relevant to this study to disclose.

References

1. Eurocat. Cases and Prevalence (Per 10,000 Births) for All Full Member Registries from 2010 to 2014. Available online: http://www.eurocat-network.eu/ACCESSPREVALENCEDATA/PrevalenceTables (accessed on 21 February 2017).
2. Santé Publique France. *Malformations Congénitales Et Anomalies Chromosomiques*; Santé Publique France: Saint-Maurice, France, 2014.
3. De-Regil, L.M.; Pena-Rosas, J.P.; Fernandez-Gaxiola, A.C.; Rayco-Solon, P. Effects and safety of periconceptional oral folate supplementation for preventing birth defects. *Cochrane Database Syst. Rev.* **2015**. [CrossRef] [PubMed]
4. MRC Vitamin Study Research Group. Prevention of neural tube defects: Results of the Medical Research Council Vitamin Study. *Lancet* **1991**, *338*, 131–137. [CrossRef]
5. Czeizel, A.E.; Dudas, I. Prevention of the first occurrence of neural-tube defects by periconceptional vitamin supplementation. *N. Engl. J. Med.* **1992**, *327*, 1832–1835. [CrossRef]
6. Cawley, S.; Mullaney, L.; McKeating, A.; Farren, M.; McCartney, D.; Turner, M.J. A review of European guidelines on periconceptional folic acid supplementation. *Eur. J. Clin. Nutr.* **2016**, *70*, 143–154. [CrossRef] [PubMed]
7. Bentley, T.G.K.; Willett, W.C.; Weinstein, M.C.; Kuntz, K.M. Population-Level Changes in Folate Intake by Age, Gender, and Race/Ethnicity after Folic Acid Fortification. *Am. J. Public Health* **2006**, *96*, 2040–2047. [CrossRef] [PubMed]

8. Mills, J.L. Strategies for Preventing Folate-Related Neural Tube Defects: Supplements, Fortified Foods, or Both? *JAMA* **2017**, *317*, 144–145. [CrossRef]
9. *Avis de l'ANSES Relatif à l'Actualisation des Repères du PNNS: Révision des Références Nutritionnelles en Vitamines et Minéraux Pour la Population Générale Adulte*; French Agency for Food Environmental and Occupational Health & Safety: Maisons-Alfort, France, 2016.
10. Sedgh, G.; Singh, S.; Hussain, R. Intended and unintended pregnancies worldwide in 2012 and recent trends. *Stud. Fam. Plan.* **2014**, *45*, 301–314. [CrossRef]
11. Guthmann, J.-P.; Célant, N.; Parent du Chatelet, I.; Duport, N.; Levy-Bruhl, D.; Rochereau, T.; Celant, N.; InVS; IRDES. Déterminants socio-économiques de vaccination et de dépistage du cancer du col par frottis cervico-utérin (FCU). In *Analyse de l'enquête santé et Protection Sociale (ESPS), 2012*; Institut de Veille Sanitaire: Saint-Maurice, France, 2016.
12. Sambamoorthi, U.; McAlpine, D.D. Racial, ethnic, socioeconomic, and access disparities in the use of preventive services among women. *Prev. Med.* **2003**, *37*, 475–484. [CrossRef]
13. Lorant, V.; Boland, B.; Humblet, P.; Deliege, D. Equity in prevention and health care. *J. Epidemiol. Community Health* **2002**, *56*, 510–516. [CrossRef]
14. Vandentorren, S.; Bois, C.; Pirus, C.; Sarter, H.; Salines, G.; Leridon, H.; Elfe Team. Rationales, design and recruitment for the Elfe longitudinal study. *BMC Pediatr.* **2009**, *9*, 58. [CrossRef] [PubMed]
15. Blondel, B.; Lelong, N.; Kermarrec, M.; Goffinet, F.; The National Coordination Group of the National Perinatal Surveys. Trends in perinatal health in France from 1995 to 2010. Results from the French National Perinatal Surveys. *J. Gynecol. Obstet. Biol. Reprod.* **2012**, *41*, e1–e15. [CrossRef] [PubMed]
16. Stockley, L.; Lund, V. Use of folic acid supplements, particularly by low-income and young women: A series of systematic reviews to inform public health policy in the UK. *Public Health Nutr.* **2008**, *11*, 807–821. [CrossRef] [PubMed]
17. Timmermans, S.; Jaddoe, V.W.; Mackenbach, J.P.; Hofman, A.; Steegers-Theunissen, R.P.; Steegers, E.A. Determinants of folic acid use in early pregnancy in a multi-ethnic urban population in The Netherlands: The Generation R study. *Prev. Med.* **2008**, *47*, 427–432. [CrossRef]
18. Nilsen, R.M.; Vollset, S.E.; Gjessing, H.K.; Magnus, P.; Meltzer, H.M.; Haugen, M.; Ueland, P.M. Patterns and predictors of folic acid supplement use among pregnant women: The Norwegian Mother and Child Cohort Study. *Am. J. Clin. Nutr.* **2006**, *84*, 1134–1141. [CrossRef] [PubMed]
19. Knudsen, V.K.; Orozova-Bekkevold, I.; Rasmussen, L.B.; Mikkelsen, T.B.; Michaelsen, K.F.; Olsen, S.F. Low compliance with recommendations on folic acid use in relation to pregnancy: Is there a need for fortification? *Public Health Nutr.* **2004**, *7*, 843–850. [CrossRef] [PubMed]
20. Blondel, B.; Kermarrec, M. *Les Naissances En 2010 Et Leur Evolution Depuis 2003*; Inserm-U 953: Paris, France, 2011.
21. Botto, L.D.; Lisi, A.; Robert-Gnansia, E.; Erickson, J.D.; Vollset, S.E.; Mastroiacovo, P.; Botting, B.; Cocchi, G.; de Vigan, C.; de Walle, H.; et al. International retrospective cohort study of neural tube defects in relation to folic acid recommendations: Are the recommendations working? *BMJ* **2005**, *330*, 571. [CrossRef] [PubMed]
22. De Walle, H.; Abramsky, L. *Prevention of Neural Tube Defects by Periconceptional Folic Acid Supplementation in Europe*; Eurocat: Newtownabbey, UK, 2009.
23. De Wals, P.; Tairou, F.; Van Allen, M.I.; Uh, S.H.; Lowry, R.B.; Sibbald, B.; Evans, J.A.; Van den Hof, M.C.; Zimmer, P.; Crowley, M.; et al. Reduction in neural-tube defects after folic acid fortification in Canada. *N. Engl. J. Med.* **2007**, *357*, 135–142. [CrossRef]
24. McGuire, M.; Cleary, B.; Sahm, L.; Murphy, D.J. Prevalence and predictors of periconceptional folic acid uptake–prospective cohort study in an Irish urban obstetric population. *Hum. Reprod.* **2010**, *25*, 535–543. [CrossRef] [PubMed]
25. Tort, J.; Lelong, N.; Prunet, C.; Khoshnood, B.; Blondel, B. Maternal and health care determinants of preconceptional use of folic acid supplementation in France: Results from the 2010 National Perinatal Survey. *BJOG* **2013**, *120*, 1661–1667. [CrossRef]
26. Ami, N.; Bernstein, M.; Boucher, F.; Rieder, M.; Parker, L.; Canadian Paediatric Society, D.T.; Hazardous Substances, C. Folate and neural tube defects: The role of supplements and food fortification. *Paediatr. Child Health* **2016**, *21*, 145–154. [CrossRef] [PubMed]

27. Mallard, S.R.; Gray, A.R.; Houghton, L.A. Delaying mandatory folic acid fortification policy perpetuates health inequalities: Results from a retrospective study of postpartum New Zealand women. *Hum. Reprod.* **2012**, *27*, 273–282. [CrossRef] [PubMed]
28. de Walle, H.E.; Cornel, M.C.; de Jong-van den Berg, L.T. Three years after the dutch folic acid campaign: Growing socioeconomic differences. *Prev. Med.* **2002**, *35*, 65–69. [CrossRef] [PubMed]
29. Eichholzer, M.; Tonz, O.; Zimmermann, R. Folic acid: A public-health challenge. *Lancet* **2006**, *367*, 1352–1361. [CrossRef]
30. de Walle, H.E.; de Jong-van den Berg, L.T. Ten years after the Dutch public health campaign on folic acid: The continuing challenge. *Eur. J. Clin. Pharmacol.* **2008**, *64*, 539–543. [CrossRef] [PubMed]
31. Mitchell, L.E. Folic Acid for the Prevention of Neural Tube Defects: The US Preventive Services Task Force Statement on Folic Acid Supplementation in the Era of Mandatory Folic Acid Fortification. *JAMA Pediatr.* **2017**, *171*, 217–218. [CrossRef] [PubMed]
32. Willet, W.C. *Nutritional Epidemiology*; Oxford University Press: New York, NY, USA, 2012.
33. Meijer, W.M.; de Walle, H.E. Differences in folic-acid policy and the prevalence of neural-tube defects in Europe; recommendations for food fortification in a EUROCAT report. *Ned. Tijdschr. Geneeskd.* **2005**, *149*, 2561–2564. [PubMed]
34. Jin, J. Folic Acid Supplementation for Prevention of Neural Tube Defects. *JAMA* **2017**, *317*, 222. [CrossRef]
35. De Benoist, B. Conclusions of a WHO Technical Consultation on folate and vitamin B12 deficiencies. *Food Nutr. Bull.* **2008**, *29*, S238–S244. [CrossRef] [PubMed]
36. Reilly, A.; Amberg-Mueller, J.; Beer, M.; Busk, L.; Castellazzi, A.-M.; Castenmiller, J.; Flynn, M.; Margaritis, I.; Lampen, A.; Parvan, C.; et al. *ESCO Report on Analysis of Risks and Benefits of Fortification of Food with Folic Acid*; EFSA European Food Safety Authority: Parma, Italy, 2009.
37. Czernichow, S.; Blacher, J.; Ducimetière, P. *Enrichissement de la Farine en Vitamines B en France Proposition d'un Programme-Pilote*; AFSSA: Maison Alfort, France, 2003.
38. World Health Organization. *WHO Recommendations on Antenatal Care for a Positive Pregnancy Experience*; World Health Organization: Geneva, Switzerland, 2016.
39. Arnaud, A.; Lioret, S.; Vandentorren, S.; Le Strat, Y. Anaemia and associated factors in homeless children in the Paris region: The ENFAMS survey. *Eur. J. Public Health* **2017**. [CrossRef]
40. Navarrete-Munoz, E.M.; Valera-Gran, D.; Garcia de la Hera, M.; Gimenez-Monzo, D.; Morales, E.; Julvez, J.; Riano, I.; Tardon, A.; Ibarluzea, J.; Santa-Marina, L.; et al. Use of high doses of folic acid supplements in pregnant women in Spain: An INMA cohort study. *BMJ Open* **2015**, *5*, e009202. [CrossRef]

© 2019 by the authors. Licensee MDPI, Basel, Switzerland. This article is an open access article distributed under the terms and conditions of the Creative Commons Attribution (CC BY) license (http://creativecommons.org/licenses/by/4.0/).

Article

Exposure to Famine During Early Life and Abdominal Obesity in Adulthood: Findings from the Great Chinese Famine During 1959–1961

Dan Liu, Dong-mei Yu, Li-yun Zhao, Hong-yun Fang, Jian Zhang, Jing-zhong Wang, Zhen-yu Yang and Wen-hua Zhao *

National Institute for Nutrition and Health, Chinese Center for Disease Control and Prevention, 27 Nanwei Road, Xicheng District, Beijing 100050, China; liudanjulie@163.com (D.L.); yudm@ninh.chinacdc.cn (D.-m.Y.); zhaoly@ninh.chinacdc.cn (L.-y.Z.); fanghy@ninh.chinacdc.cn (H.-y.F.); zhangjian@ninh.chinacdc.cn (J.Z.); wangjz@ninh.chinacdc.cn (J.-z.W.); yangzy@ninh.chinacdc.cn (Z.-y.Y.)
* Correspondence: zhaowh@chinacdc.cn; Tel.: +86-10-6623-7006

Received: 14 February 2019; Accepted: 18 April 2019; Published: 22 April 2019

Abstract: Undernutrition during early life may lead to obesity in adulthood. This study was conducted to examine the relationship between famine exposure during early life and the risk of abdominal obesity in adulthood. A total of 18,984 and 16,594 adults were surveyed in 2002 and 2010–2012 in two nationally representative cross-sectional surveys, namely China Nutrition and Health Survey, respectively. The risk of abdominal obesity was evaluated for participants born during 1956–1961 and compared with that of participants born during 1962–1964. The overall prevalence of abdominal obesity in adulthood showed a positive association with famine exposure during early life. The odds ratios of famine exposure were 1.31 (1.19–1.44) and 1.28 (1.17–1.40) in 2002 during fetal life and infancy and 1.09 (1.00–1.19) in 2012 during fetal life, respectively. The relationships between famine exposure and abdominal obesity across the famine exposure groups were distinct among females and those who lived in urban areas and were physical inactive ($p < 0.05$). Exposure to famine during early life was associated with increased risks of abdominal obesity in adulthood, which was partially alleviated by healthy lifestyle factors (e.g., physical activity).

Keywords: undernutrition; abdominal obesity; fetal; infant; adulthood

1. Introduction

Overweight and obesity, as well as abdominal obesity, in middle age are strongly related to all-cause mortality and morbidity of chronic diseases such as diabetes and hypertension and other metabolic diseases [1,2]. Several risk factors have been considered to be responsible for the development of systemic obesity and abdominal obesity, such as diet, lifestyle, and genetic background [3]. The Developmental Origins of Health and Disease hypothesis suggests that [4–6] undernutrition during early life may be associated with obesity. Nutritional status during critical window periods of early life may have long-lasting effects on health in adulthood. Famine study is a commonly used approach to test the hypothesis in humans. Undernutrition is considered as natural exposure in a famine. For example, the Dutch "hungry winter" famine and the Great Chinese Famine were used to evaluate the "fetal origins" hypothesis [7].

Previous studies have primarily focused on the relationship between famine and body mass index (BMI) and/or systemic obesity [8–12]. The relationship between early-life famine exposure and waist circumference and/or abdominal obesity in later life has been less studied. Abdominal obesity is a well-established risk factor for metabolic diseases, independent of BMI [13], and could influence the risk for disease through increased insulin resistance [14]. Therefore, it is important to explore the risk factors of abdominal obesity. The Great Chinese Famine is an opportunity to evaluate the correlations between famine exposure during early life and abdominal obesity in adulthood.

The Great Chinese Famine that occurred during 1959–1961 is one of the most disastrous catastrophes in human history, resulting in 20–30 million deaths [15–17] throughout China. The Dutch famine occurred within a well-nourished population and may have led to less severe effects on human health [18]. In contrast, the Great Chinese Famine had longer duration, and food availability was more severely curtailed nationwide. Thus, a greater impact of the Great Chinese Famine on adult health can be expected than that of the Dutch famine. Therefore, national studies on the Great Chinese Famine are good resources for investigating the relationship between famine and the subsequent effects of malnutrition in early years on abdominal obesity and other chronic diseases and also for assessing whether adult lifestyle would alter the effect of the famine.

Two nationally representative cross-sectional surveys were conducted 40 and 50 years after the Great Chinese Famine in 2002 and 2010–2012 in China, respectively. These surveys could be good sources for analyzing the relationship between famine and abdominal obesity in participants of middle age and pre-elder age.

2. Participants and Methods

2.1. Design and Participants

Data were extracted from the China Nutrition and Health Survey (CNHS), a nationally representative cross-sectional study on nutrition and chronic diseases. A stratified, multistage probability cluster sampling design was used in this survey, which has been described in detail previously [19]. In the present study, data from two surveys of CNHS conducted in 2002 and 2010–2012 were used to evaluate the long-lasting impact of famine on abdominal obesity, 40 and 50 years after the occurrence of the Great Chinese Famine, respectively. The study population consisted of participants with date of birth (DOB) between October 1, 1956, and September 30, 1964. To minimize misclassification of the exposure periods, participants with DOB between October 1, 1958, and September 30, 1959, and between October 1, 1961, and September 30, 1962, were excluded because the exact dates of the start and the end of the Chinese famine were not available in different regions. The total sample size was 18,984 adults from CNHS 2002 and 16,594 adults from CNHS 2010–2012. All procedures involving participants were approved by the Medical Ethics Committee at the National Institute for Nutrition and Health, Chinese Center for Disease Control and Prevention. All participants provided their written informed consent.

2.2. Famine Exposure

Participants were categorized into the following three predefined groups according to their DOB: (1) nonexposed, with DOB between October 1, 1962, and September 30, 1964, (2) fetal-exposed, with DOB between October 1, 1959, and September 30, 1961, and (3) infant-exposed, with DOB between October 1, 1956, and September 30, 1958. Mean ages of the participants in these three exposed groups were respectively 38.6, 41.6, and 44.6 years in 2002 and 48.2, 51.2, and 54.1 years in 2010–2012.

Although the Great Chinese Famine affected the entire mainland of China, its severity varied across provinces due to different weather conditions, population density, and local policies pertaining to food shortage [20]. The severity of the famine was determined based on the excess death rate (EDR) of each province [20]. Participants were classified into severe famine exposure group and moderate famine exposure group based on residential provinces after excluding participants without local permanent residency. The EDR was calculated as the percentage change in mortality rate from the mean level in 1956–1958 to the highest value during the period 1959–1961 [20]. The median of the EDR was used as the cutoff point, which was consistent with other studies [21]. Provinces with an EDR equal to or above the cutoff point were categorized as severe famine exposure areas, and otherwise as moderate famine exposure areas.

2.3. Anthropometric Measurements

Anthropometric measurements included body weight, height, and waist circumference. Height was measured using a stadiometer (model no. SG-210, Nantong yue kin cervix equipment Co., Ltd, Nantong, China) after removing shoes, and body weight was measured with light clothes using a beam scale (model no. RGT-14-RT, Wuxi Weighing Factory Co., Ltd, Wuxi, China). Waist circumference was measured using a waist circumference tape (model no.0403, Nanjing Kongki Commodity Co., Ltd, Nanjing, China) only after breathing out. The accuracy of the height, waist circumference, and weight measurements was 0.1 cm, 0.1 cm, and 0.1 kg, respectively. The anthropometric measurements were made according to standard anthropometric measurement methods in health surveillance [22]. All anthropometric measurement staff were trained according to the standard procedure. Standardized tests were conducted for all trainees, and only those being trained and passing an examination were given a qualification certificate for conducting anthropometric measurements. Each anthropometric measurement staff was retested partially to ensure the inter-rater reliability for each anthropometric measurement. Height and weight were used to calculate BMI, by dividing weight (kg) by height squared (m^2). Systemic obesity and abdominal obesity were defined using the Chinese criteria of weight for adults [23]. Systemic obesity was defined as BMI ≥ 28 kg/m^2. Overweight was defined as BMI ≥ 24 and < 28 kg/m^2. Abdominal obesity was defined as waist circumference >90 cm in men and ≥ 85 cm in women.

2.4. Covariates

Residential areas were classified into urban and rural. Physical activity was categorized into active and inactive level, wherein regular exercise for >20 min each time, including various activities such as running, swimming, and performing Tai Chi, was defined as active level. Current drinking status and current smoking status were considered as dichotomous variables based on the answer in the questionnaires (yes or no). Current drinking was defined as participants drinking alcohol in the past year, irrespective of the amount drunk. Current smoking was defined as smoking regularly every day or irregularly. Education level was categorized into dichotomous variable, where high school or above was considered as high educational level, otherwise as low educational level.

2.5. Statistical Analyses

The SAS version 9.4 (SAS Institute Inc., Cary, NC, USA) was used for all statistical analyses, and a two-sided p value <0.05 was considered to be statistically significant. Odds of abdominal obesity for the fetal-exposed group and the infant-exposed group, compared with the nonexposed group, were examined by the maximum likelihood method using the logistic regression model. Analyses were adjusted for sex, residential areas, education level, marital status, household income, current drinking status, current smoking status, and physical activity. To investigate whether the associations between fetal and infant exposure to famine and abdominal obesity were affected by social environment in later life, we subsequently stratified the analyses according to sex, residential areas, physical activity, and education level. The odds ratios (95% confidence interval (CI)) of abdominal obesity in the fetal- and infant-exposed groups compared with the nonexposed group were calculated within each category of the stratified variables. The stratified variables were not adjusted in the corresponding models. The odds ratios (95% CI) were plotted in a graph using Stata 13.0. Sensitivity analysis was also performed in this study. First, we selected the 75th percentile of EDR as the cutoff point, and an EDR of 150.0% was used to define the severity of famine for the purpose of distinguishing severely and moderately severely affected famine areas more significantly. Second, participants with BMI ≥ 28.0 kg/m^2 who were categorized as having systemic obesity according to the Chinese criteria were excluded to prevent the interaction between systemic obesity and abdominal obesity.

3. Results

Table 1 lists the basic characteristics of participants according to famine exposure. A total of 4352 (22.9%) and 6469 (34.1%) participants in CNHS 2002 and 4126 (24.9%) and 5975 (36.0%) participants in CNHS 2010–2012 were exposed to the Great Chinese Famine during their fetal and infant period, respectively. Compared with the nonexposed group, participants in the fetal-exposed group had a significantly greater waist circumference in both 2002 and 2010–2012 (both $p < 0.05$ after Bonferroni correction), whereas the infant-exposed group showed a greater waist circumference only in 2002. The prevalence rates of abdominal obesity in nonexposed, fetal-exposed, and infant-exposed groups were 15.5%, 19.4%, and 18.9% in 2002 and 31.6%, 33.3%, and 32.6% in 2010–2012, respectively.

Table 1. Characteristics of study population according to Chinese famine exposure.

	All	Nonexposed	Fetal-Exposed	Infant-Exposed
Birth date		1962.10–1964.9	1959.10–1961.9	1956.10–1958.9
Recruitment in 2002				
N, %	18984	8163 (43.0)	4352 (22.9)	6469 (34.1)
Moderately exposed, %	40.7	37.8	45.8	41.0
Severely exposed, %	59.3	62.2	54.3	59.0
Age, years, mean (SD)	41.3 (2.6)	38.6 (0.6)	41.6 (0.6)	44.6 (0.6)
Female (%)	55.2	55.8	55.4	54.3
Height, cm, mean (SD)	160.9 (8.2)	161.2 (8.2)	161.0 (8.1)	160.4 (8.2) *
Weight, kg, mean (SD)	60.7 (10.8)	60.6 (10.8)	61.3 (10.9) *	60.4 (10.8)
WC, cm, mean (SD)	78.2 (9.5)	77.7 (9.4)	78.8 (9.6) *	78.5 (9.6) *
BMI, kg/m^2, mean (SD)	23.4 (3.3)	23.2 (3.2)	23.6 (3.4) *	23.4 (3.4) *
Overweight (%)	38.2	36.3	41.3	38.4
Obesity (%)	8.9	8.2	9.7	9.4
Pre-central Obesity (%)	31.6	29.5	33.8	32.8
Central Obesity (%)	17.6	15.5	19.4	18.9
Recruitment in 2012				
N, %	16594	6493 (39.1)	4126 (24.9)	5975 (36.0)
Moderately exposed, %	50.8	49.5	55.5	49.0
Severely exposed, %	49.2	50.5	44.5	51.0
Age, years, mean (SD)	51.1 (2.7)	48.2 (0.9)	51.2 (0.9)	54.1 (1.0)
Female (%)	57.8	58.7	56.7	57.5
Height, cm, mean (SD)	160.4 (8.1)	160.7 (7.9)	160.5 (8.0)	160.0 (8.2) *
Weight, kg, mean (SD)	62.8 (10.7)	63.1 (10.7)	63.0 (10.7)	62.2 (10.7) *
WC, cm, mean (SD)	82.8 (9.8)	82.6 (9.8)	83.1 (9.7) *	82.7 (9.8)
BMI, kg/m^2, mean (SD)	24.3 (3.4)	24.4 (3.4)	24.4 (3.4)	24.2 (3.4) *
Overweight (%)	51.3	51.8	52.1	50.3
Obesity (%)	13.7	14.5	13.8	12.8
Pre-central Obesity (%)	52.5	51.8	53.8	52.4
Central Obesity (%)	32.4	31.6	33.3	32.6

WC: Waist circumference. BMI: Body mass index. Overweight, systemic obesity, and central obesity were defined using the Chinese criteria for adults. Moderately exposed was defined as an excess death rate lower than 50.0%. * $p < 0.05$ (Bonferroni correction); statistical significance was compared with the nonexposed group (October 1962 to September 1964).

Table 2 shows the associations of famine exposure with abdominal obesity risk and the stratified analysis according to famine severity. In general, the prevalence of abdominal obesity in 2010–2012 was higher than that in 2002 in each group. Compared with the nonexposed group (1962.10–1964.9), participants had a significantly higher prevalence of abdominal obesity in both 2002 and 2010–2012 with an odds ratio (95% CI) of 1.31 (1.19–1.44) and 1.09 (1.00–1.19) in the fetal-exposed group (1959.10–1961.9) and only in 2002 with an odds ratio (95% CI) of 1.28 (1.17–1.40) in the infant-exposed group (1956.10–1958.9). After stratification of the study areas according to famine severity, participants showed a significantly higher prevalence of abdominal obesity in severely affected famine areas in both 2002 and 2010–2012 with an odds ratio (95% CI) of 1.32 (1.14–1.52) and 1.13 (1.01–1.27) in the fetal-exposed group; in the infant-exposed group, the higher prevalence of abdominal obesity was statistically significant only in 2002 compared with that in the nonexposed group. All odds ratios were adjusted for sex, residential areas, education level, marital status, household income, current drinking status, current smoking status, and physical activity.

Table 2. Associations of famine exposure with central obesity risk in different severity of famine areas

	Nonexposed	Fetal-Exposed	Infant-Exposed
Central Obesity in 2002			
Waist circumference (cm)	77.7 (9.4)	78.8 (9.6)	78.5 (9.6)
Prevalence (%)	15.5	19.4	18.9
Odds ratio (95% CI)	1.00 (Ref)	1.31 (1.19–1.44)	1.28 (1.17–1.40)
p		<0.0001	<0.0001
Stratified by famine severity			
Moderately exposed			
Waist circumference (cm)	78.5 (9.6)	79.5 (9.9)	79.8 (9.9)
Prevalence (%)	18.0	22.4	23.2
Odds ratio (95% CI)	1.00 (Ref)	1.32 (1.14–1.52)	1.40 (1.23–1.60)
p		0.0002	0.0075
Severely exposed			
Waist circumference (cm)	77.1 (9.2)	78.2 (9.3)	77.5 (9.2)
Prevalence (%)	14.0	17.0	15.9
Odds ratio (95% CI)	1.00 (Ref)	1.24 (1.08–1.42)	1.17 (1.03–1.32)
p		0.0022	0.0139
Central Obesity in 2012			
Waist circumference (cm)	82.6 (9.8)	83.1 (9.7)	82.7 (9.8)
Prevalence (%)	31.6	33.3	32.6
Odds ratio (95% CI)	1.00 (Ref)	1.09 (1.00–1.19)	1.04 (0.97–1.13)
p		0.0493	0.2823
Stratified by famine severity			
Moderately exposed			
Waist circumference (cm)	82.9 (9.9)	83.2 (9.9)	82.9 (9.7)
Prevalence (%)	33.0	33.5	33.6
Odds ratio (95% CI)	1.00 (Ref)	1.03 (0.91–1.17)	1.02 (0.91–1.15)
p		0.6473	0.7145
Severely exposed			
Waist circumference (cm)	82.4 (9.7)	82.9 (9.5)	82.6 (9.8)
Prevalence (%)	30.6	33.2	31.9
Odds ratio (95% CI)	1.00 (Ref)	1.13 (1.01–1.27)	1.05 (0.95–1.16)
p		0.0331	0.3327

All odds ratios used the nonexposed group as the reference. Odds ratio (95% CI): adjusted for sex, residential areas, education level, marital status, household income, current drinking status, current smoking status, and physical activity.

Table 3 shows the results of the sensitivity analysis. The severely affected famine areas were defined as those with an EDR ≥150% (sensitivity analyses A). Participants who were born in severely affected famine areas had a significantly higher prevalence of abdominal obesity in both 2002 and 2010–2012 in the fetal–exposed group. In the infant-exposed group, the higher prevalence of abdominal obesity was statistically significant only in 2002.

In addition, participants with BMI ≥28 kg/m^2 were excluded in sensitivity analyses B. The results of the fetal-exposed group were similar to those in sensitivity analyses A. After excluding the effect of systemic obesity, the rate of abdominal obesity was higher in the severely affected famine areas than that in the moderately severely affected famine areas in 2010–2012.

Table 3. Associations of famine exposure with central obesity risk in different severity of famine areas: sensitivity analyses.

	Nonexposed	Fetal-Exposed	Infant-Exposed
Central Obesity in 2002			
Sensitivity analyses A			
Moderately exposed			
Prevalence (%)	17.0	20.4	20.9
Odds ratio (95% CI)	1.00 (Ref)	1.26 (1.13–1.41)	1.33 (1.20–1.47)
p		<0.0001	<0.0001
Severely exposed			
Prevalence (%)	11.7	15.0	11.3
Odds ratio (95% CI)	1.00 (Ref)	1.28 (1.01–1.62)	0.93 (0.75–1.16)
p		0.0411	0.5348
Sensitivity analyses B			
Moderately exposed			
Prevalence (%)	11.0	13.8	15.2
Odds ratio (95% CI)	1.00 (Ref)	1.26 (1.13–1.41)	1.33 (1.20–1.47)
p		<0.0001	<0.0001
Severely exposed			
Prevalence (%)	8.2	11.0	9.6
Odds ratio (95% CI)	1.00 (Ref)	1.28 (1.01–1.62)	0.93 (0.75–1.16)
p		0.0411	0.5348
Central Obesity in 2012			
Sensitivity analyses A			
Moderately exposed			
Prevalence (%)	32.3	32.7	32.9
Odds ratio (95% CI)	1.00 (Ref)	1.02 (0.93–1.12)	1.01 (0.93–1.10)
p		0.6629	0.8198
Severely exposed			
Prevalence (%)	29.5	35.8	31.7
Odds ratio (95% CI)	1.00 (Ref)	1.32 (1.11–1.58)	1.12 (0.96–1.31)
p		0.0017	0.1584
Sensitivity analyses B			
Moderately exposed			
Prevalence (%)	22.4	22.2	23.7
Odds ratio (95% CI)	1.00 (Ref)	0.98 (0.84–1.15)	1.06 (0.92–1.22)
p		0.8308	0.4090
Severely exposed			
Prevalence (%)	21.1	25.2	23.9
Odds ratio (95% CI)	1.00 (Ref)	1.27 (1.11–1.45)	1.17 (1.04–1.32)
p		0.0005	0.0102

Sensitivity analyses A: Defining severity of famine according to excess death rate 150%; Sensitivity analyses B: Excluding participants with BMI $\geq$ 28.0 kg/m^2. All odds ratios used the nonexposed group as the reference. Odds ratio (95% CI): adjusted for sex, residential areas, education level, marital status, household income, current drinking status, current smoking status, and physical activity.

In CNHS 2002 and CNHS 2010–2012, the risk of famine exposure associated with abdominal obesity was detected among female, inactive participants, those who lived in urban areas, and those with high level of education (Figure 1). The results were adjusted for sex, residential areas, education level, marital status, household income, current drinking status, current smoking status, and physical activity.

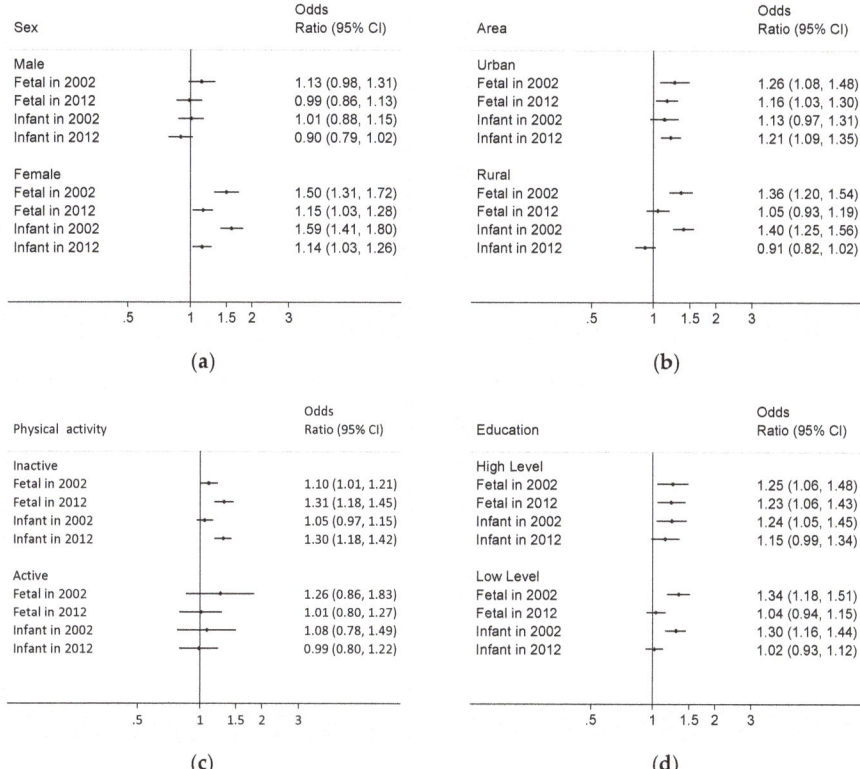

Figure 1. Subgroup analyses of the association between central obesity and famine exposure groups. (**a**) Associations of famine exposure with central obesity risk by sex; (**b**) Associations of famine exposure with central obesity risk by area; (**c**) Associations of famine exposure with central obesity risk by physical activity; (**d**) Associations of famine exposure with central obesity risk by education level. Model adjusted for sex, residential areas, education level, marital status, household income, current drinking status, current smoking status, and physical activity. Stratified variables were not adjusted in the corresponding models. CI, confidence interval. All odds ratios used the nonexposed group as the reference.

4. Discussion

Based on two nationally representative studies, the present investigation found that exposure to the Great Chinese Famine during early life increased the risk of abdominal obesity in later adulthood, especially among female, inactive participants, those who lived in urban areas, and those with high education level. The risk of abdominal obesity had a longer lasting effect on those who experienced the Great Chinese Famine in their fetal period. Current unhealthy lifestyle factors would exacerbate the effect of early-life exposure to famine on abdominal obesity.

Low birth weight has been considered to be a risk factor for obesity in adults, which indicated that exposure to famine during early life might increase the risk of obesity in adulthood [8,9]. Some studies reported that fetal or infant exposure to famine could reduce the risk of systemic obesity in adulthood [10,11]. In addition, a few studies have reported that there was no association between famine exposure during some stages of early life and later obesity [12]. However, these previous studies were different in terms of the duration of famine exposure and famine severity, and the recruitment periods were also inconsistent. For example, compared with the Dutch famine (one winter), the Chinese famine lasted longer (3 years) and was more severe. Furthermore, abdominal obesity was more harmful to metabolic diseases than systemic obesity. Because no study had evaluated the association between Chinese early-life famine exposure and abdominal obesity in adulthood during different life stages, data from CNHS 2002 and 2012 were an opportunity to fill the gap.

The fetal origin of adult disease hypothesis, also referred to as the Barker hypothesis, proposed that alterations in fetal nutrition and endocrine status lead to developmental adaptations that permanently change the body structure, physiology, and metabolism, thereby predisposing individuals to cardiovascular, metabolic, and endocrine diseases in adult life [4–6]. Several precise mechanisms might explain the associations between fetal famine exposure and risk of abdominal obesity in later life. First, early-life malnutrition might alter the neuroendocrine function, including induction of the hypothalamic–pituitary–adrenal cortical axis, which results in excessive secretion of glucocorticoids and fat accumulation in later life [24,25]. Second, a probable mechanism is that nutritional deprivation in early life affects the expression of related genes and would change the dietary behavior. Evidence from the Dutch famine study has suggested that prenatal exposure to famine would increase the preference for fatty food and might contribute to more atherogenic lipid profiles in later life, such as belly fat [26]. Third, findings from epigenetic studies have suggested that early malnutrition could lead to abnormal DNA methylation of genes, which was associated with obesity and insulin resistance in adulthood [27]. The above mentioned abnormalities could result in excessive fat deposition. Furthermore, hypertension, diabetes, and coronary heart disease were found to be associated with early-life malnutrition, indicating a relationship between famine exposure and obesity in adulthood [28].

The adaptive sex ratio adjustment hypothesis suggested that mothers who experience nutritional stress would be more likely to give birth to female babies. This is because nutrition cost was less expensive for female babies than for male babies, and female babies had a better opportunity to survive in the harsh environment [29]. Male survivors may have "acceptable" nutrition exposure during early life because vulnerable male babies were difficult to survive. On the other hand, male survivors may also be associated with the culture of son preference in China [30]. Famine might predispose female survivors to the risk of developing chronic degenerative diseases in adulthood, including obesity. Evidence has shown that undernutrition during early life had larger long-term impacts on females than on males [24,31,32]. Consistent with these findings, our study has also demonstrated a higher prevalence of abdominal obesity among female survivors in the Great Chinese Famine.

A mismatch between early life and adult life environment may explain the association between famine exposure and risk of abdominal obesity [33–35]. Nutrition restriction during early life and exposure to a "rich" environment (rich nutrition, high socioeconomic status, etc.) in later life might increase an individual's susceptibility or risk of developing obesity and other chronic diseases. Data from the CNHS 2002 showed that modern diet exacerbated the effects of famine in relation to diabetes and hypertension [21,36]. In the present study, participants who lived in urban areas or had a high education level had an increased risk for abdominal obesity across fetal and infant famine exposure groups, which might also be attributed to the modernized life. It has been suggested that people who lived in urban areas and had a high education level were more likely to have a high-energy-dense western dietary pattern in China [19,37,38]. Furthermore, the present study results have demonstrated that the adverse effects of undernutrition during early life were likely to be exacerbated by unhealthy lifestyle factors during later adulthood. Unhealthy lifestyle in the fetal- and infant-exposed groups, including inadequate levels of physical activity, imposed an increased risk of abdominal obesity compared to that in the nonexposed group. These findings further highlighted the importance of a healthy lifestyle in the prevention of adult chronic diseases.

The association between famine exposure and abdominal obesity disappeared in the 2010–2012 survey, such as in rural participants and those with low education level. The significant association existed for the female participants, but the odds ratio became smaller. Overall, the association was attenuated among participants, except for physically inactive participants in 2010–2012. This might be related to the trajectory of the human body shape over the course of the lives. Zhai et al [1] showed that the waist circumference of Chinese adults increased gradually during early age, whereas it began to decline with aging. The trajectory was slightly different between gender, wherein males began to develop a thinner body shape in their 40s and females did so in their 50s. This phenomenon could explain why the association disappeared among males in their 50s but remained in the females. The trajectory of the human body shape could also explain the change in other subgroups, such as the residential area and the education group. The risk of abdominal obesity in participants living in urban areas and having a high education level was still steady and might be related to their exposure to a "rich" environment in later life. However, rural participants with a low education level showed a declining trajectory of waist circumference in the absence of exposure to a "rich" environment.

This study has some limitations. First, the lack of birth weight data might be a concern. However, fetal programming could also occur without any marked effects on birth size [39], and hence this was not considered as a major limitation. Second, the Chinese famine affected almost the entire country. Therefore, participants had to be classified into different groups based on their birth date rather than exposure areas or nonexposure areas. Despite these limitations, our research also had irreplaceable advantages. The Great Chinese Famine lasted much longer and affected more people than other famines. This study could also demonstrate further convincing results. Data from two large, nationally representative cross-sectional surveys conducted 40 and 50 years later in 2002 and 2010–2012 after the Chinese famine were used in this study along with detailed information regarding sociodemographic characteristics, lifestyle factors, and birthplace. Therefore, our research provides valuable evidence on the hypothesis of the combined association between early-life famine exposure and later life environment and the risk of abdominal obesity in adulthood. In addition, undernutrition during early life was associated with metabolic diseases, including metabolic syndrome, diabetes mellitus, and stroke, which would occur around the age of 60 years. In future, more studies investigating the abovementioned diseases are warranted. Furthermore, China might face a high incidence of metabolic diseases in the forthcoming years.

5. Conclusions

Early-life exposure to the Great Chinese Famine exacerbated the risk of abdominal obesity, especially in females or those who lived in urban areas or were physically inactive. A healthy lifestyle might partially alleviate the adverse effects. The results suggest that promoting a healthy lifestyle should be considered as a critical strategy for the prevention of chronic diseases in adult life.

Author Contributions: D.L. participated in project design, conducted data analysis, and drafted the manuscript. D.-m.Y., L.-y.Z., and H.-y.F. participated in project design and implementation and contributed significantly to data acquisition and manuscript preparation. J.Z. and J.-z.W. conceived and designed the study and played an important role in manuscript preparation. Z.-y.Y revised the manuscript critically. W.-h.Z. conceived and designed the study, helped with interpretation of data, and revised the manuscript critically.

Funding: This research was funded by Central Finance of China, and Science & Technology Basic Resources Investigation Program of China (grant number: 2017FY101101 and 2017FY101103).

Acknowledgments: We would like to acknowledge the China Nutrition and Health Survey (CNHS, 2002 and 2010-2012) team. We thank all the participants and team members who took part in the two surveys.

Conflicts of Interest: The authors have no other funding or conflicts of interest to declare.

References

1. Zhai, Y.; Fang, H.Y.; Yu, W.T.; Yu, D.W.; Zhao, L.Y.; Liang, X.F.; Zhao, W.H. Changes in Waist Circumference and Abdominal Obesity among Chinese Adults over a Ten-year Period. *Biomed. Environ. Sci.* **2017**, *30*, 315–322.
2. Bhupathiraju, S.N.; Hu, F.B. Epidemiology of Obesity and Diabetes and Their Cardiovascular Complications. *Circ. Res.* **2016**, *118*, 1723–1735. [CrossRef]
3. Hruby, A.; Manson, J.E.; Qi, L.; Malik, V.S.; Rimm, E.B.; Sun, Q.; Willett, W.C.; Hu, F.B. Determinants and Consequences of Obesity. *Am. J. Public Health* **2016**, *106*, 1656. [CrossRef] [PubMed]

4. Barker, D.J.; Osmond, C. Infant mortality, childhood nutrition, and ischaemic heart disease in England and Wales. *Lancet* **1986**, *1*, 1077–1081. [CrossRef]
5. Barker, D.J.; Gluckman, P.D.; Godfrey, K.M.; Harding, J.E.; Owens, J.A.; Robinson, J.S. Fetal nutrition and cardiovascular disease in adult life. *Lancet* **1993**, *341*, 938–941. [CrossRef]
6. Barker, D.J.; Bull, A.R.; Osmond, C.; Simmonds, S.J. Fetal and placental size and risk of hypertension in adult life. *BMJ* **1990**, *301*, 259–262. [CrossRef]
7. Scholte, R.S.; Van Berg, G.J.D.; Lindeboom, M. Long-run effects of gestation during the Dutch Hunger Winter famine on labor market and hospitalization outcomes. *J. Health Econ.* **2015**, *39*, 17–30. [CrossRef]
8. Newby, P.K.; Dickman, P.W.; Adami, H.O.; Wolk, A. Early anthropometric measures and reproductive factors as predictors of body mass index and obesity among older women. *Int. J. Obes.* **2005**, *29*, 1084–1092. [CrossRef]
9. Minooee, S.; Tehrani, F.R.; Mirmiran, P.; Azizi, F. Low birth weight may increase body fat mass in adult women with polycystic ovarian syndrome. *Int. J. Reprod. Biomed.* **2016**, *14*, 335–340. [CrossRef]
10. Huang, C.; Li, Z.; Wang, M.; Martorell, R. Early life exposure to the 1959–1961 Chinese famine has long-term health consequences. *J. Nutr.* **2010**, *140*, 1874–1878. [CrossRef]
11. Wang, P.X.; Wang, J.J.; Lei, Y.X.; Xiao, L.; Luo, Z.C. Impact of fetal and infant exposure to the Chinese Great Famine on the risk of hypertension in adulthood. *PLoS ONE* **2012**, *7*, e49720. [CrossRef] [PubMed]
12. Hult, M.; Tornhammar, P.; Ueda, P.; Chima, C.; Bonamy, A.K.; Ozumba, B.; Norman, M. Hypertension, diabetes and overweight: Looming legacies of the Biafran famine. *PLoS ONE* **2010**, *5*, e13582. [CrossRef]
13. Lee, S.Y.; Chang, H.J.; Sung, J.; Kim, K.J.; Shin, S.; Cho, I.J.; Shim, C.Y.; Hong, G.R.; Chung, N. The impact of obesity on subclinical coronary atherosclerosis according to the risk of cardiovascular disease. *Obesity* **2014**, *22*, 1762–1768. [CrossRef] [PubMed]
14. Müller, M.J.; Lagerpusch, M.; Enderle, J.; Schautz, B.; Heller, M.; Bosy-Westphal, A. Beyond the body mass index: Tracking body composition in the pathogenesis of obesity and the metabolic syndrome. *Obes. Rev.* **2012**, *13*, 6–13. [CrossRef]
15. Smil, V. China's great famine: 40 years later. *BMJ* **1999**, *319*, 1619–1621. [CrossRef] [PubMed]
16. Li, W.; Yang, D.T. The Great Leap Forward: Anatomy of a Central Planning Disaster. *J. Political Econ.* **2005**, *113*, 840–877. [CrossRef]
17. Chen, Y.; Zhou, L.A. The long-term health and economic consequences of the 1959–1961 famine in China. *J. Health Econ.* **2007**, *26*, 659–681. [CrossRef] [PubMed]
18. He, P.; Liu, L.; Salas, J.M.I.; Guo, C.; Cheng, Y.; Chen, G.; Zheng, X. Prenatal malnutrition and adult cognitive impairment: A natural experiment from the 1959–1961 Chinese famine. *Br. J. Nutr.* **2018**, *120*, 198–203. [CrossRef]
19. Liu, D.; Zhao, L.Y.; Yu, D.M.; Ju, L.H.; Zhang, J.; Wang, J.Z.; Zhao, W.H. Dietary Patterns and Association with Obesity of Children Aged 6-17 Years in Medium and Small Cities in China: Findings from the CNHS 2010–2012. *Nutrients* **2018**, *11*, 3. [CrossRef]
20. Luo, Z.; Ren, M.U.; Zhang, X. Famine and Overweight in China. *Rev. Agric. Econ.* **2006**, *28*, 296–304. [CrossRef]
21. Li, Y.; He, Y.; Qi, L.; Jaddoe, V.W.; Feskens, E.J.; Yang, X.; Ma, G.; Hu, F.B. Exposure to the Chinese famine in early life and the risk of hyperglycemia and type 2 diabetes in adulthood. *Diabetes* **2010**, *59*, 2400–2406. [CrossRef]
22. Fu, P.; Yi, G.; Zhang, J.; Song, Y.; Wang, J. *Anthropometric Measurements Method in Health Surveillance*; National Health Commission of the People's Republic of China: Beijing, China, 2013; pp. 1–10.
23. Chen, C.; Zhao, W.; Yang, X.; Chen, J. *Criteria of Weight for Adults*; National Health Commission of the People's Republic of China: Beijing, China, 2013; pp. 1–4.
24. Barker, D.J.; Lampl, M.; Roseboom, T.; Winder, N. Resource allocation in utero and health in later life. *Placenta* **2012**, *33* (Suppl. 2), e30–e34. [CrossRef]
25. Bale, T.; Baram, T.Z.; Brown, A.S.; Goldstein, J.M.; Insel, T.R.; McCarthy, M.M.; Nemeroff, C.B.; Reyes, T.M.; Simerly, R.B.; Susser, E.S.; et al. Early life programming and neurodevelopmental disorders. *Biol. Psychiatry* **2010**, *68*, 314–319. [CrossRef]
26. Lussana, F.; Painter, R.C.; Ocke, M.C.; Buller, H.R.; Bossuyt, P.M.; Roseboom, T.J. Prenatal exposure to the Dutch famine is associated with a preference for fatty foods and a more atherogenic lipid profile. *Am. J. Clin. Nutr.* **2008**, *88*, 1648–1652. [CrossRef]

27. Tobi, E.W.; Heuvel, J.V.D.; Zwaan, B.J.; Lumey, L.H.; Heijmans, B.T.; Uller, T. Selective Survival of Embryos Can Explain DNA Methylation Signatures of Adverse Prenatal Environments. *Cell Rep.* **2018**, *25*, 2660–2667. [CrossRef] [PubMed]
28. Li, C.; Lumey, L.H. Interaction or mediation by adult obesity of the relation between fetal famine exposure and type 2 diabetes? *Int. J. Epidemiol.* **2019**, *48*, 654–656. [CrossRef] [PubMed]
29. Galante, L.; Milan, A.M. Sex-Specific Human Milk Composition: The Role of Infant Sex in Determining Early Life Nutrition. *Nutrients* **2018**, *10*, 1194. [CrossRef]
30. Zhou, C.; Wang, X.L.; Zhou, X.D.; Hesketh, T. Son preference and sex-selective abortion in China: Informing policy options. *Int. J. Public Health* **2012**, *57*, 459–465. [CrossRef] [PubMed]
31. Wang, Y.; Wang, X.; Kong, Y.; Zhang, J.H.; Zeng, Q. The Great Chinese Famine leads to shorter and overweight females in Chongqing Chinese population after 50 years. *Obesity* **2010**, *18*, 588–592. [CrossRef]
32. Song, S. Does famine influence sex ratio at birth? Evidence from the 1959-1961 Great Leap Forward Famine in China. *Proc. Biol. Sci.* **2012**, *279*, 2883–2890. [CrossRef] [PubMed]
33. Li, C.; Lumey, L.H. Exposure to the Chinese famine of 1959-61 in early life and long-term health conditions: A systematic review and meta-analysis. *Int. J. Epidemiol.* **2017**, *46*, 1157–1170. [CrossRef]
34. Meng, R.; Lv, J.; Li, L. Reply: Interaction or mediation by adult obesity of the relation between fetal famine exposure and type 2 diabetes? *Int. J. Epidemiol.* **2019**, *48*, 656–657. [CrossRef] [PubMed]
35. Meng, R.; Lv, J.; Yu, C.; Guo, Y.; Bian, Z.; Yang, L.; Chen, Y.; Zhang, H.; Chen, X.; Chen, J.; et al. Prenatal famine exposure, adulthood obesity patterns and risk of type 2 diabetes. *Int. J. Epidemiol.* **2018**, *47*, 399–408. [CrossRef] [PubMed]
36. Li, Y.; Jaddoe, V.W.; Qi, L.; He, Y.; Lai, J.; Wang, J.; Zhang, J.; Hu, Y.; Ding, E.L.; Yang, X.; et al. Exposure to the Chinese famine in early life and the risk of hypertension in adulthood. *J. Hypertens.* **2011**, *29*, 1085–1092. [CrossRef]
37. Xu, X.; Hall, J.; Byles, J.; Shi, Z. Dietary Pattern Is Associated with Obesity in Older People in China: Data from China Health and Nutrition Survey (CHNS). *Nutrients* **2015**, *7*, 8170–8188. [CrossRef] [PubMed]
38. Zhao, L.Y.; Liu, D.; Yu, D.M.; Zhang, J.; Wang, J.Z.; Zhao, W.H. Challenges Brought about by Rapid Changes in Chinese Diets: Comparison with Developed Countries and Implications for Further Improvement. *Biomed. Environ. Sci.* **2018**, *31*, 781–786. [PubMed]
39. Estampador, A.C.; Franks, P.W. Precision Medicine in Obesity and Type 2 Diabetes: The Relevance of Early-Life Exposures. *Clin. Chem.* **2018**, *64*, 130–141. [CrossRef]

© 2019 by the authors. Licensee MDPI, Basel, Switzerland. This article is an open access article distributed under the terms and conditions of the Creative Commons Attribution (CC BY) license (http://creativecommons.org/licenses/by/4.0/).

Article

Reducing Pup Litter Size Alters Early Postnatal Calcium Homeostasis and Programs Adverse Adult Cardiovascular and Bone Health in Male Rats

Jessica F. Briffa [1], Rachael O'Dowd [1], Tania Romano [1,2], Beverly S. Muhlhausler [3], Karen M. Moritz [4] and Mary E. Wlodek [1,*]

1. Department of Physiology, The University of Melbourne, Parkville 3010, Australia; jessica.griffith@unimelb.edu.au (J.F.B.); odowd.rachael.a@edumail.vic.gov.au (R.O.); t.romano@latrobe.edu.au (T.R.)
2. Department of Physiology, Anatomy and Microbiology, LaTrobe University, Bundoora 3083, Australia
3. Department of Food and Wine Science, School of Agriculture, Food and Wine, FOODplus Research Centre, The University of Adelaide, Adelaide 5064, Australia; Beverly.Muhlhausler@adelaide.edu.au
4. Child Health Research Centre and School of Biomedical Sciences, The University of Queensland, St. Lucia 4101, Australia; k.moritz1@uq.edu.au
* Correspondence: m.wlodek@unimelb.edu.au; Tel.: +61-3-8344-8801

Received: 3 December 2018; Accepted: 3 January 2019; Published: 8 January 2019

Abstract: The in utero and early postnatal environments play essential roles in offspring growth and development. Standardizing or reducing pup litter size can independently compromise long-term health likely due to altered milk quality, thus limiting translational potential. This study investigated the effect reducing litter size has on milk quality and offspring outcomes. On gestation day 18, dams underwent sham or bilateral uterine vessel ligation surgery to generate dams with normal (Control) and altered (Restricted) milk quality/composition. At birth, pups were cross-fostered onto separate dams with either an unadjusted or reduced litter size. Plasma parathyroid hormone-related protein was increased in Reduced litter pups, whereas ionic calcium and total body calcium were decreased. These data suggest Reduced litter pups have dysregulated calcium homeostasis in early postnatal life, which may impair bone mineralization decreasing adult bone bending strength. Dams suckling Reduced litter pups had increased milk long-chain monounsaturated fatty acid and omega-3 docosahexaenoic acid. Reduced litter pups suckled by Normal milk quality/composition dams had increased milk omega-6 linoleic and arachidonic acids. Reduced litter male adult offspring had elevated blood pressure. This study highlights care must be taken when interpreting data from research that alters litter size as it may mask subtle cardiometabolic health effects.

Keywords: reduced litter size; postnatal calcium homeostasis; adult bone health; milk composition

1. Introduction

It is well known that the in utero and early postnatal environments play crucial roles in offspring growth, development and long-term health. David Barker first demonstrated a causal link between size at birth and later-life cardiovascular disease [1,2], which has been since expanded to include several adverse pregnancy perturbations, including fetal growth restriction [3,4], maternal undernutrition [5,6], maternal alcohol consumption [7,8] and maternal stress [9,10]. Postnatal growth rate and development is directly proportional to the quality and quantity of milk produced and is particularly influenced by fatty acid composition [11]. Specifically, high intakes of omega-3 fatty acids in early postnatal life is associated with reduced fat deposition and improved cardiometabolic health [12,13].

More recent experimental studies have demonstrated that altered maternal nutrition during the lactation period can program adult offspring cardiometabolic disease. Specifically, pups suckled by dams fed a cafeteria diet during lactation exhibit a "thin-outside-fat-inside phenotype" (lean with increased abdominal fat) and impaired metabolic health in adulthood [14]. Whereas, male offspring cross-fostered onto a dam fed an isocaloric low-protein diet (6% protein) throughout pregnancy and lactation have increased blood pressure and renal dysfunction [15,16]. As milk is the sole source of nutrition during early postnatal life, these studies strongly support a role for altered milk quality and/or quantity as a mechanism through which maternal nutritional status during lactation influences offspring disease.

In the developmental programming field, many researchers standardize or reduce litter size at birth to normalize milk intake across pups and cohorts [17], with the degree of litter size reduction dependent on the experimental model and research question [18–25]. A previous study identified that pup body weight and development during lactation is dependent on litter size at birth [26], which the authors hypothesize is likely due to mutual maternal and offspring adjustment to a genetically determined litter size [26]. This suggests that studies that cull pups to standardize litter size have the potential to disturb this biological process and that the original litter size may continue to have some influence on offspring development. In line with this, we previously demonstrated that reducing the litter size of healthy Wistar Kyoto (WKY) rat dams at birth (from 10–14 pups per litter to 5 pups per litter) decreases offspring body weight during early life and increases adult male blood pressure likely due to mesenteric artery stiffness and compromises bone health [27–31]. Whereas, severely reducing litter size to 3 pups (inducing early postnatal overnutrition) increases offspring body weight by weaning, alters cardiac structure and function and programs poor metabolic health [32,33]. Thus, there is a need to better understand the impact of reducing litter sizes to differing extents on the subsequent outcomes of the pups.

The mechanism behind this disease programming due to reducing litter size is likely due to its effect on milk composition. Specifically, we have recently demonstrated that both the dam and pup can modulate the maternal milk composition [34]. Thus, dams suckling growth restricted pups have improved milk fatty acid composition (characterized by increased LC-polyunsaturated fatty acid (PUFA) and LC-monounsaturated fatty acid (MUFA)), which is likely to be a compensatory mechanism aimed at supporting pup growth and organ development [34]. This finding suggests that the decreased pup milk intake, rather than poor milk quality, results in the aforementioned long-term disease [27–29]. Therefore, it is possible that reducing litter size at birth to 5 pups (similarly to the reduction in litter size we observe following uteroplacental insufficiency surgery) can itself induce changes in the early postnatal environment that has consequences for adult health outcomes and may thus mask or exacerbate any offspring outcomes due to adverse pregnancy and/or lactational events.

Therefore, the aim of this study was to investigate the effect of reducing litter size at birth on pup postnatal growth and development, mammary development and maternal milk composition. We additionally characterized if there were sex-specific differences in long-term cardiometabolic and bone health outcomes due to reducing litter size at birth and whether these outcomes were exacerbated if the pup was suckled by a dam with altered milk quality/composition.

2. Materials and Methods

2.1. Animals

All experiments were approved by The University of Melbourne's animal experimentation ethics sub-committee (AEC: 02081) following the National Health and Medical Research Councils (NHMRC) Australian code for the care and use of animals for scientific purposes. Female WKY rats (9 to 13 weeks of age) were obtained from the Animal Resources Centre (Canning Vale, WA, Australia) and provided with a 12-h light/dark cycle at 19–22 °C with ad libitum access to food and water. To generate rats with 'normal' and 'altered' milk quality/composition, rats were mated and surgery performed on day 18 of gestation (term = 22 days) as described previously [35]. Briefly, F0 pregnant rats were randomly allocated to a sham (Normal milk quality/composition; Control; n = 7–8 per group) or uteroplacental insufficiency (Altered milk quality/composition; Restricted; $n = 8$ per group) group, that experience premature lactogenesis [30], and were anaesthetized with 4% isoflurane and 650 mL/min oxygen flow (reduced to 3.2% isoflurane and 250 mL/min oxygen flow when suturing to aid in the animals recovery) to reduce the duration of anesthetic exposure and aid in recovery [35]. Rats were then allowed to deliver naturally (Figure 1a). Pups from the sham (Control) operated dams were cross-fostered 1 day after birth (PN1) randomly onto separate dams (Control or Restricted) where the litter size was unaltered from the surrogate (Standard) or reduced to 5 pups (Reduced) giving rise to four experimental groups (Figure 1b); Standard litter size suckled by Normal milk quality/composition dams (mean litter size 10.38; range 9–12 pups), Reduced litter size suckled by Normal milk quality/composition (mean litter size 5.00; range 5 pups), Standard litter size suckled by Altered milk quality/composition dams (mean litter size 11.50; range 7–15 pups) and Reduced litter size suckled by Altered milk quality/composition (mean litter size 6.00; range 3–9 pups) with $n = 15$–18 dams per group [17,31,36–38]. Birth weight (3.5 ± 0.4 g vs. 4.0 ± 0.04 g for Restricted and Control dams, respectively) and litter size (8.5 ± 0.4 pups vs. 11.1 ± 0.4 pups for Restricted and Control dams, respectively) of pups born to ligation surgery (Restricted; Altered milk quality/composition) dams was reduced compared to sham-operated (Control; Normal milk quality/composition) dams.

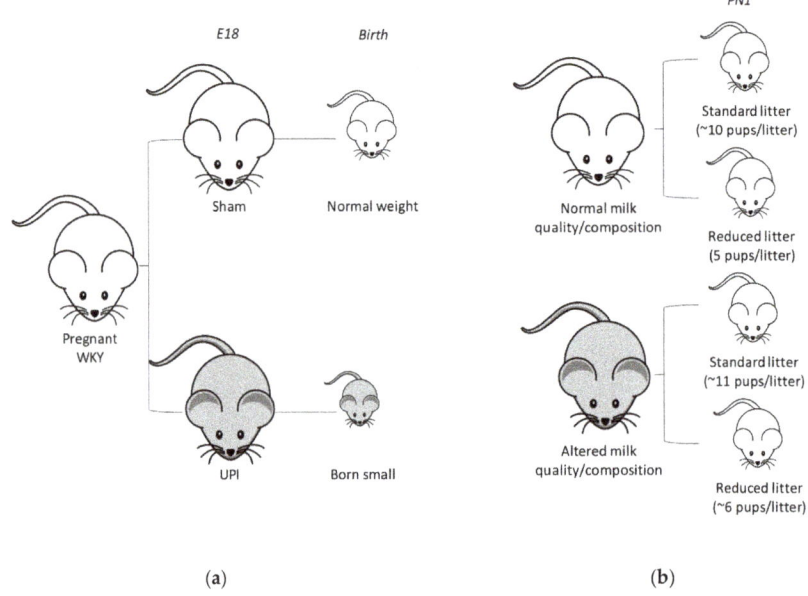

Figure 1. Study design. (**a**) Pregnant Wistar Kyoto (WKY) rats underwent Sham (white) or uteroplacental insufficiency (UPI; grey) surgery on day 18 of gestation (E18) and allowed to deliver naturally at birth. (**b**) On postnatal day 1 (PN1) pups from the Sham surgery dams were cross-fostered onto separate Sham (with Normal milk quality/composition) or UPI (with Altered milk quality/composition) dams with either an intact or reduced litter size.

2.2. Study 1: Postnatal Study

On the morning of PN6, one cohort of pups were removed from their dam and weighed, killed via decapitation, and blood was collected and pooled within litters. Aprotinin (Sigma-Aldrich; Castle Hill, NSW, Australia) was added to the tubes used for plasma PTHrP analysis [39]. One pup per litter was frozen whole and stored at -20 °C for whole body calcium analysis. Pup stomach contents were collected for fatty acid analysis. Dams were anaesthetized (Ketamine (50 mg/kg; Parnell Laboratories, Alexandria, NSW, Australia) and Ilium Xylazil-20 (10 mg/kg; Troy Laboratories, Glendenning, NSW,

Australia)) 4–6 h after removal of the pup to allow milk to accumulate and milk was collected following gentle massage of the left mammary gland and teats without the need for hormonal stimulation [40], then euthanized by cardiac puncture with blood collected for subsequent analysis. The mammary glands were dissected, weighed and the right mammary gland immediately snap frozen in liquid nitrogen or fixed in 10% neutral buffered formalin (Perrigo; Balcatta, WA, Australia) for histological analysis.

2.2.1. Mammary 'Real-Time' PCR and Histology

'Real-time' PCR was used to quantitate milk protein gene expression in mammary tissue on PN6 as described previously with $n = 3$–7 dams per group [30]. Briefly, RNA was extracted from mammary tissue using the Polytron PT 3100 (Biolab; Clayton, VIC, Australia) homogenizer and a commercially available kit (RNeasy Lipid Tissue Mini Kit from Qiagen; Chadstone, VIC, Australia). RNA was then DNase treated using the Ambion DNA free kit (Life Technologies; Mulgrave, VIC, Australia). First strand cDNA was generated from 1 µg RNA using the Superscript II Single Stranded cDNA kit (Life Technologies). qPCR was performed then conducted against the milk protein genes *Pthrp*, *b-casein* and *a-lactalbumin* with Ribosomal *18S* as the reference gene [34]. The reaction was activated by heating the mixture to 95 °C for 10 min, then 'Real-time' PCR reactions were ran for 40 cycles of 95 °C for 15 s and 60 °C for 60 s. For relative quantification of gene expression, a multiplex comparative threshold cycle method was employed [35]. *18S* values were not different between treatments.

Fixed mammary tissue was processed into paraffin blocks, sectioned at 5 µm and stained with hematoxylin and eosin ($n = 4$–5 per group). Five sections per sample were analyzed for alveolar area and number using ImagePro Software (Medai Cybernetics; Warrendale, PA, United States of America) [40].

2.2.2. PTHrP, Corticosterone, Calcium and Electrolyte Measurements

Plasma, milk and mammary tissue concentrations of PTHrP were quantified by a N-terminal radioimmunoassay with a minimum detection limit of 2 pmol/L, and intra- and inter-assay coefficients of variation of 4.8% and 13.6% respectively with $n = 7$–9 per group [41]. Plasma corticosterone was measured by enzyme immunoassay validated for direct measurements in diluted plasma following the manufacturer's protocol (Cayman Chemical; Ann Arbor, MI, United States of America) with a minimum detection limit of 30 pg/mL, and intra- and inter-assay coefficients of variation of 7.4% and 7.0% respectively ($n = 4$–8 per group). Total calcium concentrations were determined using colorimetric spectrometry using the Synchron CX-5 Clinical System (Beckman Coulter; Lane Cove, NSW, Australia) and ionic calcium (active or free calcium; regulated by PTHrP), sodium and potassium concentrations were determined using ion selective electrodes correcting for pH (Ciba-Corning model 644; Cambridge, MA, United States of America) from milk as well as from pup and maternal plasma ($n = 5$–8 per group) [39,42]. Total calcium concentration in the pup body was determined after ashing using the CX-5 Analyzer ($n = 7$–9 per group) [30,40]. Total protein and lactose concentrations were analyzed as described previously with $n = 4$–6 per group [30].

2.2.3. Fatty Acid Analysis

Total fatty acid composition of the milk was determined by the direct trans-esterification method of Lepage and Roy [43] as previously described [34] with $n = 6$–7 dams per group. Briefly, 50 to 100 µL of milk was placed into a screw-capped Teflon-lined tube containing C23:0 as an internal standard. After the 1-h trans-esterification procedure and recovery of the fatty acid methyl esters (FAMEs) in the benzene phase, the FAMEs were analyzed by capillary gas liquid chromatography. FAMEs were then separated and measured on a Shimadzu (17A) gas chromatograph with flame ionization detection. A 50 mm × 0.25 mm BPX-70 fused silica capillary column (SGE Scientific; Ringwood, VIC, Australia) with a film thickness of 0.25 µm was used in conjunction with a Shimadzu on-column auto-injector. Ultrahigh purity hydrogen was used as a carrier gas at a flow rate of 2 mL/min. A temperature gradient program was used with an initial temperature of 170 °C, increasing at 3 °C/min to 218 °C. Identification of the FAMEs was made by comparison with the retention times of chromatography reference standard mixtures (Nu-Chek Prep; Elysian, MN, United States of America) [44].

2.3. Study 2: Lactation and Adult Study

Another cohort of pups stayed with their cross-fostered dam until weaning (PN35) and were weighed from birth (PN1) to weaning (PN3, PN6, PN10, PN14, PN17, PN21, PN24, PN28 and PN35) to determine pup growth and milk intake. Pups were weighed during a 1-h maternal separation period followed by a 3-h re-feeding period. Milk intake was calculated as a percentage of pup body weight gain after feeding compared to pre-feeding [30] and the area under the milk intake curve was calculated between PN3 and PN17 as an index of milk consumption. After weaning, male and female pups were housed separately. At 6 months, offspring (1 male and 1 female per litter) underwent an intra-arterial glucose tolerance test (IAGTT) and tail-cuff blood pressure measurements as previously described [36–38]. Rats were then weighed and anaesthetized with an intraperitoneal injection (50 mg/kg Ketamine and 10 mg/kg Ilium Xylazil-20) and the right hind limbs were collected; all females underwent post-mortem when they were in estrous.

2.3.1. Bone Analyses

The right hind limb was dissected to separate the femur bone from soft tissue and femur length measured using digital calipers (n = 9–19 per group per sex). Individual femurs then underwent peripheral quantitative computed tomography (pQCT) to measure bone volumetric content, density and stress strain index using methods previously described [31,45–47]. Briefly, two slices of 1 mm thickness (voxel size 0.1000 mm^3, peel mode 20, contour mode 1) were taken at distances of 15% and 50% from the reference line to quantify both trabecular and cortical bone tissue, respectively. A tissue density of 280 mg/cm^3 or less was identified as trabecular bone, whereas a density of 710 mg/cm^3 was representative of cortical bone. Automatic density thresholding (400 mg/cm^3) was used to eliminate the effects of any soft tissue which may have remained on the femur after dissection. Bone mineral content, density, and stress strain index (index of bone bending strength) were then measured.

2.3.2. Plasma Analyses

Plasma were analyzed for glucose using enzymatic fluorometric analysis and insulin using a rat insulin radioimmunoassay kit (Merck Millipore; Bayswater, VIC, Australia) as previously described (n = 6–11 per group per sex) [38,48]. Fasting plasma glucose and insulin was taken as the average of two-time points (10 and 5 min before injection). First-phase insulin secretion was calculated as the incremental area under the insulin curve between 0 and 5 min after the intra-arterial injection of glucose. Homeostasis model assessment for insulin resistance (HOMA-IR) was then determined [38].

2.4. Statistical Analysis

Data were analyzed using a two-way ANOVA to determine differences between Pup (Standard or Reduced litter size) and Dam (Normal or Altered milk quality/composition) groups. If an interaction was present in the two-way ANOVA, a Tukey's post-hoc test was performed to identify any effect of reducing litter size within each maternal group and any effect altering maternal milk quality/composition has on pup outcomes based on litter size. Where appropriate data analysis was performed on each sex. ANOVA statistical analysis was performed using SPSS Statistics 22 (IBM; St Leonards, NSW, Australia). All data are presented as mean ± SEM and a $p < 0.05$ was assessed as being statistically significant.

3. Results

3.1. Growth to Weaning

As mentioned previously, dams that underwent uteroplacental insufficiency surgery gave birth to pups that were growth restricted. Not surprisingly, however, birth weight in the four experimental groups was not different (as only offspring from Sham-operated dams were utilized in the study) and litter size was decreased in the groups that had their litter size Reduced at birth on PN1 and PN6, with no differences in cannibalism across groups by PN6 (Table 1). From PN6 male (all ages) and female (except for PN14, PN21 and PN28) body weight was decreased (−6% and −7% for males and females, respectively) if they were suckled by a dam with Altered milk quality/composition. Body weight at weaning was not, however, different between groups (PN35; Table 1).

Table 1. Cross-fostered litter size at postnatal day (PN) 6 and body weight from birth (PN1) to weaning (PN35) (n = 12–18 litter averages per group).

Pup-on-Dam		Standard-on-Normal	Standard-on-Altered	Reduced-on-Normal	Reduced-on-Altered	Two-Way ANOVA		
						Pup	Dam	Interaction
Litter Size								
PN1		10.4 ± 0.4	11.5 ± 0.9	5.0 ± 0.0	6.0 ± 0.7	p = 0.0001	ns [1]	ns
PN6		8.9 ± 0.6	8.6 ± 0.8	4.7 ± 0.3	5.25 ± 0.8	p = 0.0001	ns	ns
Cannibalism		1.5 ± 0.5	2.9 ± 1.0	1.0 ± 0.7	0.8 ± 0.3	ns	ns	ns
Body Weight (g)								
PN1	Male	4.17 ± 0.12	4.13 ± 0.10	4.23 ± 0.15	3.99 ± 0.13	ns	ns	ns
	Female	3.87 ± 0.12	3.98 ± 0.06	3.98 ± 0.06	3.94 ± 0.11	ns	ns	ns
PN3	Male	5.62 ± 0.11	5.42 ± 0.15	5.41 ± 0.15	5.12 ± 0.15	ns	ns	ns
	Female	5.38 ± 0.12	5.16 ± 0.20	5.23 ± 0.14	4.91 ± 0.14	ns	ns	ns
PN6	Male	8.54 ± 0.21	8.31 ± 0.23	8.66 ± 0.38	7.65 ± 0.30	ns	p = 0.039	ns
	Female	8.17 ± 0.21	7.91 ± 0.37	8.29 ± 5.14	7.26 ± 0.28	ns	p = 0.049	ns
PN10	Male	14.6 ± 0.3	14.1 ± 0.4	14.6 ± 0.5	13.1 ± 0.5	ns	p = 0.037	ns
	Female	14.2 ± 0.3	13.6 ± 0.6	14.1 ± 0.5	12.7 ± 0.5	ns	p = 0.040	ns
PN14	Male	22.1 ± 0.5	21.2 ± 0.6	21.5 ± 0.7	19.6 ± 1.0	ns	p = 0.043	ns
	Female	21.7 ± 0.4	20.6 ± 0.8	20.6 ± 0.8	19.0 ± 0.9	ns	ns	ns
PN17	Male	27.4 ± 0.6	26.3 ± 0.6	26.9 ± 0.8	24.5 ± 1.2	ns	p = 0.030	ns
	Female	26.9 ± 0.5	25.6 ± 0.7	26.0 ± 0.8	24.1 ± 1.1	ns	p = 0.047	ns
PN21	Male	34.3 ± 0.6	33.6 ± 1.0	34.7 ± 0.7	31.4 ± 1.1	ns	p = 0.020	ns
	Female	33.8 ± 0.6	32.8 ± 1.0	33.5 ± 0.9	31.4 ± 1.0	ns	ns	ns
PN24	Male	43.7 ± 1.1	42.3 ± 0.9	44.3 ± 1.0	39.2 ± 1.4	ns	p = 0.004	ns
	Female	42.1 ± 0.8	40.4 ± 1.1	42.4 ± 1.1	38.5 ± 1.2	ns	p = 0.010	ns
PN28	Male	58.7 ± 1.2	57.4 ± 1.1	59.6 ± 1.2	54.0 ± 1.8	ns	p = 0.011	ns
	Female	54.9 ± 1.0	54.1 ± 1.4	55.0 ± 1.4	51.3 ± 1.3	ns	ns	ns
PN35	Male	87.7 ± 1.7	86.6 ± 1.7	87.1 ± 1.3	82.8 ± 2.8	ns	ns	ns
	Female	79.1 ± 1.2	77.8 ± 1.8	77.0 ± 1.4	75.7 ± 1.8	ns	ns	ns

Data are analysed with a two-way ANOVA reporting differences between Pup (Standard and Reduced litters) and Dam (Normal and Altered milk quality/composition) groups. Data presented as the mean ± SEM, where ns is not significant. [1] ns is not significant.

3.2. Mammary Structure and Maternal Plasma Analysis

An interaction between Dam and Pup was identified in mammary weight, where it was decreased in Reduced litter size pups suckled by Normal milk quality/composition dams compared to Standard litter size pups suckled by Normal milk quality/composition dams (−40%, Figure 2a; Tukey's post-hoc). There was, however, no difference in mammary weight between Standard litter size pups suckled by Altered or Normal milk quality/composition dams; likely due to an increased alveolar number (+25%) in Altered milk quality/composition dams, but not area (Table 2). There were no differences in maternal plasma PTHrP and total calcium between groups (Table 2). An interaction between Dam and Pup was identified in maternal ionic calcium, whereby it was increased in Reduced litter size pups suckled by Normal milk quality/composition dams compared to Standard litter size pups suckled by Normal milk quality/composition dams (+36%) and Reduced litter size pups suckled by Altered milk quality/composition dams (+97%), but was decreased in Reduced litter size pups suckled by

Altered milk quality/composition dams compared to Standard litter size pups suckled by Altered milk quality/composition dams (−40%; Figure 2b; Tukey's post-hoc). Maternal corticosterone (Table 2) along with mammary PTHrP (mRNA and protein; Table 2 and Figure 2c) were not affected by litter size or maternal treatment. A main Dam effect was identified in mammary *a-lactalbumin* gene abundance, where it was decreased in Altered milk quality/composition dams (−51%; Figure 2d). An interaction between Dam and Pup was identified in mammary *b-casein*, which was lower in Standard litter size pups suckled by Altered milk quality/composition dams (−65%) and Reduced litter size pups suckled by Normal milk quality/composition dams (−59%) compared to Standard litter size pups suckled by Normal milk quality/composition dams (Figure 2e; Tukey's post-hoc).

3.3. Pup Plasma Analysis and Milk Consumption

Milk PTHrP tended to be increased in dams suckling Reduced litter size pups (Figure 3a; $p = 0.060$, two-way ANOVA). Milk Na^+/K^+, ionic and total calcium, total protein and lactose were not affected by maternal milk quality/composition or litter size (Table 2). A main Pup effect was identified in pup plasma PTHrP, where it was increased in Reduced litter size pups (+69%; Figure 3b). This finding was in conjunction with main Pup effects in pup ionic calcium and total body calcium, where they were decreased in Reduced litter size pups (−35% and −5%, respectively; Figure 3c,d). Milk intake in both male and female pups were not affected by maternal milk quality/composition or pup litter size (Figure 3e,f).

Table 2. Maternal plasma, mammary and pooled milk composition in the four cross-foster groups on postnatal day 6.

Pup-on-Dam	Standard-on-Normal	Standard-on-Altered	Reduced-on-Normal	Reduced-on-Altered	Two-Way ANOVA		
					Pup	Dam	Interaction
Maternal (n = 4–9 per group)							
Plasma PTHrP (pmol/L)	8.2 ± 0.4	9.1 ± 1.2	10.5 ± 0.8	9.6 ± 0.7	ns[1]	ns	ns
Plasma total calcium (mmol/L)	2.6 ± 0.08	2.6 ± 0.04	2.6 ± 0.05	2.4 ± 0.04	ns	ns	ns
Plasma corticosterone (ng/mL)	593 ± 59	726 ± 106	551 ± 86	553 ± 80	ns	ns	ns
Mammary (n = 3–6 per group)							
Alveolar number	14.5 ± 1.6	15.9 ± 0.8	11.4 ± 1.2	16.5 ± 1.7	ns	$p = 0.038$	ns
Alveolar area	75.2 ± 3.0	75.1 ± 1.9	79.8 ± 6.5	78.5 ± 3.9	ns	ns	ns
PTHrP mRNA	1.0 ± 0.1	0.8 ± 0.3	1.1 ± 0.2	1.4 ± 0.3	ns	ns	ns
Milk (n = 4–8 per group)							
Na/K	7.6 ± 1.2	6.5 ± 1.6	9.8 ± 2.5	6.4 ± 0.9	ns	ns	ns
Ionic calcium (mmol/L)	8.9 ± 0.6	8.7 ± 1.0	7.4 ± 0.9	9.1 ± 0.4	ns	ns	ns
Total calcium (mmol/L)	64.7 ± 4.0	65.8 ± 3.7	60.4 ± 5.6	57.9 ± 2.8	ns	ns	ns
Total protein (mg/L)	28.2 ± 4.4	22.0 ± 3.7	25.4 ± 4.5	28.1 ± 3.6	ns	ns	ns
Lactose (mM)	30.2 ± 7.9	38.4 ± 3.8	29.5 ± 6.0	41.7 ± 10.4	ns	ns	ns

Data are analysed with a two-way ANOVA reporting differences between Pup (Standard and Reduced litters) and Dam (Normal and Altered milk quality/composition) groups. Data presented as the mean ± SEM with sex pooled per litter, where ns is not significant. [1] ns is not significant.

3.4. Milk Fatty Acid Composition

An interaction between Dam and Pup was identified in linoleic acid (LA), arachidonic acid (AA) and total n-6 fatty acids. Specifically, levels of the individual n-6 PUFAs (LA and AA), and total n-6 fatty acid content were increased in milk from Standard litter size pups suckled by Altered milk quality/composition dams (+20% and +21% for LA and total n-6) and Reduced litter size pups suckled by Normal milk quality/composition(+21%, +48% and +24%, respectively) dams compared to Standard litter size pups suckled by Normal milk quality/composition dams (Table 3; Tukey's post-hoc). There was a main Pup effect in levels of the n-3 LC-PUFA docosahexaenoic acid (DHA) which was increased in milk from dams suckling Reduced litter size pups (+31%; Table 3), but there was no difference in total n-3 fatty acid content of the milk between groups. An interaction between Dam and Pup was identified in long chain saturated fatty acid content, where they were decreased in the milk of Reduced litter size pups suckled by Normal milk quality/composition dams compared to Standard litter size pups suckled by Normal milk quality/composition dams (−24%; Table 3, Tukey's post-hoc). A main Pup effect was identified in total LC-MUFA content, where it was increased in the milk of dams suckling Reduced litter pups (+11%). The ratio of n-6:n-3 fatty acids and medium saturated fatty acids were not affected by litter size or maternal milk quality/composition (Table 3). The full list of fatty acids analyzed from the stomach contents are reported in Supplementary Table S1.

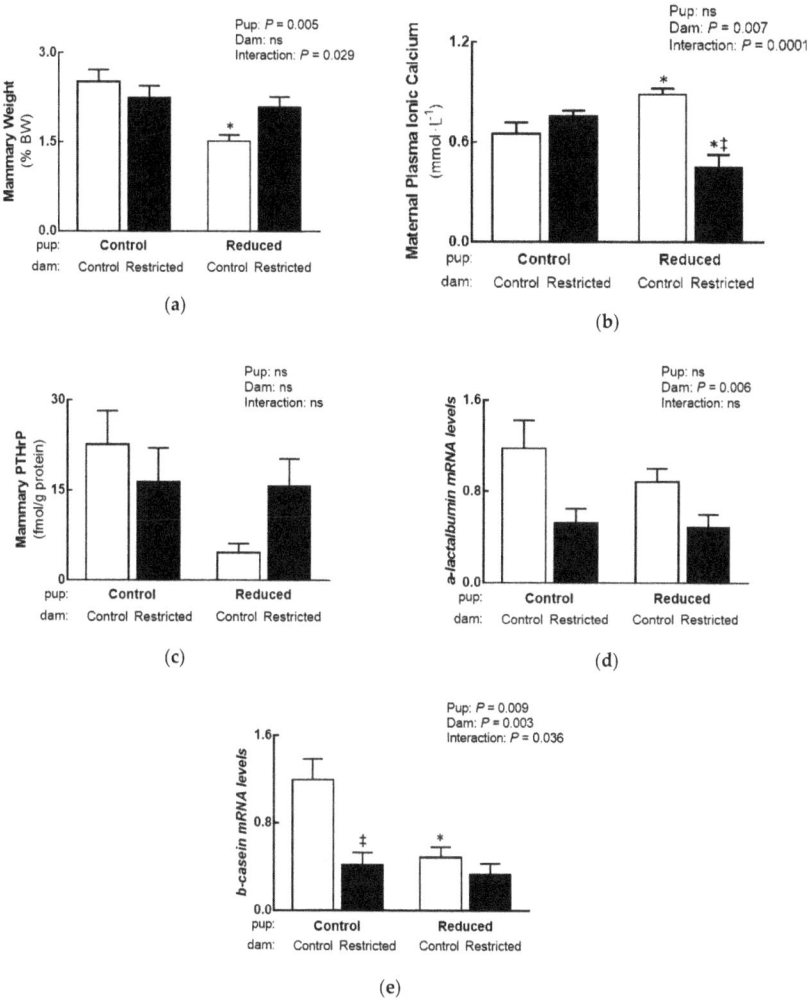

Figure 2. Effect reducing litter size has on maternal and mammary outcomes (n = 5–9 per group). (a) Mammary weight, (b) maternal ionic calcium concentrations and (c) mammary PTHrP protein concentration, (d) a-lactalbumin and (e) b-casein mRNA expression. Data are analysed with a two-way ANOVA reporting differences between Pup (Standard and Reduced litters) and Dam (Normal and Altered milk quality/composition) groups, with a Tukey's post-hoc test used to identify where interactions lie. Data presented as the mean ± SEM, where ns is not significant. Significant differences between Standard and Reduced litter pups are indicated by an asterisk (* $p < 0.05$) and differences between sham operated (Control; Normal milk quality/composition) and uteroplacental insufficiency surgery (Restricted; altered milk quality/composition) dams are indicated with a double dagger (‡ $p < 0.05$). Normal milk quality/composition dams denoted by white open bars and Altered milk quality/composition dams denoted by black closed bars.

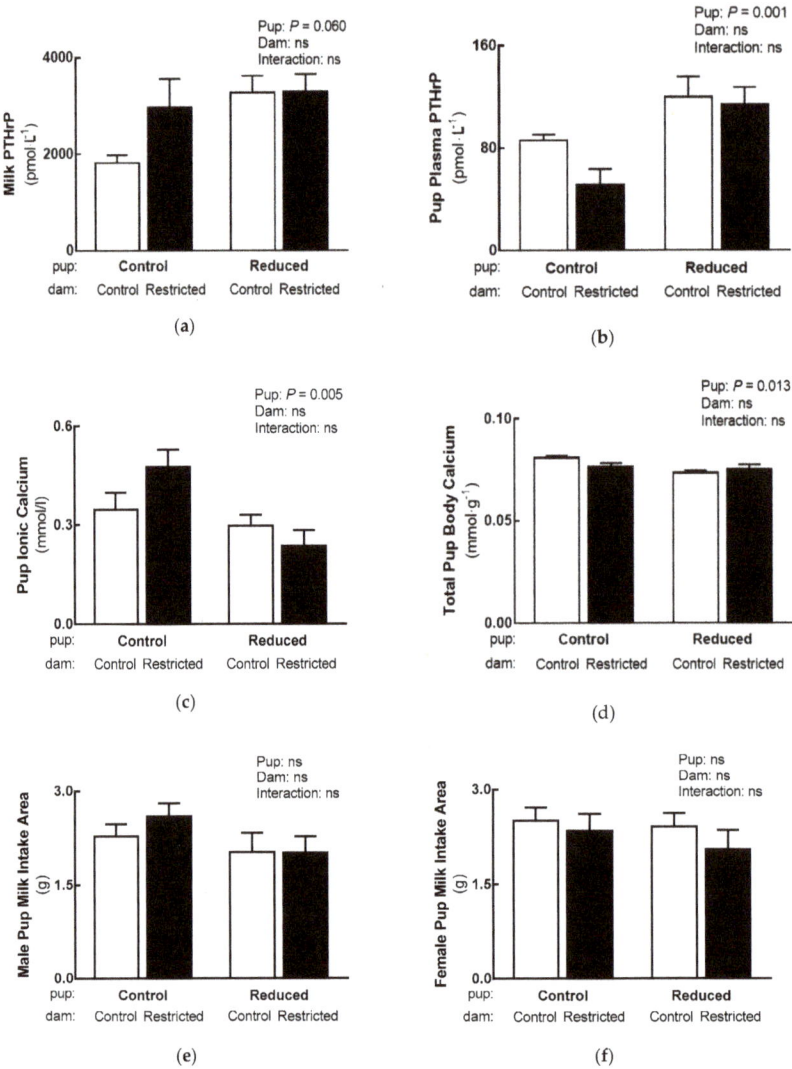

Figure 3. Effect reducing litter size has on pup calcium handling and milk intake (n = 6–10 litter averages per group, where appropriate). (**a**) Milk and (**b**) pup PTHrP concentrations, (**c**) pup ionic calcium concentrations, (**d**) pup total body calcium and (**e**,**f**) pup milk intake. Data are analysed with a two-way ANOVA reporting differences between Pup (Standard and Reduced litters) and Dam (Normal and Altered milk quality/composition) groups. Data presented as the mean ± SEM, where ns is not significant. Normal milk quality/composition dams denoted by white open bars and Altered milk quality/composition dams denoted by black closed bars.

3.5. Adult Health

Body weight at 6 months was decreased in male (−5%), but not female, offspring from Reduced litters, which was further exaggerated by Altered milk quality/composition (−5%) (Table 4). Femur length in both males and females was unaffected by litter size or maternal milk quality/composition (Table 4). An interaction between Pup and Dam was identified in male and female trabecular mineral content, whereby trabecular mineral content was decreased in male (−14%) and female (−12%) Reduced litter size pups suckled by Altered milk quality/composition dams compared to Reduced litter size pups suckled by Normal milk quality/composition counterparts (Figure 4a; Tukey's post-hoc). A main Dam effect was identified in trabecular density in male offspring, where it decreased in offspring suckled by Altered milk quality/composition dams (−6%; Figure 4b). In females, an interaction between Pup and Dam was identified trabecular density, where it was decreased in Reduced litter size pups suckled by Altered milk quality/composition dams (−8%) compared to Reduced litter size pups suckled by Normal milk quality/composition dams (Figure 4b; Tukey's post-hoc). Cortical mineral content was decreased in male offspring who were suckled by Altered milk quality/composition dams (−4%), with no change in females (Table 4). Cortical density was unaffected by litter size or maternal milk quality/composition in both male and female offspring (Table 4). A main Pup effect was identified in bone bending strength in male, but not female offspring, where it was decreased in offspring of Reduced litters (−7%; Figure 4c).

Table 3. Fatty acid composition in stomach contents consumed in milk presented as a cumulative total percentage of fatty acids on postnatal day 6 in the four cross-foster groups (n = 5–7 per group, with n = 1 representing data from 1 pooled litter).

Pup-on-Dam	Standard-on-Normal	Standard-on-Altered	Reduced-on-Normal	Reduced-on-Altered	Two-Way ANOVA		
					Pup	Dam	Interaction
Fatty Acids							
Omega-6 PUFA							
Linoleic (18:2n-6)	11.79 ± 0.25	14.12 ± 0.43 ‡[1]	14.23 ± 0.26 *[2]	13.85 ± 0.50	p = 0.010	p = 0.019	p = 0.002
Arachidonic (20:4n-6)	1.03 ± 0.06	1.33 ± 0.12	1.52 ± 0.08 *	1.35 ± 0.09	p = 0.014	ns [3]	p = 0.023
Total n-6 PUFA	14.4 ± 0.4	17.4 ± 0.7 ‡	17.9 ± 0.4 *	17.1 ± 0.7	p = 0.009	ns	p = 0.003
Omega-3 PUFA							
α-linolenic (18:3n-3)	1.40 ± 0.06	1.48 ± 0.09	1.49 ± 0.05	1.46 ± 0.11	ns	ns	ns
Eicosapentaenoic (20:5n-3)	0.21 ± 0.01	0.17 ± 0.03	0.21 ± 0.02	0.18 ± 0.04	ns	ns	ns
Docosahexaenoic (22:6n-3)	0.41 ± 0.03	0.45 ± 0.06	0.55 ± 0.04	0.58 ± 0.09	p = 0.047	ns	ns
Total n-3 PUFA	2.4 ± 0.09	2.4 ± 0.18	2.7 ± 0.15	2.7 ± 0.21	ns	ns	ns
n-6:n-3	6.07 ± 0.15	7.11 ± 0.27	6.58 ± 0.28	6.63 ± 0.54	ns	ns	ns
Total LC-MUFA	22.4 ± 0.9	22.4 ± 1.4	25.8 ± 0.9	23.9 ± 0.8	p = 0.032	ns	ns
Saturated Fatty Acids							
Medium (6–12)	19.8 ± 1.0	21.1 ± 0.7	21.1 ± 0.7	20.8 ± 0.9	ns	ns	ns
Long (14–20)	40.4 ± 1.1	35.2 ± 2.1	30.9 ± 1.4 *	34.2 ± 1.9	p = 0.006	ns	p = 0.022

Data are analysed with a two-way ANOVA reporting differences between Pup (Standard and Reduced litters) and Dam (Normal and Altered milk quality/composition) groups, with a Tukey's post-hoc test used to identify where interactions lie. Data presented as the mean ± SEM with sex pooled per litter, where ns is not significant. Significant differences between Standard and Reduced litter pups are indicated by an asterisk (* p < 0.05) and differences between sham operated (Control; Normal milk quality/composition) and uteroplacental insufficiency surgery (Restricted; Altered milk quality/composition) dams are indicated with a double dagger (‡ p < 0.05). Also see Table S1 for individual fatty acids. [1] p < 0.05 Standard-on-Altered vs. Standard-on-Normal; [2] p < 0.05 Reduced-on-Normal vs. Standard-on-Normal; [3] ns is not significant.

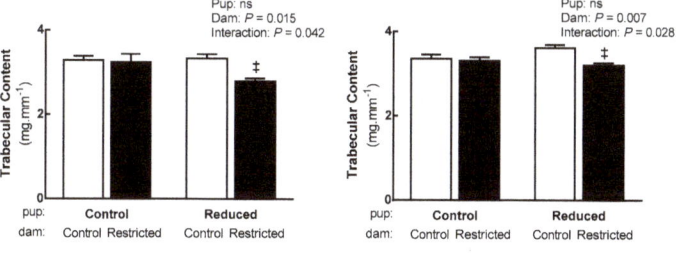

(a)

Figure 4. *Cont.*

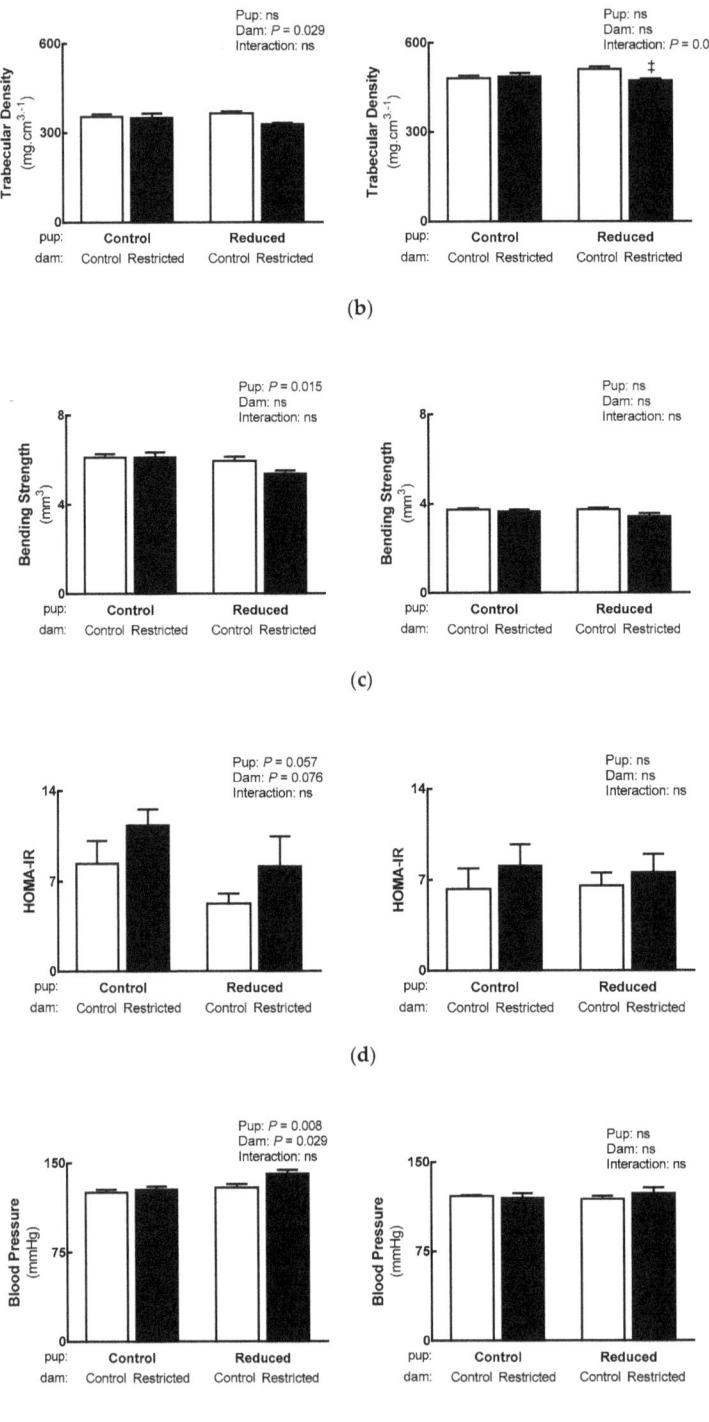

Figure 4. Effect reducing litter size has on adult offspring bone (n = 9–19 per group, with n = 1 representing 1 male and female per litter) and cardiometabolic health (n = 5–11 per group, with n = 1

representing 1 male and female per litter). (**a,b**) Trabecular mineral content and density, (**c**) bone bending strength, (**d**) homeostasis model assessment for insulin resistance (HOMA-IR) and (**e**) blood pressure at 6 months. Data are analysed with a two-way ANOVA reporting differences between Pup (Standard and Reduced litters) and Dam (Normal and Altered milk quality/composition) groups, with a Tukey's post-hoc test used to identify where interactions lie. Data presented as the mean ± SEM, where ns is not significant. Significant differences between sham operated (Control; Normal milk quality/composition) and uteroplacental insufficiency surgery (Restricted; Altered milk quality/composition) dams are indicated with a double dagger (‡ $p < 0.05$). Normal milk quality/composition dams denoted by white open bars and Altered milk quality/composition dams denoted by black closed bars, with data from males on the left-hand side and data from females on the right-hand side.

Fasting glucose at 6 months was not different between groups in either males or females (Table 4). In male, but not female offspring, a main Dam effect was identified in fasting insulin concentrations where it was higher in offspring who had been suckled by Altered milk quality/composition dams and also tended to be lower in Reduced litter size pups (Table 4; $p = 0.070$). Similarly, glucose AUC tended to be higher and first phase insulin tended to be lower in male offspring that were suckling Altered milk quality/composition dams (Table 4; $p = 0.080$ and $p = 0.069$, respectively), with no changes in females. HOMA-IR at 6 months tended to be increased in offspring who had been suckled by Altered milk quality/composition dams ($p = 0.076$) and decreased in Reduced litter size offspring ($p = 0.057$; Figure 4d). Interestingly, a main Pup and Dam effect were observed in male blood pressure, where it was increased both in Reduced litter male offspring compared to Standard litter size offspring (+7%) and in offspring who had been suckled by Altered milk quality/composition dams compared to Normal milk quality/composition dams (+5%; Figure 4e). HOMA-IR and blood pressure were not different between groups in female offspring (Figure 4d,e).

Table 4. Adult offspring physiology in the four cross-foster groups at 6 months ($n = 9$–19 per group for bone analyses and $n = 5$–11 per group for cardiometabolic analyses, with $n = 1$ representing 1 male and female per litter).

Pup-on-Dam		Standard-on-Normal	Standard-on-Altered	Reduced-on-Normal	Reduced-on-Altered	Two-Way ANOVA		
						Pup	Dam	Interaction
Body Weight (g)	Male	386.1 ± 7.0	368.9 ± 2.3	368.4 ± 7.0	351.8 ± 9.5	$p = 0.020$	$p = 0.023$	ns
	Female	239.3 ± 4.1	233.1 ± 6.5	238.3 ± 3.7	226.9 ± 4.7	ns [1]	ns	ns
Bone Parameters								
Femur Length (mm)	Male	37.01 ± 0.12	36.79 ± 0.23	36.61 ± 0.21	36.01 ± 0.16	ns	ns	ns
	Female	33.09 ± 0.14	32.75 ± 0.19	32.69 ± 0.14	32.42 ± 0.15	ns	ns	ns
Cortical content (mg·mm^{-1})	Male	10.67 ± 0.15	10.63 ± 0.28	10.74 ± 0.18	9.93 ± 0.13	ns	$p = 0.005$	ns
	Female	7.87 ± 0.09	7.79 ± 0.11	7.90 ± 0.12	7.49 ± 0.11	ns	ns	ns
Cortical density (mg (mm^3)$^{-1}$)	Male	1409.4 ± 1.8	1413.3 ± 3.7	1412.6 ± 3.8	1407.9 ± 1.9	ns	ns	ns
	Female	1407.9 ± 2.5	1412.4 ± 1.2	1404.8 ± 1.8	1399.1 ± 2.0	ns	ns	ns
Metabolic Parameters								
Fasting plasma glucose (mmol·L^{-1})	Male	5.77 ± 0.21	5.81 ± 0.19	5.52 ± 0.19	5.64 ± 0.27	ns	ns	ns
	Female	5.15 ± 0.18	5.69 ± 0.18	5.50 ± 0.20	5.64 ± 0.22	ns	ns	ns
Fasting plasma insulin (ng·ML^{-1})	Male	1.28 ± 0.24	1.82 ± 0.17	0.91 ± 0.12	1.32 ± 0.33	$p = 0.070$	$p = 0.048$	ns
	Female	1.07 ± 0.24	1.23 ± 0.23	1.09 ± 0.14	1.23 ± 0.21	ns	ns	ns
Glucose AUC	Male	757.3 ± 87.0	795.6 ± 79.8	742.2 ± 67.1	989.6 ± 62.1	ns	$p = 0.080$	ns
	Female	812.9 ± 74.1	701.9 ± 70.1	689.3 ± 52.8	635.2 ± 45.2	ns	ns	ns
First phase insulin AUC	Male	14.25 ± 1.54	10.01 ± 2.08	11.31 ± 1.92	8.53 ± 1.17	ns	$p = 0.069$	ns
	Female	13.91 ± 2.02	11.29 ± 2.05	13.81 ± 1.74	14.01 ± 1.84	ns	ns	ns

Data are analysed with a two-way ANOVA reporting differences between Pup (Standard and Reduced litters) and Dam (Normal and Altered milk quality/composition) groups. Data presented as the mean ± SEM, where ns is not significant. [1] ns is not significant.

4. Discussion

Providing offspring with adequate and appropriate nutrition during lactation is essential for long-term health, with both over- and undernutrition increasing disease susceptibility [14–16]. We have previously demonstrated that both the dam and suckling pup can influence maternal milk composition, and that these effects generally represent an attempt to improve pup growth and development [34]. The current study has demonstrated that reducing litter size at birth, regardless of whether the dam has normal or altered milk quality/composition impairs pup calcium homeostasis, which may reduce bone bending strength in male offspring. Additionally, these adult Reduced litter males appear to have increased cardiovascular disease risk and decreased diabetes risk. Furthermore, altered milk quality/composition (induced by uteroplacental insufficiency surgery) similarly programs poor bone health, high blood pressure and impairs glucose tolerance/insulin sensitivity. These finding are likely due to differential influences of the predetermined litter size at birth [26] and effects on milk quality. This highlights that caution needs to be taken when interpreting animal studies that focus on early life nutrition, as the experimental approach may independently modulate milk nutrition and quality.

4.1. Effects of Reducing Litter Size

It is important to note that as all dams underwent a surgical procedure (sham or ligation) and all litters were cross-fostered the effects reported are as direct consequences of either the maternal surgery (resulting in altered milk quality/composition) or reduced litter size, thus we have adequately accounted for all possible confounders, such as stress, across experimental groups. Additionally, growth restricted pups were not included in the study, only pups born from sham-operated dams that have intact or reduced litter sizes when cross-fostered, which would limit any changes observed in pups suckled by dams that underwent uteroplacental insufficiency surgery to the lactation environment they are exposed to in early postnatal life. Reducing litter size is commonly used in the developmental programming field, however recent studies demonstrate that this can independently program disease susceptibility [27–30]. In the current study we report dynamic changes in pup calcium homeostasis and maternal milk composition as a result of reducing the litter size to 5 pups. Interestingly, despite these changes in milk composition, no changes in postnatal body weight was observed, which suggests that other intrinsic hormonal factors may be responsible for maintaining pup growth that requires further studies. The reduction in mammary gland weight in Normal milk quality/composition dams suckled by Reduced litter pups may be due to the decreased *b-casein* gene expression, which is known to regulate mammary differentiation and is a regulator of milk protein gene expression [49]. However, this likely did not alter alveolar area or number as lactogenic differentiation is completed by PN6 [49]; although studies at earlier postnatal ages may reveal altered mammary structure. Despite no changes in maternal PTHrP, dams suckled by Reduced litter pups had increased milk PTHrP. Previous studies have demonstrated that milk PTHrP is not absorbed by the intestines into the pups circulation [50], suggesting that the increased pup PTHrP concentrations we report are independent of milk PTHrP content. This increased pup PTHrP is likely to stimulate bone resorption, and thus calcium release, in an attempt to increase plasma calcium concentrations, as shown in a previous study [51], but has also been shown to result in decreased postnatal body calcium content [52]. Interestingly, the increased maternal ionic calcium concentrations in Normal milk quality/composition dams suckled by Reduced litter pups are likely an attempt to increase milk calcium concentrations to compensate for this deficit. This, however, did not translate to increased milk calcium content at the time point investigated in this study, highlighting that milk composition needs to be assessed at additional postnatal ages. Nevertheless, the increased milk PTHrP in dams suckling Reduced litter pups would facilitate increased intestinal calcium reabsorption to increase pup plasma calcium concentrations, preventing any further bone breakdown. These data may highlight why the Reduced litter size pups suckled by Normal milk quality/quantity dams do not exhibit the same deficits in adult bone mineral content and density that are observed in the Reduced litter size pups suckled by Altered milk quality/quantity. Despite this, however, Reduced litter male offspring have decreased bending strength, indicative of increased fracture risk and thus poor bone health, suggesting that any compensatory changes in milk composition were not sufficient to fully prevent bone deficits.

In addition to alterations in pup calcium homeostasis in the Reduced litter group, we also report changes in milk fatty acid content, highlighting that Reducing litter size has profound effects on maternal milk quality. Interestingly, milk n-6 fatty acid content was increased in the Reduced litter size pups suckled by Normal milk quality/quantity dams. This is significant, since increased n-6 fatty acid intakes have been associated with increased fat deposition in early life [11]. High intakes of n-6 PUFA during lactation also influence lipid tissue status and modulate metabolic pathways that can lead to diabetes and cardiovascular disease [11]. It is important to note, however, that it is the balance of omega-6 and omega-3 PUFA in the diet that appears to be the more important determinant of physiological effects than levels of either PUFA type alone. In the current study, the n-6:n-3 ratio was not altered, at least at this age, by either litter size or maternal uteroplacental insufficiency surgery. This suggests that the observed changes in milk PUFA composition may not be a major factor contributing to the male onset cardiovascular disease we report, but may instead be due to the decreased post-weaning growth trajectory. Interestingly, reducing litter size increased milk LC-MUFA and DHA (Reduced litter size pups suckled by Normal milk quality/quantity dams only) content, which is known to have beneficial effects on metabolic function during critical periods of development [11]. The improved metabolic health in Reduced litter size pups suckled by Normal milk quality/quantity dams male offspring may also be attributed to increased milk Pentadecanoic acid (15:0) and reduced Palmitic acid (16:0) both of which are known to reduce the risk of type 2 diabetes [53]. This may explain why the Reduced litter size male offspring appear to have a decreased risk of developing diabetes.

4.2. Effects of Altered Milk Quality/Composition

Surprisingly alterations in milk quality/composition, induced by maternal uteroplacental insufficiency surgery, resulted in very few changes in milk composition compared to the large number of changes associated with reducing litter size and slowing offspring postnatal growth. In the current study, we report that dams with altered milk quality/composition have decreased mammary *α-lactalbumin* gene abundance. If this translates to decreased milk α-lactalbumin protein then this may, at least in part, contribute to the poor adult bone health we observed in male offspring reared by Altered milk quality/composition dams. Specifically, as α-lactalbumin binds calcium [54] it is possible that decreased milk α-lactalbumin impairs calcium delivery to the pup, compromising bone mineralization and development. This poor bone health was only apparent in Reduced litter size pups suckled by Altered milk quality/composition dams, likely because the effect was compounded by the decreased maternal ionic concentrations that may have compromised calcium delivery at later lactational ages. Interestingly, a recent study demonstrated that supplementing 6-week old obese diabetic Zucker rats with α-lactalbumin for 13 weeks improves metabolic function [55]; highlighting the benefits of α-lactalbumin on metabolic health. If the inverse is also true, and rats are exposed to decreased α-lactalbumin during development, it is possible that this may contribute to the increased risk of adult metabolic disease we observed in our rats. In addition, the increased n-6 fatty acid intake in Standard litter size pups suckled by Altered milk quality/quantity dams during lactation may also contribute to the increased blood pressure and impaired insulin sensitivity in male offspring [11]. As the female offspring suckled by Altered milk quality/composition dams caught up in body weight prior to weaning they may be protected against developing cardiometabolic disease, due to the benefits of early accelerated growth for long-term cardiometabolic health [36,56–58].

4.3. Study Limitations

Despite our study demonstrating that reducing the litter size from 9 pups to 5 pups alters milk quality/composition and has implications for long-term offspring health, further studies are required to investigate the effects of the more common practice of reducing litter sizes to 8–10 pups from original litter sizes of 12–20 pups, since this also has the potential to influence milk quality/quantity and thus offspring outcomes. Nevertheless, the findings of the current study highlight the potential impact reducing litter size, independent of other neonatal factors, has on programming offspring outcomes. An important factor that was not taken into consideration in the present study is the impact of offspring sex on milk composition, due to difficulties in controlling for this in our large litter bearing animal model. Indeed, several epidemiological and experimental (with smaller litter bearing animals) studies have well demonstrated differences in milk composition between male and female infants [59,60]; demonstrating that infant sex has a significant impact on milk quality. This highlights the need for well-controlled human and animal studies to identify alterations in milk composition between infant sexes in several different pregnancy complications.

A limitation of the current study is that maternal behavior was not evaluated throughout lactation. This is particularly important, as the quality and quantity of maternal care during lactation can impact on offspring behavioral, endocrine and neural development (see review by Curley and Champagne [61]). Specifically, a recent study demonstrated that dams whose litters were reduced to 3 pups (to induce postnatal overnutrition) have improved maternal care, characterize by increased time devoted to arched nursing and licking pups [62]. Not surprisingly the pups had early accelerated growth and were overweight by PN60, as indicated by increased adiposity, and were hyperglycemic and hyperleptinemic [62]. These findings are however difficult to compare directly to the results of the current study where the Reduced litter pups had decreased postnatal growth, which suggests that the extent of litter size reduction in the current study is unlikely to have resulted in improved maternal care.

5. Conclusions

This study demonstrates that reducing pup litter size and altered maternal milk quality/composition differentially program poor adult offspring health, which is likely due to altered milk quality. Specifically, reducing the litter size alters milk composition, impairs pup calcium homeostasis and programs poor bone health, which likely contributes to the lower diabetes risk and poor cardiovascular health. Alterations in maternal milk quality/composition, on the other hand, programs poor adult bone and cardiovascular health and reduces glucose tolerance/insulin sensitivity. This study highlights the need for appropriate controls in developmental research to clearly ascertain phenotypes in the model. Importantly, controls implemented to standardize outcomes across treatment groups may independently program disease susceptibility, which limit their translatability. Therefore, care must be taken in interpreting findings from studies that standardize litter size as it may mask subtle effects on cardiometabolic health.

Supplementary Materials: The following are available online at http://www.mdpi.com/2072-6643/11/1/118/s1, Table S1: Fatty acid composition in stomach contents on postnatal day 6.

Author Contributions: Conceptualization, M.E.W. and K.M.M.; Methodology, M.E.W. and K.M.M.; Validation, J.F.B., R.O., T.R., B.S.M., K.M.M. and M.E.W.; Formal analysis, J.F.B., R.O., T.R., B.S.M., K.M.M. and M.E.W.; Investigation, J.F.B., R.O. and T.R.; Resources, M.E.W. and K.M.M; Data curation, J.F.B., R.O., T.R. and M.E.W.; Writing—original draft preparation, J.F.B.; Writing—review and editing, R.O., T.R., B.S.M., K.M.M. and M.E.W.; Visualization, J.F.B., B.S.M. and M.E.W.; Supervision, M.E.W. and K.M.M.; Project administration, M.E.W.; Funding acquisition, M.E.W. and K.M.M.

Funding: This research was funded by the National Health and Medical Research Council (NHMRC) of Australia, grant numbers 208948 (M.E.W.) and 400003 (M.E.W. and K.M.M.). J.F.B. holds a Faculty of Medicine, Dentistry and Health Sciences Postdoctoral Fellowship at the University of Melbourne. K.M.M. was funded by a NHMRC Senior Research Fellowship.

Acknowledgments: We thank Ursula Lorenc for her technical assistance and Jane Moseley for her assistance in this project. We also thank Andrew Sinclair for his assistance with the fatty acid analysis, Kevin Nicholas for his assistance with mammary gland morphology, and Andrew Jefferies and Kerryn Westcott for their assistance with the animal work involved in this project.

Conflicts of Interest: The authors declare no conflict of interest. The funders had no role in the design of the study; in the collection, analyses, or interpretation of data; in the writing of the manuscript, or in the decision to publish the results.

References

1. Barker, D.J.; Winter, P.D.; Osmond, C.; Margetts, B.; Simmonds, S.J. Weight in infancy and death from ischaemic heart disease. *Lancet* **1989**, *2*, 577–580. [CrossRef]
2. Barker, D.J.P.; Godfrey, K.M.; Osmond, C.; Bull, A. The relation of fetal length, ponderal index and head circumference to blood pressure and the risk of hypertension in adult life. *Paediatr. Perinat. Epidemiol.* **1992**, *6*, 35–44. [CrossRef] [PubMed]
3. Gluckman, P.D.; Harding, J.E. Fetal growth retardation: Underlying endocrine mechanisms and postnatal consequences. *Acta Paediatr. Scand.* **1997**, *422*, 69–72. [CrossRef]
4. Wlodek, M.E.; Owens, J.A.; Siebel, A.L.; Moritz, K. Reduced nephron endowment and hypertension emerge following placental restriction in the rat. *Pediatr. Res.* **2005**, *58*, 1024.
5. Painter, R.C.; Roseboom, T.J.; Bleker, O.P. Prenatal exposure to the dutch famine and disease in later life: An overview. *Reprod. Toxicol.* **2005**, *20*, 345–352. [CrossRef]
6. Jaquiery, A.L.; Oliver, M.H.; Honeyfield-Ross, M.; Harding, J.E.; Bloomfield, F.H. Periconceptional undernutrition in sheep affects adult phenotype only in males. *J. Nutr. Metab.* **2012**, *2012*, 123610. [CrossRef]
7. Sokol, R.J.; Delaney-Black, V.; Nordstrom, B. Fetal alcohol spectrum disorder. *JAMA* **2003**, *290*, 2996–2999. [CrossRef]
8. Gray, S.P.; Denton, K.M.; Cullen-McEwen, L.; Bertram, J.F.; Moritz, K.M. Prenatal exposure to alcohol reduces nephron number and raises blood pressure in progeny. *J. Am. Soc. Nephrol.* **2010**, *21*, 1891–1902. [CrossRef]
9. Virk, J.; Li, J.; Vestergaard, M.; Obel, C.; Kristensen, J.K.; Olsen, J. Prenatal exposure to bereavement and type-2 diabetes: A danish longitudinal population based study. *PLoS ONE* **2012**, *7*, e43508.
10. Singh, R.R.; Cuffe, J.S.; Moritz, K.M. Short- and long-term effects of exposure to natural and synthetic glucocorticoids during development. *Clin. Exp. Pharmacol. Physiol.* **2012**, *39*, 979–989. [CrossRef]
11. Mennitti, L.V.; Oliveira, J.L.; Morais, C.A.; Estadella, D.; Oyama, L.M.; Oller do Nascimento, C.M.; Pisani, L.P. Type of fatty acids in maternal diets during pregnancy and/or lactation and metabolic consequences of the offspring. *J. Nutr. Biochem.* **2015**, *26*, 99–111. [CrossRef] [PubMed]
12. Ailhaud, G.; Massiera, F.; Weill, P.; Legrand, P.; Alessandri, J.M.; Guesnet, P. Temporal changes in dietary fats: Role of n-6 polyunsaturated fatty acids in excessive adipose tissue development and relationship to obesity. *Prog. Lipid Res.* **2006**, *45*, 203–236. [CrossRef] [PubMed]
13. Gallo, L.A.; Tran, M.; Moritz, K.M.; Wlodek, M.E. Developmental programming: Variations in early growth and adult disease. *Clin. Exp. Pharmacol. Physiol.* **2013**, *40*, 795–802. [CrossRef] [PubMed]
14. Pomar, C.A.; van, N.R.; Sanchez, J.; Pico, C.; Keijer, J.; Palou, A. Maternal consumption of a cafeteria diet during lactation in rats leads the offspring to a thin-outside-fat-inside phenotype. *Int. J. Obes. (Lond.)* **2017**, *41*, 1279–1287. [CrossRef] [PubMed]
15. Lozano, G.; Elmaghrabi, A.; Salley, J.; Siddique, K.; Gattineni, J.; Baum, M. Effect of prenatal programming and postnatal rearing on glomerular filtration rate in adult rats. *Am. J. Physiol. Ren. Physiol.* **2015**, *308*, F411–F419. [CrossRef] [PubMed]
16. Siddique, K.; Guzman, G.L.; Gattineni, J.; Baum, M. Effect of postnatal maternal protein intake on prenatal programming of hypertension. *Reprod. Sci.* **2014**, *21*, 1499–1507. [CrossRef] [PubMed]
17. Dickinson, H.; Moss, T.J.; Gatford, K.L.; Moritz, K.M.; Akison, L.; Fullston, T.; Hryciw, D.H.; Maloney, C.A.; Morris, M.J.; Wooldridge, A.L.; et al. A review of fundamental principles for animal models of dohad research: An australian perspective. *J. Dev. Orig. Health Dis.* **2016**, *7*, 449–472. [CrossRef] [PubMed]
18. De Almeida, F.J.; Duque-Guimaraes, D.; Carpenter, A.A.; Loche, E.; Ozanne, S.E. A post-weaning obesogenic diet exacerbates the detrimental effects of maternal obesity on offspring insulin signaling in adipose tissue. *Sci. Rep.* **2017**, *7*, 44949. [CrossRef]
19. Galyon, K.D.; Farshidi, F.; Han, G.; Ross, M.G.; Desai, M.; Jellyman, J.K. Maternal bisphenol a exposure alters rat offspring hepatic and skeletal muscle insulin signaling protein abundance. *Am. J. Obstet. Gynecol.* **2017**, *216*, 290.e1–290.e9. [CrossRef]
20. Maniam, J.; Antoniadis, C.P.; Wang, K.W.; Morris, M.J. Early life stress induced by limited nesting material produces metabolic resilience in response to a high-fat and high-sugar diet in male rats. *Front. Endocrinol. (Lausanne)* **2015**, *6*, 138. [CrossRef]
21. Firth, E.C.; Gamble, G.D.; Cornish, J.; Vickers, M.H. Neonatal leptin treatment reverses the bone-suppressive effects of maternal undernutrition in adult rat offspring. *Sci. Rep.* **2017**, *7*, 7686. [CrossRef] [PubMed]

22. Dasinger, J.H.; Intapad, S.; Backstrom, M.A.; Carter, A.J.; Alexander, B.T. Intrauterine growth restriction programs an accelerated age-related increase in cardiovascular risk in male offspring. *Am. J. Physiol. Ren. Physiol.* **2016**, *311*, F312–F319. [CrossRef] [PubMed]
23. Zambrano, E.; Sosa-Larios, T.; Calzada, L.; Ibanez, C.A.; Mendoza-Rodriguez, C.A.; Morales, A.; Morimoto, S. Decreased basal insulin secretion from pancreatic islets of pups in a rat model of maternal obesity. *J. Endocrinol.* **2016**, *231*, 49–57. [CrossRef] [PubMed]
24. Niu, Y.; Herrera, E.A.; Evans, R.D.; Giussani, D.A. Antioxidant treatment improves neonatal survival and prevents impaired cardiac function at adulthood following neonatal glucocorticoid therapy. *J. Physiol.* **2013**, *591*, 5083–5093. [CrossRef] [PubMed]
25. Shah, A.; Reyes, L.M.; Morton, J.S.; Fung, D.; Schneider, J.; Davidge, S.T. Effect of resveratrol on metabolic and cardiovascular function in male and female adult offspring exposed to prenatal hypoxia and a high-fat diet. *J. Physiol.* **2016**, *594*, 1465–1482. [CrossRef] [PubMed]
26. Chahoud, I.; Paumgartten, F.J. Influence of litter size on the postnatal growth of rat pups: Is there a rationale for litter-size standardization in toxicity studies? *Environ. Res.* **2009**, *109*, 1021–1027. [CrossRef] [PubMed]
27. Wlodek, M.E.; Westcott, K.; Siebel, A.L.; Owens, J.A.; Moritz, K.M. Growth restriction before or after birth reduces nephron number and increases blood pressure in male rats. *Kidney Int.* **2008**, *74*, 187–195. [CrossRef] [PubMed]
28. Tare, M.; Parkington, H.C.; Bubb, K.J.; Wlodek, M.E. Uteroplacental insufficiency and lactational environment separately influence arterial stiffness and vascular function in adult male rats. *Hypertension* **2012**, *60*, 378–386. [CrossRef]
29. Tran, M.; Young, M.E.; Jefferies, A.J.; Hryciw, D.H.; Ward, M.M.; Fletcher, E.L.; Wlodek, M.E.; Wadley, G.D. Uteroplacental insufficiency leads to hypertension, but not glucose intolerance or impaired skeletal muscle mitochondrial biogenesis, in 12-month-old rats. *Physiol. Rep.* **2015**, *3*, e12556. [CrossRef]
30. O'Dowd, R.; Kent, J.C.; Moseley, J.M.; Wlodek, M.E. Effects of uteroplacental insufficiency and reducing litter size on maternal mammary function and postnatal offspring growth. *Am. J. Physiol. Regul. Integr. Comp. Physiol.* **2008**, *294*, R539–R548. [CrossRef]
31. Romano, T.; Wark, J.D.; Owens, J.A.; Wlodek, M.E. Prenatal growth restriction and postnatal growth restriction followed by accelerated growth independently program reduced bone growth and strength. *Bone* **2009**, *45*, 132–141. [CrossRef] [PubMed]
32. Habbout, A.; Guenancia, C.; Lorin, J.; Rigal, E.; Fassot, C.; Rochette, L.; Vergely, C. Postnatal overfeeding causes early shifts in gene expression in the heart and long-term alterations in cardiometabolic and oxidative parameters. *PLoS ONE* **2013**, *8*, e56981. [CrossRef] [PubMed]
33. Habbout, A.; Delemasure, S.; Goirand, F.; Guilland, J.C.; Chabod, F.; Sediki, M.; Rochette, L.; Vergely, C. Postnatal overfeeding in rats leads to moderate overweight and to cardiometabolic and oxidative alterations in adulthood. *Biochimie* **2012**, *94*, 117–124. [CrossRef] [PubMed]
34. Briffa, J.F.; O'Dowd, R.; Moritz, K.M.; Romano, T.; Jedwab, L.R.; McAinch, A.J.; Hryciw, D.H.; Wlodek, M.E. Uteroplacental insufficiency reduces rat plasma leptin concentrations and alters placental leptin transporters: Ameliorated with enhanced milk intake and nutrition. *J. Physiol.* **2017**, *595*, 3389–3407. [CrossRef] [PubMed]
35. Wlodek, M.E.; Westcott, K.T.; O'Dowd, R.; Serruto, A.; Wassef, L.; Moritz, K.M.; Moseley, J.M. Uteroplacental restriction in the rat impairs fetal growth in association with alterations in placental growth factors including pthrp. *Am. J. Physiol. Regul. Integr. Comp. Physiol.* **2005**, *288*, R1620–R1627. [CrossRef] [PubMed]
36. Wlodek, M.E.; Mibus, A.; Tan, A.; Siebel, A.L.; Owens, J.A.; Moritz, K.M. Normal lactational environment restores nephron endowment and prevents hypertension after placental restriction in the rat. *J. Am. Soc. Nephrol.* **2007**, *18*, 1688–1696. [CrossRef] [PubMed]
37. Moritz, K.M.; Mazzuca, M.Q.; Siebel, A.L.; Mibus, A.; Arena, D.; Tare, M.; Owens, J.A.; Wlodek, M.E. Uteroplacental insufficiency causes a nephron deficit, modest renal insufficiency but no hypertension with ageing in female rats. *J. Physiol.* **2009**, *587*, 2635–2646. [CrossRef]
38. Siebel, A.L.; Mibus, A.; De Blasio, M.J.; Westcott, K.T.; Morris, M.J.; Prior, L.; Owens, J.A.; Wlodek, M.E. Improved lactational nutrition and postnatal growth ameliorates impairment of glucose tolerance by uteroplacental insufficiency in male rat offspring. *Endocrinology* **2008**, *149*, 3067–3076. [CrossRef]
39. Wlodek, M.E.; Westcott, K.T.; Ho, P.W.M.; Serruto, A.; Di Nicolantonio, R.; Farrugia, W.; Moseley, J.M. Reduced fetal, placental, and amniotic fluid pthrp in the growth-restricted spontaneously hypertensive rat. *Am. J. Physiol. Regul. Integr. Comp. Physiol.* **2000**, *279*, R31–R38. [CrossRef]

40. Wlodek, M.E.; Westcott, K.T.; Serruto, A.; O'Dowd, R.; Wassef, L.; Ho, P.W.M.; Moseley, J.M. Impaired mammary function and parathyroid hormone-related protein during lactation in growth-restricted spontaneously hypertensive rats. *J. Endocrinol.* **2003**, *177*, 233–245. [CrossRef]
41. Wlodek, M.E.; Ho, P.W.M.; Rice, G.E.; Moseley, J.M.; Martin, T.J.; Brennecke, S.P. Parathyroid hormone-related protein (pthrp) concentrations in human amniotic fluid during gestation and at the time of labour. *Reprod. Fertil. Dev.* **1995**, *7*, 1509–1513. [CrossRef] [PubMed]
42. Wlodek, M.E.; Ceranic, V.; O'Dowd, R.; Westcott, K.T.; Siebel, A.L. Maternal progesterone treatment rescues the mammary impairment following uteroplacental insufficiency and improves postnatal pup growth in the rat. *Reprod. Sci.* **2009**, *16*, 380–390. [CrossRef]
43. LePage, G.; Roy, C.C. Direct transesterification of all lipid classes in a one-step reaction. *J. Lipid Res.* **1986**, *27*, 114–120. [PubMed]
44. Weisinger, H.S.; Vingrys, A.; Bui, B.; Sinclair, A.J. Effects of dietary n-3 fatty acid deficiency and repletion in the guinea pig retina. *Investig. Ophthalmol. Vis. Sci.* **1999**, *40*, 327–338.
45. Anevska, K.; Gallo, L.A.; Tran, M.; Jefferies, A.J.; Wark, J.D.; Wlodek, M.E.; Romano, T. Pregnant growth restricted female rats have bone gains during late gestation which contributes to second generation adolescent and adult offspring having normal bone health. *Bone* **2015**, *74*, 199–207. [CrossRef] [PubMed]
46. Romano, T.; Wark, J.D.; Wlodek, M.E. Developmental programming of bone deficits in growth-restricted offspring. *Reprod. Fertil. Dev.* **2014**, *27*, 823–833. [CrossRef]
47. Anevska, K.; Cheong, J.N.; Wark, J.D.; Wlodek, M.E.; Romano, T. Maternal stress does not exacerbate long-term bone deficits in female rats born growth restricted, with differential effects on offspring bone health. *Am. J. Physiol. Regul. Integr. Comp. Physiol.* **2018**, *314*, R161–R170. [CrossRef]
48. Wadley, G.D.; Siebel, A.L.; Cooney, G.J.; McConell, G.K.; Wlodek, M.E.; Owens, J.A. Uteroplacental insufficiency and reducing litter size alters skeletal muscle mitochondrial biogenesis in a sex specific manner in the adult rat. *Am. J. Physiol. Endocrinol. Metab.* **2008**, *294*, E861–E869. [CrossRef]
49. Rijnkels, M.; Kabotyanski, E.; Montazer-Torbati, M.B.; Hue, B.C.; Vassetzky, Y.; Rosen, J.M.; Devinoy, E. The epigenetic landscape of mammary gland development and functional differentiation. *J. Mammary Gland Biol. Neoplasia* **2010**, *15*, 85–100. [CrossRef]
50. Kukreja, S.C.; D'Anza, J.J.; Melton, M.E.; Wimbiscus, S.A.; Grill, V.; Martin, T.J. Lack of effects of neutralisation of parathyroid hormone-related protein on calcium homeostasis in neonatal mice. *J. Bone Miner. Res.* **1991**, *6*, 1197–1201. [CrossRef]
51. Kovacs, C.S. Calcium metabolism during pregnancy and lactation. *J. Mammary Gland Biol. Neoplasia* **2005**, *10*, 105–118. [CrossRef] [PubMed]
52. Kovacs, C.S. Bone development and mineral homeostasis in the fetus and neonate: Roles of the calciotropic and phosphotropic hormones. *Physiol. Rev.* **2014**, *94*, 1143–1218. [CrossRef] [PubMed]
53. Hirahatake, K.M.; Slavin, J.L.; Maki, K.C.; Adams, S.H. Associations between dairy foods, diabetes, and metabolic health: Potential mechanisms and future directions. *Metabolism* **2014**, *63*, 618–627. [CrossRef]
54. Permyakov, E.A.; Berliner, L.J. Alpha-lactalbumin: Structure and function. *FEBS Lett.* **2000**, *473*, 269–274. [CrossRef]
55. Gregersen, S.; Bystrup, S.; Overgaard, A.; Jeppesen, P.B.; Sonderstgaard Thorup, A.C.; Jensen, E.; Hermansen, K. Effects of whey proteins on glucose metabolism in normal wistar rats and zucker diabetic fatty (zdf) rats. *Rev. Diabet. Stud.* **2013**, *10*, 252–269. [CrossRef] [PubMed]
56. Eriksson, J.; Forsen, T.; Tuomilehto, J.; Osmond, C.; Barker, D. Fetal and childhood growth and hypertension in adult life. *Hypertension* **2000**, *36*, 790–794. [CrossRef]
57. Lucas, A.; Fewtrell, M.S.; Davies, P.S.; Bishop, N.J.; Clough, H.; Cole, T.J. Breastfeeding and catch-up growth in infants born small for gestational age. *Acta Paediatr.* **1997**, *86*, 564–569. [CrossRef]
58. Eriksson, J.G.; Forsén, T.; Tuomilehto, J.; Osmond, C.; Barker, D.J.P. Early growth and coronary heart disease in later life: Longitudinal study. *BMJ* **2001**, *322*, 948–953. [CrossRef]
59. Galante, L.; Milan, A.M.; Reynolds, C.M.; Cameron-Smith, D.; Vickers, M.H.; Pundir, S. Sex-specific human milk composition: The role of infant sex in determining early life nutrition. *Nutrients* **2018**, *10*, 1194. [CrossRef]
60. Thakkar, S.K.; Giuffrida, F.; Cristina, C.H.; De Castro, C.A.; Mukherjee, R.; Tran, L.A.; Steenhout, P.; Lee le, Y.; Destaillats, F. Dynamics of human milk nutrient composition of women from singapore with a special focus on lipids. *Am. J. Hum. Biol.* **2013**, *25*, 770–779. [CrossRef]

61. Curley, J.P.; Champagne, F.A. Influence of maternal care on the developing brain: Mechanisms, temporal dynamics and sensitive periods. *Front. Neuroendocrinol.* **2016**, *40*, 52–66. [CrossRef] [PubMed]
62. Enes-Marques, S.; Giusti-Paiva, A. Litter size reduction accentuates maternal care and alters behavioral and physiological phenotypes in rat adult offspring. *J. Physiol. Sci.* **2018**, *68*, 789–798. [CrossRef] [PubMed]

© 2019 by the authors. Licensee MDPI, Basel, Switzerland. This article is an open access article distributed under the terms and conditions of the Creative Commons Attribution (CC BY) license (http://creativecommons.org/licenses/by/4.0/).

Review

The Good, the Bad, and the Ugly of Pregnancy Nutrients and Developmental Programming of Adult Disease

Chien-Ning Hsu [1,2] and You-Lin Tain [3,4,]*

1. Department of Pharmacy, Kaohsiung Chang Gung Memorial Hospital, Kaohsiung 833, Taiwan; chien_ning_hsu@hotmail.com
2. School of Pharmacy, Kaohsiung Medical University, Kaohsiung 807, Taiwan
3. Department of Pediatrics, Kaohsiung Chang Gung Memorial Hospital and Chang Gung University College of Medicine, Kaohsiung 833, Taiwan
4. Institute for Translational Research in Biomedicine, Kaohsiung Chang Gung Memorial Hospital and Chang Gung University College of Medicine, Kaohsiung 833, Taiwan
* Correspondence: tainyl@hotmail.com; Tel.: +886-975-056-995; Fax: +886-7733-8009

Received: 21 March 2019; Accepted: 17 April 2019; Published: 20 April 2019

Abstract: Maternal nutrition plays a decisive role in developmental programming of many non-communicable diseases (NCDs). A variety of nutritional insults during gestation can cause programming and contribute to the development of adult-onset diseases. Nutritional interventions during pregnancy may serve as reprogramming strategies to reverse programming processes and prevent NCDs. In this review, firstly we summarize epidemiological evidence for nutritional programming of human disease. It will also discuss evidence from animal models, for the common mechanisms underlying nutritional programming, and potential nutritional interventions used as reprogramming strategies.

Keywords: developmental origins of health and disease (DOHaD); gut microbiota; non-communicable disease; nutrient-sensing signal; nutrition; oxidative stress; pregnancy; reprogramming

---

1. Introduction

Maternal nutrition plays an essential role in fetal growth and development. It has long been known that adverse nutritional conditions during pregnancy may permanently change the structure and function of specific organs in the offspring, leading to many adult diseases. This concept is now currently referred to as the Developmental Origins of Health and Disease (DOHaD) [1]. Conversely, the DOHaD concept provides a strategy to reverse programming processes by shifting the therapeutic interventions from adulthood to fetal or infantile stage, before clinical phenotype becomes evident [2,3]. Nutritional interventions during pregnancy have started to gain importance as a reprogramming strategy to prevent DOHaD-associated diseases [4–6].

Nutritional exposure at early life is particularly important, as the plasticity of developing organs that shape the way in which the body reacts to challenges in later life. This review highlights evidence for the impact of nutritional programming on offspring health and the role of nutritional interventions as a reprogramming strategy in the emerging area of DOHaD research.

2. Nutritional Programming of Health and Disease: Good, Bad, or Ugly?

The term nutritional programming describes the process through which exposure to early-life nutritional stimuli brings about morphological changes or functional adaption of the offspring [7,8]. The long-term consequences of nutritional insults are variable and depend on several factors such as the type of nutrient, exposure duration and intensity, species, sex, and the critical time-windows of development during which it is applied. Nutritional programming is emerging as a critical risk factor for a number of non-communicable diseases (NCDs), including hypertension, cardiovascular disease, diabetes, obesity, allergic diseases, kidney disease, neurocognitive impairments, nonalcoholic fatty liver disease (NAFLD), and metabolic syndrome [3–10]. NCDs constitute the main cause of death all over the world. Although NCDs are generally preventable, current approaches are obviously insufficient.

Conversely, nutritional programming can also be advantageous. From an evolutionary perspective, developmental plasticity seems to represent an adaptive process. Developmental plasticity is beneficial for fitness, despite such benefits may come at a cost to health outcomes. Developmental plasticity can offer a survival advantage to the offspring based on evolutionary grounding [11]. Also, developmental plasticity is beneficial when environmental conditions change within generations [11]. Importantly, several reports suggest, at least in animal models, that developmental programming of adult disease is potentially reversible by nutritional interventions during the period of developmental plasticity. In genetic models of hypertension, early-life nutritional interventions can improve cardiovascular outcomes in adult offspring [12,13]. The identification of critical time-windows and specific nutritional interventions is thus a promising way to explore in order to offer novel reprogramming strategies.

Considering the good and bad sides of nutritional programming during pregnancy, there is still an ugly side of it. Currently, little information exists on certain nutrients for women with pre-existing deficiencies, as the dietary reference intakes are established for healthy individuals. Almost all dietary reference intake reports lack long-term offspring outcome data to accurately inform recommendations for women in the pregnancy stage. However, currently in human studies it is difficult to confirm causation linking maternal nutrition status with phenotypes in offspring. Also, these cohorts do not identify molecular mechanisms by which the phenotype is generated and aid in developing specific nutritional interventions for a wide spectrum of adult diseases. As a consequence of ethical considerations concerning what is achievable or not in human studies, animal models are essential. It is for this reason that much of our knowledge of the type of nutritional insults driving the programming process, the critical time-window of vulnerability for nutritional insults, potential mechanisms underpinning nutritional programming, and reprogramming strategy largely come from studies in animal models.

On the contrary, there are some limitations of animal models when translating into replications in human trials. Possible problems include different models with varying similarity to the human conditions, variability in animals for study, follow-up duration may not correspond to disease latency in humans, and outcome measures have uncertain relevance to the human conditions [14].

Despite interesting results using nutritional interventions to prevent various adult diseases have been obtained from animal models, many challenges still lie ahead for successful translation of promising animal therapies to humans. As nutrients do not drive their programming effect independently from each other, key questions for future research include what are the nutrient–nutrient interactions, nutrient–drug interactions, and nutrient–environment interactions that can affect the programming power on pregnancy and offspring outcomes.

A schematic summarizing the good, the bad, and the ugly sides of nutritional programming involved in the developmental programming of adult diseases is presented in Figure 1.

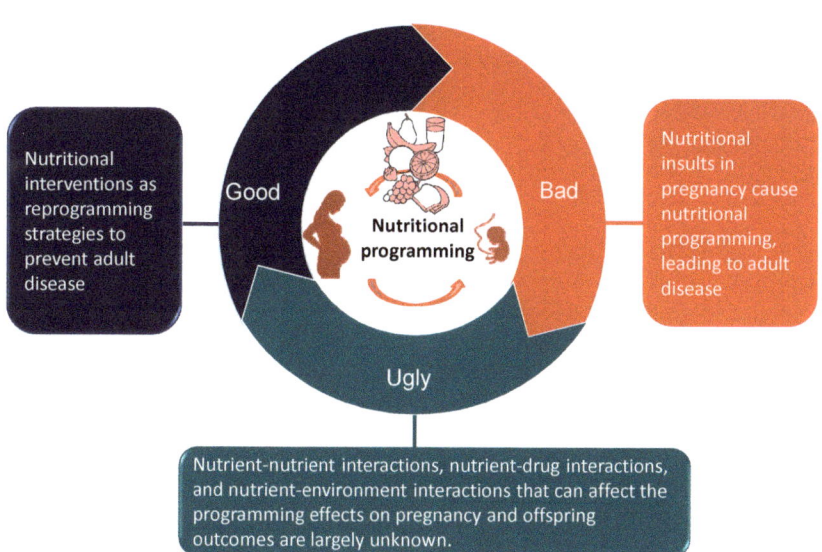

Figure 1. Schematic illustration of the good, the bad, and the ugly sides of nutritional programming involved in developmental programming of health and disease.

3. Epidemiological Evidence for Nutritional Programming of Human Disease

Excessive or insufficient consumption of a specific nutrient has been linked to developmental programming of a variety of NCDs [15–27]. Table 1 mainly summarizes cohort studies documenting adverse offspring outcomes in response to both undernutrition and overnutrition during pregnancy [15–27]. Although low birth weight is considered as a surrogate marker of early-life nutrition and numerous studies of maternal under- and over-nutrition have revealed its association with offspring health [28], we have restricted this review only to adverse outcomes starting from childhood.

First, several famine cohort studies consistently showed that offspring exposed to gestational famine are more vulnerable to metabolic syndrome-related disorders [15–18]. One famous example is the Dutch famine study, which demonstrated that undernutrition in pregnancy is associated with increased risk of adult offspring developing coronary heart disease, hyperlipidemia, obesity, obstructive airways disease, kidney disease, and hypertension [16,18]. The Dutch famine study also showed that offspring exposed to famine in early gestation are prone to developing coronary heart disease, hyperlipidemia, and obesity. Exposure in mid gestation is associated with obstructive airways disease and microalbuminuria. These findings suggest that the timing of undernutrition in pregnancy predicts which organ system is reprogrammed and leads to the development of adult diseases [16]. However, no such consequence was observed in the Leningrad siege famine study [29]. The dissimilar effects of exposure to the two famines suggest the importance of the timing during gestation and the organs and systems developing during that critical time-window. A recent meta-analysis of 20 studies reported that fetal famine exposure may increase the risk of overweight and obesity [30].

Overnutrition is a form of malnutrition in which the intake of nutrients or a specific nutrient is oversupplied, especially in unbalanced proportions. Only a small number of studies have examined the contribution of excessive intake of a specific macronutrient during pregnancy to the adverse offspring outcomes [23–27]. It has been demonstrated that high-free-sugar intake in pregnancy is associated with childhood atopy and asthma [23]. There is also an association between high-protein intake in pregnancy and the risk for high blood pressure (BP) in adult offspring [24–26]. Additionally, consumption of a high-fat diet during pregnancy has been reported to increase the risk of overweight in male adult offspring [27]. It is noteworthy that most studies mainly focused on cardiovascular outcomes and metabolic outcomes, rather less attention has been paid to other programming effects, including neurobehavioral, endocrine, and respiratory outcomes.

Table 1. Effects of malnutrition during pregnancy on adverse offspring outcomes in human cohort studies.

Nutritional Risk Factors	Outcome	Offspring Number	Age at Measure (Year)	Country	Cohort Study
Undernutrition					
Undernutrition	Hypertension, impaired glucose tolerance, and overweight	1339	40	Nigeria	Biafran famine study [15]
Undernutrition	Coronary heart disease, hyperlipidemia, obesity, obstructive airways disease, and microalbuminuria	741	50	Netherlands	Dutch famine study [16]
Undernutrition	Overweight, type 2 diabetes, hyperglycemia, and metabolic syndrome	1029–8973	55	China	China great leap forward famine study [17]
Undernutrition	Elevation of blood pressure	359	59	Netherlands	Dutch famine study [18]
Low vitamin B12	Insulin resistance	653	6	India	PMNS [19]
Low vitamin B12	Impaired cognition function	118	9	India	PMNS [20]
Vitamin D deficiency	Elevation of blood pressure	1834	5–6	Netherlands	ABCD [21]
Vitamin D deficiency	Elevation of blood pressure	3525	9.9	United Kingdom	ALSPAC [22]
Overnutrition					
Free sugar intake	Atopy and asthma	8956	7–9	United Kingdom	ALSPAC [23]
High-protein, low-carbohydrate diet	Elevation of blood pressure	253	40	Scotland	Aberdeen maternity hospital study [24]
High-protein, low-carbohydrate diet	Elevation of blood pressure	626	30	Scotland	Motherwell study [25]
High-protein intake	Elevation of blood pressure	434	20	Denmark	DaFO88 [26]
High-fat diet	Obesity	965	20	Denmark	Aarhus birth cohort study [27]

Studies tabulated according to nutritional risk factor and age at measure. ABCD, Amsterdam born children and their development; ALSPAC, the Avon longitudinal study of parents and children; CHDS, child health and development studies; CPP, the collaborative perinatal project pregnancy cohort; DaFO88 = Danish fetal origins cohort; MUSP, Mater University study of pregnancy and its outcomes; PMNS, Pune maternal nutrition study.

Since epidemiological studies do not dissect the physiological and molecular mechanisms by which the disorder is created, animal models with full control over the dietary manipulations are essential in the discovery of mechanisms underlying nutritional programming and the development of ideal nutritional interventions before they are implemented in humans.

4. Insights from Animal Models of Nutritional Programming

4.1. Animal Models for Nutitional Programming

Animal models particularly aid in inferring relevant information for possible translation to clinical practice in the field of nutritional programming. Using animal model with proven construct validity can increase the probability of successful translation of preclinical findings to clinical application. The translatability may further be increased by using a broader spectrum of relevant animal models. The choice of one animal model over another mainly depends on its similarity to the human diseases. For example, rabbits offer a better model of human atherosclerosis than rodents as they are prone to develop fatty streak lesions in studies of high-fat intake [31]. Although non-human primate is considered as the gold standard because of their similarity to humans, the most common species used are rodents in the DOHaD field [32]. Other species such as rabbits, sheep, pigs have been used depending on the experimental approaches, to study nutritional programming related to offspring outcomes [32]. Mice models provide a low-cost and easy-to-handle option, allowing genetic modification. Rabbits are suitable for studies on the blastocyst due to their large size. Additionally, their placenta structure and lipid metabolism are close to those in humans. Sheep are a monotocous species with a long gestational period as in humans. Pigs are considered as a good model for studying early stages of fertilization and development. Thus, animal studies play a pivotal role for us to clarify programming effects, identify critical developmental windows, and develop ideal reprogramming strategies for nutritional programming. The range of nutritional insults that have been utilized can be grouped into models that seek to restrict calorie intake, manipulate macronutrient intake, or restrict micronutrient intake.

4.2. Animal Models of Maternal Caloric rRstriction

Rodent studies of calorie restriction ranging from 30–70% to pregnant dams showed hypertension in the adult offspring [33–35]. Offspring that experienced a 50% calorie restriction during the last week of gestation showed impairment of beta cell development [36]. In rodents, severe 70% caloric restriction during pregnancy resulted in obesity, hypertension, hyperleptinemia, and hyperinsulinism in adult offspring [37]. Similar adverse cardiovascular outcomes have been observed in cows and sheep [38,39]. Pups exposed to a more severe degree of caloric restriction were likely to develop hypertension earlier [5]. Additionally, the severity of adverse offspring outcome seems relevant to the timing of exposure. In a maternal 50% caloric restriction sheep model, no major change in glucose–insulin homeostasis was observed in adult offspring born of undernourished ewes during early pregnancy [40]. However, maternal undernutrition during late pregnancy caused glucose intolerance and insulin resistance in adult offspring. These findings suggest that undernutrition in late pregnancy, during the period of maximal fetal growth, impacts negatively on the subsequent adult offspring's glucose–insulin homeostasis.

4.3. Animal Models of Macronutrients Imbalance

Macronutrients include carbohydrates, proteins, and fats. Sugar consumption, particularly fructose, has grown over the past several decades and its growth has been paralleled by an increase in obesity, diabetes, and hypertension [41]. Several rodent studies showed that maternal diets consisting of 75% simple carbohydrates (dextrose and maltodextrin) led to a significantly higher body weight gain in offspring [42–44]. Fructose is a monosaccharide naturally present in fruits. However, most of the increase in fructose consumption now is derived from refined sugars and high-fructose corn syrup [45]. As we reviewed elsewhere [46], consumption of high-fructose alone or as a part of diet by rodent mothers induces several features of metabolic syndrome in adult offspring, including hypertension, insulin resistance, obesity, hepatic steatosis, and dyslipidemia.

Of note, adverse effects of fructose feeding depend on the amount and duration of fructose consumption [47]. Despite being viewed as far in excess of a relevant load, most rodent studies have been performed using diets containing 50–60% fructose [48,49]. However, a recent study indicated that maternal consumption of 10% w/v fructose significantly increased BP in mice offspring after 1 year [50]. On the other hand, several studies used fructose as a part of maternal diet along with fat and salt to induce hypertension in adult offspring [51,52]. As Western diet is characterized by the intake of high-sugar drinks, high-fat products, and excess salt, it is important to dissect the interplay between fructose, fat, and salt on the nutritional programming. Indeed, animal studies examining the combined effects of key components of the Western diet have shown their synergistic effects of fructose, fat, and salt on the elevation of BP in adult offspring [53,54].

Protein is another macronutrient. The low-protein model has been extensively used to study the mechanisms of nutritional programming. In rodents, protein restriction during pregnancy leads to intrauterine growth retardation (IUGR) with subsequent hyperglycemia, glucose intolerance, insulin resistance, obesity, and adipocyte hypertrophy [10]. Additionally, rodent studies of protein restriction ranging from 6–9% to pregnant dams showed hypertension in the adult offspring [55–59]. Similar to caloric restriction, a greater degree of protein restriction shows a higher propensity to develop hypertension earlier in offspring [5]. Similar observations in several model species have confirmed the effects of maternal low-protein diet on the above-mentioned metabolic syndrome-related phenotypes in mammals [59]. Programming effects of protein restriction occur not only during pregnancy but also through breastfeeding. Mice offspring born to dams fed with low-protein diet during lactation have impaired cardiac structure and function that limit exercise/functional capacity [60].

Maternal high-fat diet is a widely used animal model to induce developmental programming of obesity and related disorders [61]. Adult offspring exposed to a maternal high-fat consumption in utero developed insulin resistance, increased lipid profile, hepatic steatosis, increased visceral fat mass, and adipocyte hypertrophy [61–63]. Nevertheless, maternal high-fat-diet-induced responses of BP include an increase [64,65] or no change [66,67], mainly depending on strain, sex, age, and diverse fatty acid compositions.

Additionally, alteration of the protein-to-carbohydrate ratio of the maternal diet was related to an adverse outcome in offspring. Male offspring rats born of dams exposed to a diet with high-protein/carbohydrate ratio were characterized by elevated BP and a greater degree of glomerulosclerosis [68]. These data are in agreement with human studies showing that a maternal high-protein, low-carbohydrate diet is linked to elevated BP in the offspring [24,25].

4.4. Animal Models of Micronutrients Deficiency

There is also evidence to support that deficiencies in micronutrients, including minerals, trace elements and vitamins, in pregnant mothers are associated with the development of NCDs. Evidence in rats showed both maternal low- and high-salt diets lead to elevated BP in male adult offspring [69]. Additionally, maternal calcium-deficient diet increased BP and insulin resistance in adult rat offspring [70,71]. Furthermore, magnesium-deficient diet in pregnancy caused impaired kidney function but had no effect on BP in rat offspring [72]. Deficiencies in trace elements and vitamins in pregnant mothers are associated with the development of hypertension, impaired glucose tolerance, and insulin resistance in their adult offspring [73–78]. These micronutrients include iron [73,74], zinc [75,76], and vitamin D [77,78]. Additionally, maternal folate and vitamin B12 restriction increased visceral adiposity and altered lipid metabolism in rat offspring [79].

The vitamins folic acid, B12, B6, and B2 have a key role as coenzymes in one-carbon metabolism, serving as methyl-donor nutrients. Because one-carbon metabolism is required for epigenetic regulation of gene expression, pregnant women are recommended to eat methyl-donor food to reduce adverse birth outcomes [80]. These foods contain nutrients that directly provide methyl-donors or serve as cofactors, such as methionine, folic acid, vitamins B2, B6, B12, and choline. Although methyl-donor supplementation may be of benefit in preventing disease, we recently found that pregnant rats fed with high-methyl-donor diet led to hypertension in male adult offspring [81]. Thus, additional studies should be taken to assess the programming effects of these methyl-donor nutrients and to solve the current conflicting data in regards to maternal methyl-donor diet.

Vitamin C, E, B6, and coenzyme Q-10 have shown beneficial effects on lowering BP [82]. However, whether deficiencies of these micronutrients on pregnant mothers lead to programmed hypertension in their offspring remains unclear. So far, no studies have been conducted examining the role of deficient non-essential nutrients on developmental programming of adult disease.

5. Common Mechanisms Underlying Nutritional Programming of Disease

In view of the fact that diversely nutritional insults in gestation create very similar outcome in adult offspring, there might be some common mechanisms contributing to the pathogenesis of nutritional programming of disease. Currently, possible mechanisms underlying nutritional programming include nutrient-sensing signals, oxidative stress, tissue remodeling, epigenetic regulation, gut microbiota, and sex differences [5–10,59]. Each mechanism will be discussed in turn.

5.1. Nutrient-Sensing Signals

Fetal metabolism and development are determined by maternal nutritional status via nutrient-sensing signals. Sirtuin 1 (SIRT1), cyclic adenosine monophosphate (AMP)-activated protein kinase (AMPK), mammalian target of rapamycin (mTOR), and peroxisome proliferator-activated receptors (PPARs) are well-known nutrient-sensing signals [83]. In excessive nutrient conditions, the mTOR is activated by increases in glucose, amino acid, and insulin levels, while in nutrient-depleted conditions, AMPK and SIRT1 are activated by increases in intracellular AMP and NAD^+ levels, respectively. Dysregulated nutrient-sensing signals are present in several metabolic diseases [83]. The interplay between AMPK and SIRT1 driven by maternal nutritional insults was reported to mediate PPARs and their target genes, thereby generating developmental programming of adult disease [84]. Maternal high-fat diet increased fetal fat mass accompanied by elevated PPARγ but decreased SIRT1 expression [85]. As we mentioned earlier, maternal high-fructose diet induced several phenotypes of metabolic syndrome in adult offspring [46]. Using the maternal high-fructose rat model, we found PPAR signaling pathway is significantly regulated in the offspring liver, heart, and kidney using next-generation RNA sequencing (NGS) analysis [48,86].

On the other hand, pharmacological interventions targeting AMPK or PPARs signaling have been reported to prevent the development of hypertension and metabolic syndrome in a variety of fetal programming models [84,87]. Although nutrient-sensing signaling are not likely to be the sole mechanism that increases susceptibility to disease in later life, it is required to elucidate their interplay with other mechanisms of nutritional programming in determining its impact on adult diseases of developmental origins.

5.2. Oxidative Stress

The embryo and fetus have low-antioxidant capacity. An overproduction of reactive oxygen species (ROS) under suboptimal intrauterine conditions would prevail over antioxidants, leading to oxidative stress damage and thus compromising fetal development [88]. These damages include oxidation of biological molecules, like lipids, proteins, and DNA. Excessive pro-oxidant molecules can overpower the antioxidant defense, resulting in biological damage, namely, oxidative stress. There are several types of nutritionally mediated oxidative stress sources, including diets high in carbohydrates, animal-based protein, and saturated fat [89]. Oxidative stress during pregnancy may lead to lifelong effects on vulnerable organs leading to adverse offspring outcomes in later life [90]. As reviewed elsewhere [91], several nutritional insults in pregnancy have been linked to oxidative stress in adult offspring, including maternal caloric restriction [33,35], high-fructose diet [48], low-protein diet [57], high-fat diet [64,65], zinc-deficiency diet [75], iron-deficiency diet [92,93], and methyl-donor diet [81]. Detailed mechanisms that underlie the interactions between maternal nutrition and oxidative stress and their roles in the programming process toward adult diseases of developmental origins, however, remain to be determined.

The major antioxidant nutrients in our body are copper, zinc, manganese, selenium, vitamins E, C, and A, and the glutathione system. Although some prospective assessment of the effect of supplemental antioxidants on human health suggests benefit, there are conflicting results in this area [94]. Although antioxidant nutrients therapy in pregnancy have been shown to prevent adult

NCDs [33], there is a lack of data on when and how to use antioxidant nutrients to reprogram oxidative stress-related adult diseases.

5.3. Tissue Remodeling

In the first trimester, there is rapid differentiation and development of organs and systems. The second trimester is characterized by the development of structures and the start of functional activities. The third trimester is a period of the most rapid growth during the gestation. Most organ development lasts to birth, while some others continue to develop after birth, like the liver and nervous system [95].

If nutritional insults in gestation impact on critical periods of rapid growth, then it would be expected that the organs will mature at a smaller size and with a reduced functional capacity. For example, maternal low-protein diet caused fewer islets, smaller islets, and reduced islet vascularization in rat offspring pancreas [96]. The same maternal insult also influenced the thymic growth in a low-protein diet mice model [97].

Another example is the kidney. Low nephron number plays a key role in the developmental programming of cardiovascular disease, hypertension, and kidney disease [8]. There is an increasing body of literature demonstrating the relationship between maternal malnutrition and low nephron number, as reviewed elsewhere [98,99]. In rodent models, reported nutritional insults include maternal caloric restriction [35], low-protein diet [56], high-salt diet [69], low-salt diet [69], zinc deficiency [75], vitamin A restriction [100], and iron restriction [101]. In low-protein diet model, adult male offspring developed low nephron number and hypertension, which is related to renal hyperfiltration and activation of the renin–angiotensin system (RAS) [56]. On the other hand, nutritional programming can also have no effect [102], or even augment nephron number in adult offspring [103]. These observations highlight that the nutritional programming might be not driven by a single factor (i.e., low nephron number) and other organ- or phenotype-specific mechanisms remain to be identified.

5.4. Epigenetic Regulation

Persistent epigenetic modification of genes linked to disease is considered as a crucial mechanism for developmental programming [104]. Epigenetics is defined as alterations in gene expression that are not explained by changes in the DNA sequence. Epigenetic mechanisms include DNA methylation, histone modification, and non-coding RNAs, all of which are involved in translating nutritional insults in pregnancy to the developmental programming of adult diseases [104].

In rats, low-protein diet in pregnancy was shown to alter methylation and expression levels of *PPARα* gene and glucocorticoid receptor (GR) genes in the liver of the offspring [105]. Of note is that such alterations can be reversed by folic acid supplementation. Using DNA microarray technology, a previous study showed maternal low-protein diet caused a global upregulation of genes implicated in adipocyte differentiation, as well as in protein, carbohydrate, and lipid metabolism in the visceral adipose tissue [106]. Additionally, global methylation patterns in the offspring liver and brain have been studied in two different programming models, low-protein diet model [107], and micronutrient deficiency model [108], respectively. Several genes belong to the renin–angiotensin system have been reported to be altered by epigenetic regulation, culminating to the development of hypertension [109]. Using a mouse model, a report showed that maternal low-protein diet caused significant increases in the mRNA and protein expression of angiotensin converting enzyme-1 (ACE1) in offspring brain, with changes in DNA methylation and miRNA [110]. Furthermore, maternal undernutrition resulted in epigenetic changes in hypothalamic genes, GR, neuropeptide Y, and proopiomelanocortin, which predispose offspring to altered regulation of food intake and glucose homeostasis in later life [111].

Another important point to note is the impact of methyl-donor nutrients involved in the DNA methylation. It has been reported that maternal serum vitamin B12 levels inversely correlated with offspring's global methylation status at birth [112]. Another report showed that folic acid supplementation directly impacted the methylation status of the IGF2 gene in the infants up to 17 months of age [113]. Moreover, maternal blood concentrations of methyl-donor nutrients measured around the time of conception can predict the methylation patterns at metastable epialleles in their offspring [114].

Overall, these studies support that malnutrition in pregnancy can epigenetically program development of adult diseases in later life. Nevertheless, the detailed mechanisms underlying the epigenetic modulation of common genes by different maternal nutritional insults still await for further study in diverse models of developmental programming.

5.5. Gut Microbiota

There is now growing evidence suggesting the role of early-life gut microbiota in developmental programming of metabolic syndrome-related disorders [115,116]. Maternal nutritional insults may cause a microbial imbalance, namely dysbiosis. Dysbiosis of gut microbiota may link early-life nutritional insults and later risk of adult diseases [117]. The gut microbiota produces a variety of metabolites detectable in host circulation, including short-chain fatty acids, small organic acids, bile acids, vitamins, and choline metabolites [118]. During pregnancy, the diet–gut microbiota interactions can mediate changes in global histone acetylation and methylation not only in mother but also in the fetus via the contact with their metabolites [119]. Emerging evidence shows that gut microbiota dysbiosis in early-life might be correlated with adverse offspring outcomes including obesity [120], metabolic syndrome [120], diabetes [120], hypertension [121], allergy [122], and neurological disorder [123]. On the other hand, non-essential nutrients are substances within foods that can have a significant impact on health, for example, dietary fiber. Of note, consumption of dietary fiber has become one dietary strategy for modulating the gut microbiota. A recent study from our laboratory examined the maternal high-fructose-induced hypertension model and found that modulation of gut microbiota by inulin is able to protect adult offspring against programmed hypertension [124]. Although recent studies demonstrating that microbiota-targeted therapies can be applied to a variety of diseases [125], their roles on nutritional programming-related disorders, especially the use of prebiotics in pregnancy, demand further exploration.

5.6. Sex Differences

There is a growing body of evidence that indicates sex differences exist in nutrient intake, bioavailability, and metabolism [126,127]. Several above-mentioned mechanisms related to nutritional programming, such as nutrient-sensing signal [128], oxidative stress [129], and epigenetic regulation [130] have been reported to respond to maternal nutritional insults in a sex-specific manner.

Same maternal nutritional insults, such as caloric restriction [34], high-fructose diet [49], or high-fat diet [67], can induce different phenotypes on male and female offspring. For example, male offspring are more prone to hypertension than female offspring [5]. This difference has led many studies to target their efforts totally to one sex, especially to males [35,56,57,69,75].

Evidence from animal models of maternal overnutrition has suggested that females are more vulnerable to programming of glucose homeostasis, whereas males are more susceptible to changes in adiposity and body weight [131–133]. On the other hand, male offspring of pregnant rats with caloric restriction during early gestation displayed obesity later in life, but female offspring were unaffected [134].

A few studies have investigated the sex-specific programming response to maternal diet on transcriptome profiles of the offspring [49,135]. A previous study has shown that more genes in the placenta were affected in females than in males under different maternal diets [135]. This is in accord with a report revealing that sex-specific placental adaptations are often associated with male offspring developing adult disease while females are minimally affected [136]. In a maternal high-fructose diet model [49], we found a sex-specific alteration of renal transcriptome response to maternal high-fructose intake and female offspring are more fructose-sensitive. However, whether higher sensitivity to nutritional insults is beneficial or harmful for programming effects in female offspring awaits further evaluation. As males and females likely respond differentially to nutritional programming, future research is required to clarify whether nutritional interventions also result in sex-specific reprograming effects.

6. Nutritional Interventions as Reprogramming Strategies

Reprogramming strategies aim to reverse the programmed development and achieve normal development. These strategies could be nutritional intervention, exercise, lifestyle modification, or pharmacological therapy. Nutritional programming is theoretically bidirectional. Inadequate (e.g., folate deficiency) or excessive intake (e.g., excess of macronutrients) of certain nutrients in pregnancy leads to adult diseases in later life. Conversely, adverse programmed processes during a compromised pregnancy can be prevented or at least reduced by appropriate nutritional interventions. It is well known that a balanced, diverse, and nutritious diet is universally recommended to meet nutritional needs to improve maternal and birth outcomes [137,138]; however, little is known whether supplementing with certain classes of nutrients in gestation can be beneficial on DOHaD-related disorders induced by various early-life insults in humans. In the current review, we only restrict to nutritional interventions during pregnancy and/or lactation periods as reprogramming strategies to prevent adult diseases of developmental origins in all sorts of animal models, some of which are listed in Table 2 [33,35,124,139–150]. Rats are the most commonly used species for developmental programming research. Rats become sexually mature at approximately 5–6 weeks of age. In adulthood, one rat month is comparable to three human years [151]. Accordingly, Table 2 lists the ages of reprogramming effects measured in rats ranging from 4 to 14 months, which can be translated to humans of specific age groups, from childhood to young adulthood. However, limited information is available about the long-term effects of nutritional interventions on older adult offspring. Additionally, developmental window is not uniform across different organ systems or species. For example, renal development in rodents, unlike in humans, continues up to postnatal week 1–2 [99]. Therefore, nutritional interventions during pregnancy and early lactation period may preserve nephrogenesis and increase nephron numbers to prevent the developmental programming of kidney disease [35,140].

Table 2. Reprogramming strategies aimed at nutritional interventions to prevent developmental programming of adult diseases in animal models.

Treatments	Animal Models	Species/Gender	Period of Treatment	Reprogramming Effects	Age at Measure	Reference
Macronutrients						
Glycine	Maternal low-protein diet, 9%	Wistar/M	Pregnancy	Prevented hypertension	4 weeks	[139]
Citrulline	Maternal caloric restriction, 50%	SD/M	Pregnancy and lactation	Prevented reduced nephron numbr and kidney injury	12 weeks	[35]
Citrulline	Streptozotocin-induced diabetes	SD/M	Pregnancy and lactation	Prevented reduced nephron numbr and hypertension	12 weeks	[140]
Citrulline	Prenatal dexamethasone exposure	SD/M	Pregnancy and lactation	Prevented reduced nephron number and hypertension	12 weeks	[141]
Citrulline	Maternal nitric oxide deficiency	SD/M	Pregnancy and lactation	Prevented hypertension	12 weeks	[142]
Branched-chain amino acid	Maternal 70% caloric restriction	SD/M	Pregnancy	Prevented hypertension	16 weeks	[143]
Taurine	Streptozotocin-induced diabetes	Wistar/M + F	Pregnancy and lactation	Prevented hypertension, hyperglycemia, and dyslipidemia	16 weeks	[144]

Table 2. Cont.

Treatments	Animal Models	Species/Gender	Period of Treatment	Reprogramming Effects	Age at Measure	Reference
Conjugated linoleic acid	Maternal high-fat diet	SD/M	Pregnancy and lactation	Prevented hypertension and endothelial dysfunction	18 weeks	[145]
Omega-3 polyunsaturated fatty acids	Maternal low-protein diet, 5%	Wistar/M + F	Pregnancy and lactation	Attenuated hypertension and cardiac remodeling	6 months	[146]
Omega-3 polyunsaturated fatty acids	Maternal caferteria diet	SD/M + F	Pregnancy and lactation	Prevented liver steatosis	14 months	[147]
Micronutrients						
Vitamin C, E, selenium and folic acid	Maternal 50% caloric restriction	Wistar/M + F	Pregnancy	Prevented hypertension and endothelial dysfunction	14–16 weeks	[33]
Folic acid	Maternal low-protein diet, 9%	Wistar/M	Pregnancy	Prevented hypertension and cardiovascular dysfunction	15 weeks	[148]
Vitamin E	Cholesterol-enriched diet	Rabbit	Pregnancy	Prevented hypertension and atherosclerosis	6 and 12 months	[149]
Non-essential nutrients						
Long-chain inulin	Maternal high-fructose diet	SD/M	Pregnancy and lactation	Prevented hypertension	12 weeks	[124]
Oligofructose	Maternal high-fat/-sucrose diet	SD/M	Pregnancy and lactation	Attenuated hepatic steatosis and insulin resistance	21 weeks	[150]

Studies tabulated according to types of treatments and age at measure. SD, Sprague–Dawley rat. M, male; F, female.

Most importantly, some nutritional interventions are employed in non-nutritional-insults-induced programming models [140–142,144]. Nutritional interventions are grouped into macronutrients, micronutrients, and non-essential nutrients.

6.1. Macronutrieents

Macronutrients used as reprogramming strategies are mainly directed at amino acids. Within the body, amino acids are used for a wide variety of structural proteins and, hence, they play critical roles in organogenesis and fetal development. Among them, the most commonly used for reprogramming is citrulline. Maternal citrulline supplementation has been reported to protect adult offspring against hypertension and kidney disease in several models of developmental programming, including maternal caloric restriction [35], streptozotocin-induced diabetes [140], prenatal dexamethasone exposure [141], and maternal nitric oxide (NO) deficiency [142]. Citrulline is a precursor to arginine, and it is involved in the generation of NO [152]. Despite post-weaning arginine supplementation being reported to improve hypertension, insulin sensitivity, and beta cell function in offspring rats [153,154], the reprogramming effects of arginine supplementation in pregnancy are yet to be examined. Supplemental citrulline is more efficient than arginine to produce NO as citrulline can bypass hepatic metabolism and allow renal conversion to arginine. Thus, a better understanding of maternal citrulline supplementation in the prevention of diverse animal models of developmental programming is warranted before it is implemented in humans.

Another three amino acids, glycine, branched-chain amino acid, and taurine have also been used as reprogramming interventions in models of maternal low-protein diet [139], maternal caloric restriction [143], and streptozotocin-induced diabetes [144], respectively. Glycine is a simple amino acid not essential to the human diet. Glycine and vitamins (folic acid, vitamin B2, B6, and B12) take part in one-carbon metabolism and DNA methylation. Thus, glycine supplementation may have important implications for fetal programming through epigenetic mechanisms. Additionally, maternal branched-chain amino acid supplementation protects adult rat offspring against hypertension in a maternal caloric restriction model [143]. Taurine is an abundant semi-essential, sulfur-containing amino acid. Taurine supplementation during gestation prevented maternal diabetes-induced programmed hypertension [144]. Taurine supplementation can lower BP and increase hydrogen sulfide simultaneously [155]. Cumulative evidence supports hydrogen sulfide as a reprogramming strategy for long-term protection against programmed hypertension [156]. Additional studies are required to clarify whether the protective effects of maternal taurine supplementation on programmed hypertension is related to hydrogen sulfide pathway in other programming models.

So far, three reports showed maternal polyunsaturated fatty acids' (PUFAs) supplementation has reprograming effects against hypertension [145,146], cardiac remodeling [146], and liver steatosis [147] in adult rat offspring. Although PUFAs have been recommended for pregnant and breastfeeding women [138], a recent meta-analysis that included 3644 children showed maternal supplementation with omega-3 PUFA during pregnancy does not have a beneficial effect on obesity risk [157]. Thus, whether there are differential reprogramming effects of individual PUFAs used as dietary supplement during pregnancy on diverse adult diseases remain to be determined.

6.2. Micronutrients

Micronutrients include vitamins and minerals. Relatively few reports demonstrated reprogramming effects of maternal micronutrients supplementation on offspring outcome [33,148,149]. One previous report demonstrated that combined micronutrients vitamin C, E, selenium, and folic acid supplementation in gestation prevented maternal caloric restriction-induced hypertension and endothelial dysfunction [33]. Vitamin C, E, and selenium have antioxidant properties. These micronutrients were shown to prevent programmed hypertension by reducing oxidative stress [158]. Folic acid is known to prevent neural tube defects, leading to global recommendations for folic acid supplementation before and during early pregnancy. Besides, folic acid, combined with other vitamins B12, B6, and B2, are the source of coenzymes which participate in one-carbon metabolism. In our body, one-carbon metabolites serve as methyl-donors that are required for DNA methylation. Although folic-acid supplementation during pregnancy was reported to improve offspring cardiovascular dysfunction induced by maternal low-protein diet [148], whether its reprogramming effects is related to epigenetic regulation remains undiscovered. On the other hand, our recent study showed that pregnant rats fed with high-methyl-donor diet or methyl-deficient diet both resulted in programmed hypertension in their male adult offspring [81]. Additional studies are required to approve whether methyl-donor food intake in pregnant women is beneficial to reduce adverse offspring and not just birth outcomes [80]. In a rabbit model, maternal vitamin E supplementation protected adult offspring against hypertension and hypercholesterolemia [135]. Many of the physiological functions of vitamin E, including its antioxidative effects, have been well studied. However, there still exists an ongoing debate regarding its beneficial effects on human health [159]. Despite recent studies demonstrating that antioxidant vitamins and minerals can be applied to a variety of diseases [94], their reprogramming effects on different adult diseases for dietary supplement use during pregnancy remain to be identified.

6.3. Non-Essential Nutrients

Dietary fiber, one of the non-essential nutrients, has been used as reprogramming strategies to prevent developmental programming of hypertension [124], hepatic steatosis, and insulin resistance [150]. Maternal supplementation with inulin protected adult rat offspring born to dams fed with high-fructose diet against the developmental programming of hypertension [124]. Another report demonstrated that maternal oligofructose supplementation attenuated hepatic steatosis and insulin resistance in adult rat offspring in a high-fat/-sucrose diet model [150]. Inulin and oligofructose are the most studied forms of prebiotic foods for beneficial gut microorganisms. As gut microbiota can influence development of sensitization and allergy, the World Allergy Organization (WAO) experts issued their guidelines on the use of prebiotics for all formula-fed infants in allergy prevention [160]. Nevertheless, whether prebiotics can be used in pregnant women for prevention of allergy in their offspring required further preclinical and clinical evidence.

Additionally, combined supplementations with high-fiber diet and acetate prevented hypertension in deoxycorticosterone acetate (DOCA)–salt hypertensive mice [161]. Acetate is the most abundant short-chain fatty acids, the major metabolites of gut microbiota. Considering that the gut microbiota dysbiosis is a key mechanism underlying nutritional programming, maternal supplementation with prebiotics or other microbiota-targeting nutritional interventions (e.g., probiotics and postbiotics) might be used as reprogramming strategies in different adult diseases of developmental origins.

It is noteworthy that reprogramming strategies can be created based on the above-mentioned mechanisms leading to a variety of adult diseases. For example, maternal inulin supplementation prevented adult rat offspring against high-fructose diet-induced programmed hypertension associated with nutrient-sensing signals [124]. The beneficial effects of vitamin C and E, selenium, and folic acid [33], citrulline [140,142], conjugated linoleic acid [145], and folic acid [148] on developmental programming of hypertension are relevant to the reduction of oxidative stress. Moreover, low nephron number was restored by citrulline supplementation in the maternal caloric restriction model [35], streptozotocin-induced diabetes model [140], and prenatal dexamethasone exposure model [141]. Thus, nutritional interventions in pregnancy may reprogram common mechanisms of both nutritional and non-nutritional in utero insults to prevent offspring against adult diseases of developmental origins. This effect requires more efforts to bridge gaps between basic animal research and clinical practice.

7. Conclusions

Almost all NCDs can originate in early-life. It stressed the importance of taking a DOHaD approach to prevent and control global burden of NCDs. All nutrients in pregnancy play a pivotal role in fetal growth and development. Without a doubt, maternal malnutrition is definitely bad for offspring health as it induces nutritional programming, leading to the developmental programming of numerous NCDs. Conversely, nutritional programming can also be advantageous. Through animal models, several nutritional interventions have been proven effective as a reprogramming strategy to prevent the development of various NCDs. Although there is a great opportunity that reprogramming strategy is good for offspring health, there is also an ugly side to it.

Given that the complexities of nutritional programming in pregnancy lead to both beneficial and deleterious effects in mother and offspring health, there remains a long road ahead to determine the "right" nutritional intervention for the "right" person (mother or offspring) at the "right" time, to prevent the DOHaD-related diseases. It is clear that better understanding of the type of nutrient, dose of supplement, and therapeutic duration for nutritional interventions are needed before patients could benefit from this reprogramming strategy. We expect that studies utilizing animal models of nutritional programming will provide the much-needed scientific basis for the development of ideal reprogramming strategies for clinical practice that will improve pregnancy and offspring outcomes in humans.

Author Contributions: C.-N.H. contributed to concept generation, data interpretation, drafting of the manuscript, critical revision of the manuscript, and approval of the article; Y.-L.T. contributed to concept generation, data interpretation, drafting of the manuscript, critical revision of the manuscript, and approval of the article.

Funding: This work was supported by grant MOST 107-2314-B-182-045-MY3 from the Ministry of Science and Technology, Taiwan, and the grants CMRPG8H0831 and CMRPG8J0251 from Chang Gung Memorial Hospital, Kaohsiung, Taiwan.

Conflicts of Interest: The authors declare no conflict of interest.

References

1. Hanson, M. The birth and future health of DOHaD. *J. Dev. Orig. Health Dis.* **2015**, *6*, 434–437. [CrossRef]

2. Tain, Y.L.; Joles, J.A. Reprogramming: A preventive strategy in hypertension focusing on the kidney. *Int. J. Mol. Sci.* **2016**, *17*, 23. [CrossRef]
3. Paauw, N.D.; van Rijn, B.B.; Lely, A.T.; Joles, J.A. Pregnancy as a critical window for blood pressure regulation in mother and child: Programming and reprogramming. *Acta Physiol.* **2017**, *219*, 241–259.
4. Mathias, P.C.; Elmhiri, G.; de Oliveira, J.C.; Delayre-Orthez, C.; Barella, L.F.; Tófolo, L.P.; Fabricio, G.S.; Chango, A.; Abdennebi-Najar, L. Maternal diet, bioactive molecules, and exercising as reprogramming tools of metabolic programming. *Eur. J. Nutr.* **2014**, *53*, 711–722. [CrossRef]
5. Hsu, C.N.; Tain, Y.L. The Double-Edged Sword Effects of Maternal Nutrition in the Developmental Programming of Hypertension. *Nutrients* **2018**, *10*, 1917. [CrossRef]
6. Nüsken, E.; Dötsch, J.; Weber, L.T.; Nüsken, K.D. Developmental Programming of Renal Function and Re-Programming Approaches. *Front. Pediatr.* **2018**, *6*, 36. [CrossRef]
7. Langley-Evans, S.C. Nutritional programming of disease: Unravelling the mechanism. *J. Anat.* **2009**, *215*, 36–51. [CrossRef] [PubMed]
8. Bagby, S.P. Maternal nutrition, low nephron number, and hypertension in later life: Pathways of nutritional programming. *J. Nutr.* **2007**, *137*, 1066–1072. [CrossRef] [PubMed]
9. Hoffman, D.J.; Reynolds, R.M.; Hardy, D.B. Developmental origins of health and disease: Current knowledge and potential mechanisms. *Nutr. Rev.* **2017**, *75*, 951–970. [CrossRef] [PubMed]
10. Remacle, C.; Bieswal, F.; Bol, V.; Reusens, B. Developmental programming of adult obesity and cardiovascular disease in rodents by maternal nutrition imbalance. *Am. J. Clin. Nutr.* **2011**, *94*, 1846S–1852S. [CrossRef] [PubMed]
11. Lea, A.J.; Tung, J.; Archie, E.A.; Alberts, S.C. Developmental plasticity: Bridging research in evolution and human health. *Evol. Med. Public Health* **2018**, *2017*, 162–175. [CrossRef]
12. Noyan-Ashraf, M.H.; Wu, L.; Wang, R.; Juurlink, B.H. Dietary approaches to positively influence fetal determinants of adult health. *FASEB J.* **2006**, *20*, 371–373. [CrossRef] [PubMed]
13. Care, A.S.; Sung, M.M.; Panahi, S.; Gragasin, F.S.; Dyck, J.R.; Davidge, S.T.; Bourque, S.L. Perinatal Resveratrol Supplementation to Spontaneously Hypertensive Rat Dams Mitigates the Development of Hypertension in Adult Offspring. *Hypertension* **2016**, *67*, 1038–1044. [CrossRef] [PubMed]
14. Perel, P.; Roberts, I.; Sena, E.; Wheble, P.; Briscoe, C.; Sandercock, P.; Macleod, M.; Mignini, L.E.; Jayaram, P.; Khan, K.S. Comparison of treatment effects between animal experiments and clinical trials: Systematic review. *BMJ* **2007**, *334*, 197. [CrossRef] [PubMed]
15. Hult, M.; Tornhammar, P.; Ueda, P.; Chima, C.; Bonamy, A.K.; Ozumba, B.; Norman, M. Hypertension, diabetes and overweight: Looming legacies of the Biafran famine. *PLoS ONE* **2010**, *5*, e13582. [CrossRef] [PubMed]
16. Painter, R.C.; Roseboom, T.J.; Bleker, O.P. Prenatal exposure to the Dutch famine and disease in later life: An overview. *Reprod. Toxicol.* **2005**, *20*, 345–352. [CrossRef]
17. Li, C.; Lumey, L.H. Exposure to the Chinese famine of 1959-61 in early life and long-term health conditions: A systematic review and meta-analysis. *Int. J. Epidemiol.* **2017**, *46*, 1157–1170. [CrossRef] [PubMed]
18. Stein, A.D.; Zybert, P.A.; van der Pal-de Bruin, K.; Lumey, L.H. Exposure to famine during gestation, size at birth, and blood pressure at age 59 y: Evidence from the Dutch Famine. *Eur. J. Epidemiol.* **2006**, *21*, 759–765. [CrossRef] [PubMed]
19. Yajnik, C.S.; Deshpande, S.S.; Jackson, A.A.; Refsum, H.; Rao, S.; Fisher, D.J.; Bhat, D.S.; Naik, S.S.; Coyaji, K.J.; Joglekar, C.V.; et al. Vitamin B12 and folate concentrations during pregnancy and insulin resistance in the offspring: The Pune Maternal Nutrition Study. *Diabetologia* **2008**, *51*, 29–38. [CrossRef]
20. Bhate, V.; Deshpande, S.; Bhat, D.; Ladkat, R.; Watve, S.; Yajnik, C.; Joshi, N.; Fall, C.; De Jager, C.A.; Refsum, H. Vitamin B12 status of pregnant Indian women and cognitive function in their 9-year-old children. *Food Nutr. Bull.* **2008**, *29*, 249–254. [CrossRef]
21. Hrudey, E.J.; Reynolds, R.M.; Oostvogels, A.J.; Brouwer, I.A.; Vrijkotte, T.G. The association between maternal 25-hydroxyvitamin D concentration during gestation and early childhood cardio-metabolic outcomes: Is there interaction with pre-pregnancy BMI? *PLoS ONE* **2015**, *10*, e0133313.
22. Williams, D.M.; Fraser, A.; Fraser, W.D.; Hyppönen, E.; Davey Smith, G.; Deanfield, J.; Hingorani, A.; Sattar, N.; Lawlor, D.A. Associations of maternal 25-hydroxyvitamin D in pregnancy with offspring cardiovascular risk factors in childhood and adolescence: Findings from the Avon Longitudinal Study of Parents and Children. *Heart* **2013**, *99*, 1849–1856. [CrossRef]

23. Bédard, A.; Northstone, K.; Henderson, A.J.; Shaheen, S.O. Maternal intake of sugar during pregnancy and childhood respiratory and atopic outcomes. *Eur. Respir. J.* **2017**, *50*, 1700073. [CrossRef]
24. Campbell, D.M.; Hall, M.H.; Barker, D.J.; Cross, J.; Shiell, A.W.; Godfrey, K.M. Diet in pregnancy and the offspring's blood pressure 40 years later. *Br. J. Obstet. Gynaecol.* **1996**, *103*, 273–280. [CrossRef]
25. Shiell, A.W.; Campbell-Brown, M.; Haselden, S.; Robinson, S.; Godfrey, K.M.; Barker, D.J. High-meat, low-carbohydrate diet in pregnancy: Relation to adult blood pressure in the offspring. *Hypertension* **2001**, *38*, 1282–1288. [CrossRef] [PubMed]
26. Hrolfsdottir, L.; Halldorsson, T.I.; Rytter, D.; Bech, B.H.; Birgisdottir, B.E.; Gunnarsdottir, I.; Granström, C.; Henriksen, T.B.; Olsen, S.F.; Maslova, E. Maternal Macronutrient Intake and Offspring Blood Pressure 20 Years Later. *J. Am. Heart Assoc.* **2017**, *6*, e005808. [CrossRef] [PubMed]
27. Maslova, E.; Rytter, D.; Bech, B.H.; Henriksen, T.B.; Olsen, S.F.; Halldorsson, T.I. Maternal intake of fat in pregnancy and offspring metabolic health—A prospective study with 20 years of follow-up. *Clin. Nutr.* **2016**, *35*, 475–483. [CrossRef] [PubMed]
28. Barker, D.J. The origins of the developmental origins theory. *J. Intern. Med.* **2007**, *261*, 412–417. [CrossRef]
29. Stanner, S.A.; Yudkin, J.S. Fetal programming and the Leningrad Siege study. *Twin Res.* **2001**, *4*, 287–292. [CrossRef]
30. Zhou, J.; Zhang, L.; Xuan, P.; Fan, Y.; Yang, L.; Hu, C.; Bo, Q.; Wang, G.; Sheng, J.; Wang, S. The relationship between famine exposure during early life and body mass index in adulthood: A systematic review and meta-analysis. *PLoS ONE* **2018**, *13*, e0192212. [CrossRef]
31. Fan, J.; Kitajima, S.; Watanabe, T.; Xu, J.; Zhang, J.; Liu, E.; Chen, Y.E. Rabbit models for the study of human atherosclerosis: From pathophysiological mechanisms to translational medicine. *Pharmacol. Ther.* **2015**, *146*, 104–119. [CrossRef]
32. Chavatte-Palmer, P.; Tarrade, A.; Rousseau-Ralliard, D. Diet before and during Pregnancy and Offspring Health: The Importance of Animal Models and What Can Be Learned from Them. *Int. J. Environ. Res. Public Health* **2016**, *13*, 586. [CrossRef]
33. Franco Mdo, C.; Ponzio, B.F.; Gomes, G.N.; Gil, F.Z.; Tostes, R.; Carvalho, M.H.; Fortes, Z.B. Micronutrient prenatal supplementation prevents the development of hypertension and vascular endothelial damage induced by intrauterine malnutrition. *Life Sci.* **2009**, *85*, 327–333. [CrossRef]
34. Ozaki, T.; Nishina, H.; Hanson, M.A.; Poston, L. Dietary restriction in pregnant rats causes gender-related hypertension and vascular dysfunction in offspring. *J. Physiol.* **2001**, *530*, 141–152. [CrossRef]
35. Tain, Y.L.; Hsieh, C.S.; Lin, I.C.; Chen, C.C.; Sheen, J.M.; Huang, L.T. Effects of maternal L-citrulline supplementation on renal function and blood pressure in offspring exposed to maternal caloric restriction: The impact of nitric oxide pathway. *Nitric. Oxide* **2010**, *23*, 34–41. [CrossRef]
36. Garofano, A.; Czernichow, P.; Bréant, B. In utero undernutrition impairs rat beta-cell development. *Diabetologia* **1997**, *40*, 1231–1234. [CrossRef]
37. Vickers, M.H.; Reddy, S.; Ikenasio, B.A.; Breier, B.H. Dysregulation of the adipoinsular axis—A mechanism for the pathogenesis of hyperleptinemia and adipogenic diabetes induced by fetal programming. *J. Endocrinol.* **2001**, *170*, 323–332. [CrossRef]
38. Mossa, F.; Carter, F.; Walsh, S.W.; Kenny, D.A.; Smith, G.W.; Ireland, J.L.; Hildebrandt, T.B.; Lonergan, P.; Ireland, J.J.; Evans, A.C. Maternal undernutrition in cows impairs ovarian and cardiovascular systems in their offspring. *Biol. Reprod.* **2013**, *88*, 92. [CrossRef]
39. Gilbert, J.S.; Lang, A.L.; Grant, A.R.; Nijland, M.J. Maternal nutrient restriction in sheep: Hypertension and decreased nephron number in offspring at 9 months of age. *J. Physiol.* **2005**, *565*, 137–147. [CrossRef]
40. Gardner, D.S.; Tingey, K.; Van Bon, B.W.; Ozanne, S.E.; Wilson, V.; Dandrea, J.; Keisler, D.H.; Stephenson, T.; Symonds, M.E. Programming of glucose-insulin metabolism in adult sheep after maternal undernutrition. *Am. J. Physiol. Regul. Integr. Comp. Physiol.* **2005**, *289*, R947–R954. [CrossRef]
41. Johnson, R.J.; Segal, M.S.; Sautin, Y.; Nakagawa, T.; Feig, D.I.; Kang, D.H.; Gersch, M.S.; Benner, S.; Sánchez-Lozada, L.G. Potential role of sugar (fructose) in the epidemic of hypertension, obesity and the metabolic syndrome, diabetes, kidney disease, and cardiovascular disease. *Am. J. Clin. Nutr.* **2007**, *86*, 899–906. [PubMed]
42. Shankar, K.; Harrell, A.; Liu, X.; Gilchrist, J.M.; Ronis, M.J.J.; Badger, T.M. Maternal obesity at conception programs obesity in the offspring. *Am. J. Physiol. Regul. Integr. Comp. Physiol.* **2008**, *294*, R528–R538. [CrossRef]

43. Shankar, K.; Harrell, A.; Kang, P.; Singhal, R.; Ronis, M.J.J.; Badger, T.M. Carbohydrate-responsive gene expression in the adipose tissue of rats. *Endocrinology* **2010**, *151*, 153–164. [CrossRef]
44. Borengasser, S.J.; Zhong, Y.; Kang, P.; Lindsey, F.; Ronis, M.J.J.; Badger, T.M.; Gomez-Acevedo, H.; Shankar, K. Maternal obesity enhances white adipose tissue differentiation and alters genome-scale DNA methylation in male rat offspring. *Endocrinology* **2013**, *154*, 4113–4125. [CrossRef]
45. Havel, P.J. Dietary fructose: Implications for dysregulation of energy homeostasis and lipid/carbohydrate metabolism. *Nutr. Rev.* **2005**, *63*, 133–157. [PubMed]
46. Lee, W.C.; Wu, K.L.H.; Leu, S.; Tain, Y.L. Translational insights on developmental origins of metabolic syndrome: Focus on fructose consumption. *Biomed. J.* **2018**, *41*, 96–101. [CrossRef]
47. Dai, S.; McNeill, J.H. Fructose-induced hypertension in rats is concentration- and duration-dependent. *J. Pharmacol. Toxicol. Methods* **1995**, *33*, 101–107. [CrossRef]
48. Tain, Y.L.; Wu, K.L.; Lee, W.C.; Leu, S.; Chan, J.Y. Maternal fructose-intake-induced renal programming in adult male offspring. *J. Nutr. Biochem.* **2015**, *26*, 642–650. [CrossRef] [PubMed]
49. Hsu, C.N.; Wu, K.L.; Lee, W.C.; Leu, S.; Chan, J.Y.; Tain, Y.L. Aliskiren administration during early postnatal life sex-specifically alleviates hypertension programmed by maternal high fructose consumption. *Front. Physiol.* **2016**, *7*, 299. [CrossRef] [PubMed]
50. Saad, A.F.; Dickerson, J.; Kechichian, T.B.; Yin, H.; Gamble, P.; Salazar, A.; Patrikeev, I.; Motamedi, M.; Saade, G.R.; Costantine, M.M. High-fructose diet in pregnancy leads to fetal programming of hypertension, insulin resistance, and obesity in adult offspring. *Am. J. Obstet. Gynecol.* **2016**, *215*, e1–e6. [CrossRef]
51. Gray, C.; Gardiner, S.M.; Elmes, M.; Gardner, D.S. Excess maternal salt or fructose intake programmes sex-specific, stress- and fructose-sensitive hypertension in the offspring. *Br. J. Nutr.* **2016**, *115*, 594–604. [CrossRef] [PubMed]
52. Yamada-Obara, N.; Yamagishi, S.I.; Taguchi, K.; Kaida, Y.; Yokoro, M.; Nakayama, Y.; Ando, R.; Asanuma, K.; Matsui, T.; Ueda, S.; et al. Maternal exposure to high-fat and high-fructose diet evokes hypoadiponectinemia and kidney injury in rat offspring. *Clin. Exp. Nephrol.* **2016**, *20*, 853–886. [CrossRef]
53. Tain, Y.L.; Lee, W.C.; Leu, S.; Wu, K.; Chan, J. High salt exacerbates programmed hypertension in maternal fructose-fed male offspring. *Nutr. Metab. Cardiovasc. Dis.* **2015**, *25*, 1146–1151. [CrossRef] [PubMed]
54. Tain, Y.L.; Lee, W.C.; Wu, K.; Leu, S.; Chan, J.Y.H. Maternal high fructose intake increases the vulnerability to post-weaning high-fat diet induced programmed hypertension in male offspring. *Nutrients* **2018**, *10*, 56. [CrossRef] [PubMed]
55. Sathishkumar, K.; Elkins, R.; Yallampalli, U.; Yallampalli, C. Protein restriction during pregnancy induces hypertension and impairs endothelium-dependent vascular function in adult female offspring. *J. Vasc. Res.* **2009**, *46*, 229–239. [CrossRef] [PubMed]
56. Woods, L.L.; Ingelfinger, J.R.; Nyengaard, J.R.; Rasch, R. Maternal protein restriction suppresses the newborn renin-angiotensin system and programs adult hypertension in rats. *Pediatr. Res.* **2001**, *49*, 460–467. [CrossRef]
57. Cambonie, G.; Comte, B.; Yzydorczyk, C.; Ntimbane, T.; Germain, N.; Lê, N.L.; Pladys, P.; Gauthier, C.; Lahaie, I.; Abran, D.; et al. Antenatal antioxidant prevents adult hypertension, vascular dysfunction, and microvascular rarefaction associated with in utero exposure to a low-protein diet. *Am. J. Physiol. Regul. Integr. Comp. Physiol.* **2007**, *292*, R1236–R1245. [CrossRef] [PubMed]
58. Bai, S.Y.; Briggs, D.I.; Vickers, M.H. Increased systolic blood pressure in rat offspring following a maternal low-protein diet is normalized by maternal dietary choline supplementation. *J. Dev. Orig. Health Dis.* **2012**, *3*, 342–349. [CrossRef]
59. McMillen, I.C.; Robinson, J.S. Developmental origins of the metabolic syndrome: Prediction, plasticity, and programming. *Physiol. Rev.* **2005**, *85*, 571–633.
60. Ferguson, D.P.; Monroe, T.O.; Heredia, C.P.; Fleischmann, R.; Rodney, G.G.; Taffet, G.E.; Fiorotto, M.L. Postnatal undernutrition alters adult female mouse cardiac structure and function leading to limited exercise capacity. *J. Physiol.* **2019**, *597*, 1855–1872. [CrossRef]
61. Williams, L.; Seki, Y.; Vuguin, P.M.; Charron, M.J. Animal models of in utero exposure to a high fat diet: A review. *Biochim. Biophys. Acta* **2014**, *1842*, 507–519. [CrossRef] [PubMed]
62. Ashino, N.G.; Saito, K.N.; Souza, F.D.; Nakutz, F.S.; Roman, E.A.; Velloso, L.A.; Torsoni, A.S.; Torsoni, M.A. Maternal high-fat feeding through pregnancy and lactation predisposes mouse offspring to molecular insulin resistance and fatty liver. *J. Nutr. Biochem.* **2012**, *23*, 341–348. [CrossRef] [PubMed]

63. Connor, K.L.; Vickers, M.H.; Beltrand, J.; Meaney, M.J.; Sloboda, D.M. Nature, nurture or nutrition? Impact of maternal nutrition on maternal care, offspring development and reproductive function. *J. Physiol.* **2012**, *590*, 2167–2180. [CrossRef]
64. Resende, A.C.; Emiliano, A.F.; Cordeiro, V.S.; de Bem, G.F.; de Cavalho, L.C.; de Oliveira, P.R.; Neto, M.L.; Costa, C.A.; Boaventura, G.T.; de Moura, R.S. Grape skin extract protects against programmed changes in the adult rat offspring caused by maternal high-fat diet during lactation. *J. Nutr. Biochem.* **2013**, *24*, 2119–2126. [CrossRef] [PubMed]
65. Torrens, C.; Ethirajan, P.; Bruce, K.D.; Cagampang, F.R.; Siow, R.C.; Hanson, M.A.; Byrne, C.D.; Mann, G.E.; Clough, G.F. Interaction between maternal and offspring diet to impair vascular function and oxidative balance in high fat fed male mice. *PLoS ONE* **2012**, *7*, e50671. [CrossRef] [PubMed]
66. Khan, I.Y.; Taylor, P.D.; Dekou, V.; Seed, P.T.; Lakasing, L.; Graham, D.; Dominiczak, A.F.; Hanson, M.A.; Poston, L. Gender-linked hypertension in offspring of lard-fed pregnant rats. *Hypertension* **2003**, *41*, 168–175. [CrossRef] [PubMed]
67. Tain, Y.L.; Lin, Y.J.; Sheen, J.M.; Yu, H.R.; Tiao, M.M.; Chen, C.C.; Tsai, C.C.; Huang, L.T.; Hsu, C.N. High Fat Diets Sex-Specifically Affect the Renal Transcriptome and Program Obesity, Kidney Injury, and Hypertension in the Offspring. *Nutrients* **2017**, *9*, 357. [CrossRef]
68. Thone-Reineke, C.; Kalk, P.; Dorn, M.; Klaus, S.; Simon, K.; Pfab, T.; Godes, M.; Persson, P.; Unger, T.; Hocher, B. High-protein nutrition during pregnancy and lactation programs blood pressure, food efficiency, and body weight of the offspring in a sex-dependent manner. *Am. J. Physiol. Regul. Integr. Comp. Physiol.* **2006**, *291*, R1025–R1030. [CrossRef]
69. Koleganova, N.; Piecha, G.; Ritz, E.; Becker, L.E.; Müller, A.; Weckbach, M.; Nyengaard, J.R.; Schirmacher, P.; Gross-Weissmann, M.L. Both high and low maternal salt intake in pregnancy alter kidney development in the offspring. *Am. J. Physiol. Renal Physiol.* **2011**, *301*, F344–F354. [CrossRef] [PubMed]
70. Bergel, E.; Belizán, J.M. A deficient maternal calcium intake during pregnancy increases blood pressure of the offspring in adult rats. *BJOG* **2002**, *109*, 540–545. [CrossRef]
71. Takaya, J.; Yamanouchi, S.; Kino, J.; Tanabe, Y.; Kaneko, K. A Calcium-Deficient Diet in Dams during Gestation Increases Insulin Resistance in Male Offspring. *Nutrients* **2018**, *10*, 1745. [CrossRef] [PubMed]
72. Schlegel, R.N.; Moritz, K.M.; Paravicini, T.M. Maternal hypomagnesemia alters renal function but does not program changes in the cardiovascular physiology of adult offspring. *J. Dev. Orig. Health Dis.* **2016**, *7*, 473–480. [CrossRef] [PubMed]
73. Gambling, L.; Dunford, S.; Wallace, D.I.; Zuur, G.; Solanky, N.; Srai, K.S.; McArdle, H.J. Iron deficiency during pregnancy affects post-natal blood pressure in the rat. *J. Physiol.* **2003**, *552*, 603–610. [CrossRef]
74. Lewis, R.M.; Petry, C.J.; Ozanne, S.E.; Hales, C.N. Effects of maternal iron restriction in the rat on blood pressure, glucose tolerance, and serum lipids in the 3-month-old offspring. *Metabolism* **2001**, *50*, 562–567. [CrossRef] [PubMed]
75. Tomat, A.; Elesgaray, R.; Zago, V.; Fasoli, H.; Fellet, A.; Balaszczuk, A.M.; Schreier, L.; Costa, M.A.; Arranz, C. Exposure to zinc deficiency in fetal and postnatal life determines nitric oxide system activity and arterial blood pressure levels in adult rats. *Br. J. Nutr.* **2010**, *104*, 382–389. [CrossRef] [PubMed]
76. Padmavathi, I.J.; Kishore, Y.D.; Venu, L.; Ganeshan, M.; Harishankar, N.; Giridharan, N.V.; Raghunath, M. Prenatal and perinatal zinc restriction: Effects on body composition, glucose tolerance and insulin response in rat offspring. *Exp. Physiol.* **2009**, *94*, 761–769. [CrossRef] [PubMed]
77. Tare, M.; Emmett, S.J.; Coleman, H.A.; Skordilis, C.; Eyles, D.W.; Morley, R.; Parkington, H.C. Vitamin D insufficiency is associated with impaired vascular endothelial and smooth muscle function and hypertension in young rats. *J. Physiol.* **2011**, *589*, 4777–4786. [CrossRef]
78. Zhang, H.; Chu, X.; Huang, Y.; Li, G.; Wang, Y.; Li, Y.; Sun, C. Maternal vitamin D deficiency during pregnancy results in insulin resistance in rat offspring, which is associated with inflammation and Iκbα methylation. *Diabetologia* **2014**, *57*, 2165–2172. [CrossRef] [PubMed]
79. Kumar, K.A.; Lalitha, A.; Pavithra, D.; Padmavathi, I.J.; Ganeshan, M.; Rao, K.R.; Venu, L.; Balakrishna, N.; Shanker, N.H.; Reddy, S.U.; et al. Maternal dietary folate and/or vitamin B12 restrictions alter body composition (adiposity) and lipid metabolism in Wistar rat offspring. *J. Nutr. Biochem.* **2013**, *24*, 25–31. [CrossRef] [PubMed]
80. O'Neill, R.J.; Vrana, P.B.; Rosenfeld, C.S. Maternal methyl supplemented diets and effects on offspring health. *Front. Genet.* **2014**, *5*, 289. [CrossRef] [PubMed]

81. Tain, Y.L.; Chan, J.Y.H.; Lee, C.T.; Hsu, C.N. Maternal melatonin therapy attenuates methyl-donor diet-induced programmed hypertension in male adult rat offspring. *Nutrients* **2018**, *10*, 1407. [CrossRef] [PubMed]
82. Houston, M.C. The role of cellular micronutrient analysis, nutraceuticals, vitamins, antioxidants and minerals in the prevention and treatment of hypertension and cardiovascular disease. *Ther. Adv. Cardiovasc. Dis.* **2010**, *4*, 165–183. [CrossRef]
83. Efeyan, A.; Comb, W.C.; Sabatini, D.M. Nutrient-sensing mechanisms and pathways. *Nature* **2015**, *517*, 302–310. [CrossRef] [PubMed]
84. Tain, Y.L.; Hsu, C.N.; Chan, J.Y. PPARs Link Early Life Nutritional insults to later programmed hypertension and metabolic syndrome. *Int. J. Mol. Sci.* **2015**, *17*, 20. [CrossRef] [PubMed]
85. Qiao, L.; Guo, Z.; Bosco, C.; Guidotti, S.; Wang, Y.; Wang, M.; Parast, M.; Schaack, J.; Hay, W.W., Jr.; Moore, T.R.; et al. Maternal High-Fat Feeding Increases Placental Lipoprotein Lipase Activity by Reducing SIRT1 Expression in Mice. *Diabetes* **2015**, *64*, 3111–3120. [CrossRef] [PubMed]
86. Tain, Y.L.; Chan, J.Y.; Hsu, C.N. Maternal Fructose Intake Affects Transcriptome Changes and Programmed Hypertension in Offspring in Later Life. *Nutrients* **2016**, *8*, 757. [CrossRef] [PubMed]
87. Tain, Y.L.; Hsu, C.N. AMP-Activated Protein Kinase as a Reprogramming Strategy for Hypertension and Kidney Disease of Developmental Origin. *Int. J. Mol. Sci.* **2018**, *19*, 1744. [CrossRef] [PubMed]
88. Dennery, P.A. Oxidative stress in development: Nature or nurture? *Free Radic. Biol. Med.* **2010**, *49*, 1147–1151. [CrossRef] [PubMed]
89. Tan, B.L.; Norhaizan, M.E.; Liew, W.P. Nutrients and Oxidative Stress: Friend or Foe? *Oxid. Med. Cell Longev.* **2018**, *2018*, 9719584. [CrossRef] [PubMed]
90. Avila, J.G.; Echeverri, I.; de Plata, C.A.; Castillo, A. Impact of oxidative stress during pregnancy on fetal epigenetic patterns and early origin of vascular diseases. *Nutr. Rev.* **2015**, *73*, 12–21. [CrossRef] [PubMed]
91. Tain, Y.L.; Hsu, C.N. Interplay between oxidative stress and nutrient sensing signaling in the developmental origins of cardiovascular disease. *Int. J. Mol. Sci.* **2017**, *18*, 841. [CrossRef] [PubMed]
92. Woodman, A.G.; Mah, R.; Keddie, D.; Noble, R.M.N.; Panahi, S.; Gragasin, F.S.; Lemieux, H.; Bourque, S.L. Prenatal iron deficiency causes sex-dependent mitochondrial dysfunction and oxidative stress in fetal rat kidneys and liver. *FASEB J.* **2018**, *32*, 3254–3263. [CrossRef] [PubMed]
93. Woodman, A.G.; Mah, R.; Keddie, D.L.; Noble, R.M.N.; Holody, C.D.; Panahi, S.; Gragasin, F.S.; Lemieux, H.; Bourque, S.L. Perinatal iron deficiency and a high salt diet cause long-term kidney mitochondrial dysfunction and oxidative stress. *Cardiovasc. Res.* **2019**. [CrossRef]
94. Yeung, A.W.K.; Tzvetkov, N.T.; El-Tawil, O.S.; Bungău, S.G.; Abdel-Daim, M.M.; Atanasov, A.G. Antioxidants: Scientific Literature Landscape Analysis. *Oxid. Med. Cell Longev.* **2019**, *2019*, 8278454. [CrossRef]
95. Gilbert, S.F. *Developmental Biology*, 6th ed.; Sinauer Associates, Inc.: Sunderland, MA, USA, 2000.
96. Snoeck, A.; Remacle, C.; Reusens, B.; Hoet, J.J. Effect of a low protein diet during pregnancy on the fetal rat endocrine pancreas. *Biol. Neonate* **1990**, *57*, 107–118. [CrossRef] [PubMed]
97. Chen, J.H.; Tarry-Adkins, J.L.; Heppolette, C.A.; Palmer, D.B.; Ozanne, S.E. Early-life nutrition influences thymic growth in male mice that may be related to the regulation of longevity. *Clin. Sci.* **2009**, *118*, 429–438. [CrossRef]
98. Wood-Bradley, R.J.; Barrand, S.; Giot, A.; Armitage, J.A. Understanding the role of maternal diet on kidney development; an opportunity to improve cardiovascular and renal health for future generations. *Nutrients* **2015**, *7*, 1881–1905. [CrossRef]
99. Tain, Y.L.; Hsu, C.N. Developmental origins of chronic kidney disease: Should we focus on early life? *Int. J. Mol. Sci.* **2017**, *18*, 381. [CrossRef] [PubMed]
100. Lelièvre-Pégorier, M.; Vilar, J.; Ferrier, M.L.; Moreau, E.; Freund, N.; Gilbert, T.; Merlet-Bénichou, C. Mild vitamin A deficiency leads to inborn nephron deficit in the rat. *Kidney Int.* **1998**, *54*, 1455–1462. [CrossRef]
101. Lisle, S.J.; Lewis, R.M.; Petry, C.J.; Ozanne, S.E.; Hales, C.N.; Forhead, A.J. Effect of maternal iron restriction during pregnancy on renal morphology in the adult rat offspring. *Br. J. Nutr.* **2003**, *90*, 33–39. [CrossRef] [PubMed]
102. Woods, L.L.; Morgan, T.K.; Resko, J.A. Castration fails to prevent prenatally programmed hypertension in male rats. *Am. J. Physiol. Regul. Integr. Comp. Physiol.* **2010**, *298*, R1111–R1116. [CrossRef] [PubMed]
103. Hokke, S.; Puelles, V.G.; Armitage, J.A.; Fong, K.; Bertram, J.F.; Cullen-McEwen, L.A. Maternal fat feeding augments offspring nephron endowment in mice. *PLoS ONE* **2016**, *11*, e0161578. [CrossRef]

104. Bianco-Miotto, T.; Craig, J.M.; Gasser, Y.P.; van Dijk, S.J.; Ozanne, S.E. Epigenetics and DOHaD: From basics to birth and beyond. *J. Dev. Orig. Health Dis.* **2017**, *8*, 513–519. [CrossRef] [PubMed]
105. Lillycrop, K.A.; Phillips, E.S.; Jackson, A.A.; Hanson, M.A.; Burdge, G.C. Dietary protein restriction of pregnant rats induces and folic acid supplementation prevents epigenetic modification of hepatic gene expression in the offspring. *J. Nutr.* **2005**, *135*, 1382–1386. [CrossRef] [PubMed]
106. Guan, H.; Arany, E.; van Beek, J.P.; Chamson-Reig, A.; Thyssen, S.; Hill, D.J.; Yang, K. Adipose tissue gene expression profiling reveals distinct molecular pathways that define visceral adiposity in offspring of maternal protein-restricted rats. *Am. J. Physiol. Endocrinol. Metab.* **2005**, *288*, E663–E673. [CrossRef]
107. Rees, W.D.; Hay, S.M.; Brown, D.S.; Antipatis, C.; Palmer, R.M. Maternal protein deficiency causes hypermethylation of DNA in the livers of rat fetuses. *J. Nutr.* **2000**, *130*, 1821–1826. [CrossRef] [PubMed]
108. Sable, P.; Randhir, K.; Kale, A.; Chavan-Gautam, P.; Joshi, S. Maternal micronutrients and brain global methylation patterns in the offspring. *Nutr. Neurosci.* **2015**, *18*, 30–36. [CrossRef]
109. Bogdarina, I.; Welham, S.; King, P.J.; Burns, S.P.; Clark, A.J. Epigenetic modification of the renin-angiotensin system in the fetal programming of hypertension. *Circ. Res.* **2007**, *100*, 520–526. [CrossRef]
110. Goyal, R.; Goyal, D.; Leitzke, A.; Gheorghe, C.P.; Longo, L.D. Brain renin-angiotensin system: Fetal epigenetic programming by maternal protein restriction during pregnancy. *Reprod. Sci.* **2010**, *17*, 227–238. [CrossRef]
111. Stevens, A.; Begum, G.; Cook, A.; Connor, K.; Rumball, C.; Oliver, M.; Challis, J.; Bloomfield, F.; White, A. Epigenetic changes in the hypothalamic proopiomelanocortin and glucocorticoid receptor genes in the ovine fetus after periconceptional undernutrition. *Endocrinology* **2010**, *151*, 3652–3664. [CrossRef]
112. McKay, J.; Groom, A.; Potter, C.; Coneyworth, L.J.; Ford, D.; Mathers, J.C.; Relton, C.L. Genetic and non-genetic influences during pregnancy on infant global and site specific DNA methylation: Role for folate gene variants and vitamin B 12. *PLoS ONE* **2012**, *7*, e33290. [CrossRef]
113. Steegers-Theunissen, R.; Obermann-Borst, S.; Kremer, D.; Lindemans, L.; Siebel, C.; Steegers, E.A.; Slagboom, P.E.; Heijmans, B.T. Periconceptional maternal folic acid use of 400 µg per day is related to increased methylation of the IGF2 gene in the very young child. *PLoS ONE* **2009**, *4*, 1–5. [CrossRef]
114. Dominguez-Salas, P.; Moore, S.E.; Baker, M.S.; Bergen, A.W.; Cox, S.E.; Dyer, R.A.; Fulford, A.J.; Guan, Y.; Laritsky, E.; Silver, M.J.; et al. Maternal nutrition at conception modulates DNA methylation of human metastable epialleles. *Nat. Commun.* **2014**, *5*, 3746. [CrossRef] [PubMed]
115. Tamburini, S.; Shen, N.; Wu, H.C.; Clemente, J.C. The microbiome in early life: Implications for health outcomes. *Nat. Med.* **2016**, *22*, 713–722. [CrossRef] [PubMed]
116. Stiemsma, L.T.; Michels, K.B. The role of the microbiome in the developmental origins of health and disease. *Pediatrics* **2018**, *141*, e20172437. [CrossRef] [PubMed]
117. Chu, D.M.; Meyer, K.M.; Prince, A.L.; Aagaard, K.M. Impact of maternal nutrition in pregnancy and lactation on offspring gut microbial composition and function. *Gut Microbes* **2016**, *7*, 459–470. [CrossRef]
118. Nicholson, J.K.; Holmes, E.; Kinross, J.; Burcelin, R.; Gibson, G.; Jia, W.; Pettersson, S. Host-gut microbiota metabolic interactions. *Science* **2012**, *336*, 1262–1267. [CrossRef]
119. Krautkramer, K.A.; Kreznar, J.H.; Romano, K.A.; Vivas, E.I.; Barrett-Wilt, G.A.; Rabaglia, M.E.; Keller, M.P.; Attie, A.D.; Rey, F.E.; Denu, J.M. Diet-Microbiota Interactions Mediate Global Epigenetic Programming in Multiple Host Tissues. *Mol. Cell* **2016**, *64*, 982–992. [CrossRef]
120. Mulligan, C.M.; Friedman, J.E. Maternal modifiers of the infant gut microbiota: Metabolic consequences. *J. Endocrinol.* **2017**, *235*, R1–R12.
121. Al Khodor, S.; Reichert, B.; Shatat, I.F. The Microbiome and Blood Pressure: Can Microbes Regulate Our Blood Pressure? *Front. Pediatr.* **2017**, *5*, 138. [CrossRef]
122. Cukrowska, B. Microbial and Nutritional Programming-The Importance of the Microbiome and Early Exposure to Potential Food Allergens in the Development of Allergies. *Nutrients* **2018**, *10*, 1541. [CrossRef]
123. Codagnone, M.G.; Spichak, S.; O'Mahony, S.M.; O'Leary, O.F.; Clarke, G.; Stanton, C.; Dinan, T.G.; Cryan, J.F. Programming Bugs: Microbiota and the Developmental Origins of Brain Health and Disease. *Biol. Psychiatry* **2019**, *85*, 150–163. [CrossRef]
124. Hsu, C.N.; Lin, Y.J.; Hou, C.Y.; Tain, Y.L. Maternal Administration of Probiotic or Prebiotic Prevents Male Adult Rat Offspring against Developmental Programming of Hypertension Induced by High Fructose Consumption in Pregnancy and Lactation. *Nutrients* **2018**, *10*, 1229. [CrossRef]
125. Lankelma, J.M.; Nieuwdorp, M.; de Vos, W.M.; Wiersinga, W.J. The gut microbiota in internal medicine: Implications for health and disease. *Neth. J. Med.* **2015**, *73*, 61–68.

126. Marino, M.; Masella, R.; Bulzomi, P.; Campesi, I.; Malorni, W.; Franconi, F. Nutrition and human health from a sex-gender perspective. *Mol. Aspects Med.* **2011**, *32*, 1–70. [CrossRef] [PubMed]
127. Clegg, D.J.; Riedy, C.A.; Smith, K.A.; Benoit, S.C.; Woods, S.C. Differential sensitivity to central leptin and insulin in male and female rats. *Diabetes* **2003**, *52*, 682–687. [CrossRef]
128. Loganathan, N.; Belsham, D.D. Nutrient-sensing mechanisms in hypothalamic cell models: Neuropeptide regulation and neuroinflammation in male- and female-derived cell lines. *Am. J. Physiol. Regul. Integr. Comp. Physiol.* **2016**, *311*, R217–R221. [CrossRef]
129. Vina, J.; Gambini, J.; Lopez-Grueso, R.; Abdelaziz, K.M.; Jove, M.; Borras, C. Females live longer than males: Role of oxidative stress. *Curr. Pharm. Des.* **2011**, *17*, 3959–3965. [CrossRef]
130. Hartman, R.J.G.; Huisman, S.E.; den Ruijter, H.M. Sex differences in cardiovascular epigenetics-a systematic review. *Biol. Sex Differ.* **2018**, *9*, 19. [CrossRef]
131. Pankey, C.L.; Walton, M.W.; Odhiambo, J.F.; Smith, A.M.; Ghenis, A.B.; Nathanielsz, P.W.; Ford, S.P. Intergenerational impact of maternal overnutrition and obesity throughout pregnancy in sheep on metabolic syndrome in grandsons and granddaughters. *Domest. Anim. Endocrinol.* **2017**, *60*, 67–74. [CrossRef] [PubMed]
132. Fuente-Martín, E.; Granado, M.; García-Cáceres, C.; Sanchez-Garrido, M.A.; Frago, L.M.; Tena-Sempere, M.; Argente, J.; Chowen, J.A. Early nutritional changes induce sexually dimorphic long-term effects on body weight gain and the response to sucrose intake in adult rats. *Metabolism* **2012**, *61*, 812–822. [CrossRef] [PubMed]
133. Dearden, L.; Balthasar, N. Sexual dimorphism in offspring glucosesensitive hypothalamic gene expression and physiological responses to maternal high-fat diet feeding. *Endocrinology* **2014**, *155*, 2144–2154. [CrossRef]
134. Jones, A.P.; Friedman, M.I. Obesity and adipocyte abnormalities in offspring of rats undernourished during pregnancy. *Science* **1982**, *215*, 1518–1519. [CrossRef] [PubMed]
135. Mao, J.; Zhang, X.; Sieli, P.T.; Falduto, M.T.; Torres, K.E.; Rosenfeld, C.S. Contrasting effects of different maternal diets on sexually dimorphic gene expression in the murine placenta. *Proc. Natl. Acad. Sci. USA* **2010**, *107*, 5557–5562. [CrossRef] [PubMed]
136. Cheong, J.N.; Wlodek, M.E.; Moritz, K.M.; Cuffe, J.S. Programming of maternal and offspring disease: Impact of growth restriction, fetal sex and transmission across generations. *J. Physiol.* **2016**, *594*, 4727–4740. [CrossRef] [PubMed]
137. Haider, B.A.; Bhutta, Z.A. Multiple-micronutrient supplementation for women during pregnancy. *Cochrane Database Syst. Rev.* **2017**, *4*, CD004905. [CrossRef]
138. Schwarzenberg, S.J.; Georgieff, M.K.; COMMITTEE ON NUTRITION. Advocacy for Improving Nutrition in the First 1000 Days to Support Childhood Development and Adult Health. *Pediatrics* **2018**, *141*, e20173716. [CrossRef] [PubMed]
139. Jackson, A.A.; Dunn, R.L.; Marchand, M.C.; Langley-Evans, S.C. Increased systolic blood pressure in rats induced by a maternal low-protein diet is reversed by dietary supplementation with glycine. *Clin. Sci.* **2002**, *103*, 633–639. [CrossRef] [PubMed]
140. Tain, Y.L.; Lee, W.C.; Hsu, C.N.; Lee, W.C.; Huang, L.T.; Lee, C.T.; Lin, C.Y. Asymmetric dimethylarginine is associated with developmental programming of adult kidney disease and hypertension in offspring of streptozotocin-treated mothers. *PLoS ONE* **2013**, *8*, e55420. [CrossRef] [PubMed]
141. Tain, Y.L.; Sheen, J.M.; Chen, C.C.; Yu, H.R.; Tiao, M.M.; Kuo, H.C.; Huang, L.T. Maternal citrulline supplementation prevents prenatal dexamethasone-induced programmed hypertension. *Free Radic. Res.* **2014**, *48*, 580–586. [CrossRef] [PubMed]
142. Tain, Y.L.; Huang, L.T.; Lee, C.T.; Chan, J.Y.; Hsu, C.N. Maternal citrulline supplementation prevents prenatal N^G-nitro-l-arginine-methyl ester (L-NAME)-induced programmed hypertension in rats. *Biol. Reprod.* **2015**, *92*, 7. [CrossRef]
143. Fujii, T.; Yura, S.; Tatsumi, K.; Kondoh, E.; Mogami, H.; Fujita, K.; Kakui, K.; Aoe, S.; Itoh, H.; Sagawa, N.; et al. Branched-chain amino acid supplemented diet during maternal food restriction prevents developmental hypertension in adult rat offspring. *J. Dev. Orig. Health Dis.* **2011**, *2*, 176–183. [CrossRef] [PubMed]
144. Thaeomor, A.; Teangphuck, P.; Chaisakul, J.; Seanthaweesuk, S.; Somparn, N.; Roysommuti, S. Perinatal Taurine Supplementation Prevents Metabolic and Cardiovascular Effects of Maternal Diabetes in Adult Rat Offspring. *Adv. Exp. Med. Biol.* **2017**, *975*, 295–305. [PubMed]

145. Gray, C.; Vickers, M.H.; Segovia, S.A.; Zhang, X.D.; Reynolds, C.M. A maternal high fat diet programmes endothelial function and cardiovascular status in adult male offspring independent of body weight, which is reversed by maternal conjugated linoleic acid (CLA) supplementation. *PLoS ONE* **2015**, *10*, e0115994.
146. Gregório, B.M.; Souza-Mello, V.; Mandarim-de-Lacerda, C.A.; Aguila, M.B. Maternal fish oil supplementation benefits programmed offspring from rat dams fed low-protein diet. *Am. J. Obstet. Gynecol.* **2008**, *199*, e1–e7. [CrossRef] [PubMed]
147. Sánchez-Blanco, C.; Amusquivar, E.; Bispo, K.; Herrera, E. Dietary fish oil supplementation during early pregnancy in rats on a cafeteria-diet prevents fatty liver in adult male offspring. *Food Chem. Toxicol.* **2019**, *123*, 546–552. [CrossRef]
148. Torrens, C.; Brawley, L.; Anthony, F.W.; Dance, C.S.; Dunn, R.; Jackson, A.A.; Poston, L.; Hanson, M.A. Folate supplementation during pregnancy improves offspring cardiovascular dysfunction induced by protein restriction. *Hypertension* **2006**, *47*, 982–987. [CrossRef] [PubMed]
149. Palinski, W.; D'Armiento, F.P.; Witztum, J.L.; de Nigris, F.; Casanada, F.; Condorelli, M.; Silvestre, M.; Napoli, C. Maternal hypercholesterolemia and treatment during pregnancy influence the long-term progression of atherosclerosis in offspring of rabbits. *Circ. Res.* **2001**, *89*, 991–996. [CrossRef] [PubMed]
150. Paul, H.A.; Collins, K.H.; Nicolucci, A.C.; Urbanski, S.J.; Hart, D.A.; Vogel, H.J.; Reimer, R.A. Maternal prebiotic supplementation reduces fatty liver development in offspring through altered microbial and metabolomic profiles in rats. *FASEB J.* **2019**. [CrossRef]
151. Sengupta, P. The Laboratory Rat: Relating Its Age with Human's. *Int. J. Prev. Med.* **2013**, *4*, 624–630.
152. Romero, M.J.; Platt, D.H.; Caldwell, R.B.; Caldwell, R.W. Therapeutic use of citrulline in cardiovascular disease. *Cardiovasc. Drug Rev.* **2006**, *24*, 275–290. [CrossRef]
153. Alves, G.M.; Barão, M.A.; Odo, L.N.; Nascimento Gomes, G.; do Carmo Franco, M.; Nigro, D.; Lucas, S.R.; Laurindo, F.R.; Brandizzi, L.I.; Zaladek Gil, F. L-Arginine effects on blood pressure and renal function of intrauterine restricted rats. *Pediatr. Nephrol.* **2002**, *17*, 856–862. [CrossRef] [PubMed]
154. Carvalho, D.S.; Diniz, M.M.; Haidar, A.A.; Cavanal, M.F.; da Silva Alves, E.; Carpinelli, A.R.; Gil, F.Z.; Hirata, A.E. L-Arginine supplementation improves insulin sensitivity and beta cell function in the offspring of diabetic rats through AKT and PDX-1 activation. *Eur. J. Pharmacol.* **2016**, *791*, 780–787. [CrossRef]
155. DiNicolantonio, J.J.; OKeefe, J.H.; McCarty, M.F. Boosting endogenous production of vasoprotective hydrogen sulfide via supplementation with taurine and N-acetylcysteine: A novel way to promote cardiovascular health. *Open Heart* **2017**, *4*, e000600. [CrossRef] [PubMed]
156. Hsu, C.N.; Tain, Y.L. Hydrogen Sulfide in Hypertension and Kidney Disease of Developmental Origins. *Int. J. Mol. Sci.* **2018**, *19*, 1438. [CrossRef] [PubMed]
157. Vahdaninia, M.; Mackenzie, H.; Dean, T.; Helps, S. The effectiveness of ω-3 polyunsaturated fatty acid interventions during pregnancy on obesity measures in the offspring: An up-to-date systematic review and meta-analysis. *Eur. J. Nutr.* **2018**. [CrossRef] [PubMed]
158. Ji, Y.; Wu, Z.; Dai, Z.; Sun, K.; Wang, J.; Wu, G. Nutritional epigenetics with a focus on amino acids: Implications for the development and treatment of metabolic syndrome. *J. Nutr. Biochem.* **2016**, *27*, 1–8. [CrossRef]
159. Miyazawa, T.; Burdeos, G.C.; Itaya, M.; Nakagawa, K.; Miyazawa, T. Vitamin E: Regulatory Redox Interactions. *IUBMB Life* **2019**. [CrossRef] [PubMed]
160. Cuello-Garcia, C.A.; Fiocchi, A.; Pawankar, R.; Yepes-Nuñez, J.J.; Morgano, G.P.; Zhang, Y.; Ahn, K.; Al-Hammadi, S.; Agarwal, A.; Gandhi, S.; et al. World Allergy Organization- McMaster University Guidelines for Allergic Disease Prevention (GLAD-P.): Prebiotics. *World Allergy Organ J.* **2016**, *9*, 10. [CrossRef] [PubMed]
161. Marques, F.Z.; Nelson, E.; Chu, P.Y.; Horlock, D.; Fiedler, A.; Ziemann, M.; Tan, J.K.; Kuruppu, S.; Rajapakse, N.W.; El-Osta, A.; et al. High-Fiber Diet and Acetate Supplementation Change the Gut Microbiota and Prevent the Development of Hypertension and Heart Failure in Hypertensive Mice. *Circulation* **2017**, *135*, 964–977. [CrossRef] [PubMed]

© 2019 by the authors. Licensee MDPI, Basel, Switzerland. This article is an open access article distributed under the terms and conditions of the Creative Commons Attribution (CC BY) license (http://creativecommons.org/licenses/by/4.0/).

MDPI
St. Alban-Anlage 66
4052 Basel
Switzerland
Tel. +41 61 683 77 34
Fax +41 61 302 89 18
www.mdpi.com

Nutrients Editorial Office
E-mail: nutrients@mdpi.com
www.mdpi.com/journal/nutrients

www.ingramcontent.com/pod-product-compliance
Lightning Source LLC
LaVergne TN
LVHW071951080526
838202LV00064B/6719